Texts in Theoretical Computer Science
An EATCS Series

Editors: W. Brauer G. Rozenberg A. Salomaa
On behalf of the European Association
for Theoretical Computer Science (EATCS)

Advisory Board: G. Ausiello M. Broy C.S. Calude
S. Even J. Hartmanis J. Hromkovič N. Jones
T. Leighton M. Nivat C. Papadimitriou D. Scott

Springer
Berlin
Heidelberg
New York
Hong Kong
London
Milan
Paris
Tokyo

Leonid Libkin

Elements of Finite Model Theory

With 24 Figures

Author

Prof. Leonid Libkin
Department of Computer Science
University of Toronto
Toronto ON M5S 3H5
Canada
libkin@cs.toronto.edu
www.cs.toronto.edu/~libkin

Series Editors

Prof. Dr. Wilfried Brauer
Institut für Informatik der TUM
Boltzmannstr. 3, 85748 Garching, Germany
Brauer@informatik.tu-muenchen.de

Prof. Dr. Grzegorz Rozenberg
Leiden Institute of Advanced Computer Science
University of Leiden
Niels Bohrweg 1, 2333 CA Leiden, The Netherlands
rozenber@liacs.nl

Prof. Dr. Arto Salomaa
Turku Centre for Computer Science
Lemminkäisenkatu 14 A, 20520 Turku, Finland
asalomaa@utu.fi

ACM Computing Classification (1998): F.4, F.1, H.2, F.3

ISBN 978-3-642-05948-3

This work is subject to copyright. All rights are reserved, whether the whole or part of the material is concerned, specifically the rights of translation, reprinting, reuse of illustrations, recitation, broadcasting, reproduction on microfilm or in any other way, and storage in data banks. Duplication of this publication or parts thereof is permitted only under the provisions of the German Copyright Law of September 9, 1965, in its current version, and permission for use must always be obtained from Springer-Verlag. Violations are liable for prosecution under the German Copyright Law.

Springer-Verlag is a part of Springer Science+Business Media

springeronline.com

© Springer-Verlag Berlin Heidelberg 2010
Printed in Germany

The use of general descriptive names, trademarks, etc. in this publication does not imply, even in the absence of a specific statement, that such names are exempt from the relevant protective laws and therefore free for general use.

Cover Design: KünkelLopka, Heidelberg

Printed on acid-free paper 45/3142/GF - 5 4 3 2 1 0

To Helen and Daniel
Алёне и Даниле

Preface

Finite model theory is an area of mathematical logic that grew out of computer science applications.

The main sources of motivational examples for finite model theory are found in database theory, computational complexity, and formal languages, although in recent years connections with other areas, such as formal methods and verification, and artificial intelligence, have been discovered.

The birth of finite model theory is often identified with Trakhtenbrot's result from 1950 stating that validity over finite models is not recursively enumerable; in other words, completeness fails over finite models. The technique of the proof, based on encoding Turing machine computations as finite structures, was reused by Fagin almost a quarter century later to prove his celebrated result that put the equality sign between the class NP and existential second-order logic, thereby providing a machine-independent characterization of an important complexity class. In 1982, Immerman and Vardi showed that over ordered structures, a fixed point extension of first-order logic captures the complexity class PTIME of polynomial time computable properties. Shortly thereafter, logical characterizations of other important complexity classes were obtained. This line of work is often referred to as descriptive complexity.

A different line of finite model theory research is associated with the development of relational databases. By the late 1970s, the relational database model had replaced others, and all the basic query languages for it were essentially first-order predicate calculus or its minor extensions. In 1974, Fagin showed that first-order logic cannot express the transitive closure query over finite relations. In 1979, Aho and Ullman rediscovered this result and brought it to the attention of the computer science community. Following this, Chandra and Harel proposed a fixed-point extension of first-order logic on finite relational structures as a query language capable of expressing queries such as the transitive closure. Logics over finite models have become the standard starting point for developing database query languages, and finite model theory techniques are used for proving results about their expressiveness and complexity.

Yet another line of work on logics over finite models originated with Büchi's work from the early 1960s: he showed that regular languages are precisely those definable in monadic second-order logic over strings. This line of work is the automata-theoretic counterpart of descriptive complexity: instead of logical characterizations of time/space restrictions of Turing machines, one provides such characterizations for weaker devices, such as automata. More recently, connections between database query languages and automata have been explored too, as the field of databases started moving away from relations to more complex data models.

In general, finite model theory studies the behavior of logics on finite structures. The reason this is a separate subject, and not a tiny chapter in classical model theory, is that most standard model-theoretic tools (most notably, compactness) fail over finite models. Over the past 25–30 years, many tools have been developed to study logics over finite structures, and these tools helped answer many questions about complexity theory, databases, formal languages, etc.

This book is an introduction to finite model theory, geared towards theoretical computer scientists. It grew out of my finite model theory course, taught to computer science graduate students at the University of Toronto. While teaching that course, I realized that there is no single source that covers all the main areas of finite model theory, and yet is suitable for computer science students. There are a number of excellent books on the subject. *Finite Model Theory* by Ebbinghaus and Flum was the first standard reference and heavily influenced the development of the field, but it is a book written for mathematicians, not computer scientists. There is also a nice set of notes by Väänänen, available on the web. Immerman's *Descriptive Complexity* deals extensively with complexity-theoretic aspects of finite model theory, but does not address other applications. *Foundations of Databases* by Abiteboul, Hull, and Vianu covers many database applications, and Thomas's chapter "Languages, automata, and logic" in the *Handbook of Formal Languages* describes connections between logic and formal languages. Given the absence of a single source for all the subjects, I decided to write course notes, which eventually became this book.

The reader is assumed to have only the most basic computer science and logic background: some discrete mathematics, theory of computation, complexity, propositional and predicate logic. The book also includes a background chapter, covering logic, computability theory, and computational complexity. In general, the book should be accessible to senior undergraduate students in computer science.

A note on exercises: there are three kinds of these. Some are the usual exercises that the reader should be able to do easily after reading each chapter. If I indicate that an exercise comes from a paper, it means that its level could range from moderately to extremely difficult: depending on the exact level, such an "exercise" could be a question on a take-home exam, or even a course project, whose main goal is to understand the paper where the result is proven. Such exercises also gave me the opportunity to mention a number of interesting results that otherwise could not have been included in the book. There are also exercises marked with an asterisk: for these, I do not know solutions.

It gives me the great pleasure to thank my colleagues and students for their help. I received many comments from Marcelo Arenas, Pablo Barceló, Michael Benedikt, Ari Brodsky, Anuj Dawar, Ron Fagin, Arthur Fischer, Lauri Hella, Christoph Koch, Janos Makowsky, Frank Neven, Juha Nurmonen, Ben Rossman, Luc Segoufin, Thomas Schwentick, Jan Van den Bussche, Victor Vianu, and Igor Walukiewicz. Ron Fagin, as well as Yuri Gurevich, Alexander Livchak, Michael Taitslin, and Vladimir Sazonov, were also very helpful with historical comments. I taught two courses based on this book, and students in both classes provided very useful feedback; in addition to those I already thanked, I would like to acknowledge Antonina Kolokolova, Shiva Nejati, Ken Pu, Joseph Rideout, Mehrdad Sabetzadeh, Ramona Truta, and Zheng Zhang. Despite their great effort, mistakes undoubtedly remain in the book; if you find one, please let me know. My email is `libkin@cs.toronto.edu`.

Many people in the finite model theory community influenced my view of the field; it is impossible to thank them all, but I want to mention Scott Weinstein, from whom I learned finite model theory, and immediately became fascinated with the subject.

Finally, I thank Ingeborg Mayer, Alfred Hofmann, and Frank Holzwarth at Springer-Verlag for editorial assistance, and Denis Thérien for providing ideal conditions for the final proofreading of the book.

This book is dedicated to my wife, Helen, and my son, Daniel. Daniel was born one week after I finished teaching a finite model theory course in Toronto, and after several sleepless nights I decided that perhaps writing a book is the type of activity that goes well with the lack of sleep. By the time I was writing Chap. 6, Daniel had started sleeping through the night, but at that point it was too late to turn back. And without Helen's help and support I certainly would not have finished this book in only two years.

Toronto, Ontario, Canada
May 2004

Leonid Libkin

Contents

1

Introduction

Finite model theory studies the expressive power of logics on finite models. Classical model theory, on the other hand, concentrates on infinite structures: its origins are in mathematics, and most objects of interest in mathematics are infinite, e.g., the sets of natural numbers, real numbers, etc. Typical examples of interest to a model-theorist would be algebraically closed fields (e.g., $\langle \mathbb{C}, +, \cdot \rangle$), real closed fields (e.g., $\langle \mathbb{R}, +, \cdot, < \rangle$), various models of arithmetic (e.g., $\langle \mathbb{N}, +, \cdot \rangle$ or $\langle \mathbb{N}, + \rangle$), and other structures such as Boolean algebras or random graphs.

The origins of finite model theory are in computer science where most objects of interest are finite. One is interested in the expressiveness of logics over finite graphs, or finite strings, other finite relational structures, and sometimes restrictions of arithmetic structures to an initial segment of natural numbers.

The areas of computer science that served as a primary source of examples, as well as the main consumers of techniques from finite model theory, are databases, complexity theory, and formal languages (although finite model theory found applications in other areas such as AI and verification). In this chapter, we give three examples that illustrate the need for studying logics over finite structures.

1.1 A Database Example

While early database systems used rather ad hoc data models, from the early 1970s the world switched to the relational model. In that model, a database stores tables, or relations, and is queried by a logic-based declarative language. The most standard such language, *relational calculus*, has precisely the power of first-order predicate calculus. In real life, it comes equipped with a specialized programming syntax (e.g., the `select-from-where` statement of SQL).

Suppose that we have a company database, and one of its relations is the `Reports_To` relation: it stores pairs (x, y), where x is an employee, and y is his/her immediate manager. Organizational hierarchies tend to be quite complicated and often result in many layers of management, so one may want to skip the immediate manager level and instead look for the manager's manager.

In SQL, this would be done by the following query:

```
select R1.employee, R2.manager
from   Reports_To R1, Reports_To R2
where  R1.manager=R2.employee
```

This is simply a different way of writing the following first-order logic formula:

$$\varphi(x,y) \;\equiv\; \exists z \left(\texttt{Reports\_To}(x,z) \wedge \texttt{Reports\_To}(z,y)\right) .$$

Continuing, we may ask for someone's manager's manager's manager:

$$\exists z_1 \exists z_2 \left(\texttt{Reports\_To}(x,z_1) \wedge \texttt{Reports\_To}(z_1,z_2) \wedge \texttt{Reports\_To}(z_2,y)\right) ,$$

and so on.

But what if we want to find everyone who is higher in the hierarchy than a given employee? Speaking graph-theoretically, if we associate a pair (x,y) in the Reports_To relation with a directed edge from x to y in a graph, then we want to find, for a given node, all the nodes reachable from it. This does not seem possible in first-order logic, but how can one prove this?

There are other queries naturally related to this reachability property. Suppose that once in a while, the company wants to make sure that its management hierarchy is logically consistent; that is, we cannot have cycles in the Reports_To relation. In graph-theoretic terms, it means that Reports_To is acyclic. Again, if one thinks about it for a while, it seems that first-order logic does not have enough power to express this query.

We now consider a different kind of query. Suppose we have two managers, x and y, and let X be the set of all the employees directly managed by x (i.e., all x' such that (x',x) is in Reports_To), and likewise let Y be the set of all the employees directly managed by y. Can we write a query asking whether $|X| = |Y|$; that is, a query asking whether x and y have the same number of people reporting to them?

It turns out that first-order logic is again not sufficiently expressive for this kind of query, but since queries like those described above are so common in practice, SQL adds special features to the language to perform them. That is, SQL can count: it can apply the cardinality function (and more complex functions as well) to entire columns in relations. For example, in SQL one can write a query that finds all pairs of managers x and y who have the same number of people reporting to them:

```
select R1.manager, R2.manager
from   Reports_To R1, Reports_To R2
where   (select count(Reports_To.employee)
         from Reports_To
         where Reports_To.manager = R1.manager)
      = (select count(Reports_To.employee)
         from Reports_To
         where Reports_To.manager = R2.manager)
```

Since this cannot be done in first-order logic, but can be done in SQL (and, in fact, in some rather simple extensions of first-order logic with counting), it is natural to ask whether counting provides enough expressiveness to define queries such as reachability (can node x be reached from node y in a given graph?) and acyclicity.

Typical applications of finite model theory in databases have to deal with questions of this sort: what can, and, more importantly, what cannot, be expressed in various query languages.

Let us now give intuitive reasons why reachability queries are not expressible in first-order logic. Consider a different example. Suppose that we have an airline database, with a binary relation R (for routes), such that an entry (A, B) in R indicates that there is a flight from A to B. Now suppose we want to find all pairs of cities A, B such that there is a direct flight between them; this is done by the following query:

$$q_0(x, y) \equiv R(x, y),$$

which is simply a first-order formula with two free variables. Next, suppose we want to know if one can get from x to y with exactly one change of plane; then we write

$$q_1(x, y) \equiv \exists z\ R(x, z) \wedge R(z, y).$$

Doing "with at most one change" means having a disjunction

$$Q_1(x, y) \equiv q_1(x, y) \vee q_0(x, y).$$

Clearly, for each fixed k we can write a formula stating that one can get from x to y with exactly k stops:

$$q_k(x, y) = \exists z_1 \ldots \exists z_k\ R(x, z_1) \wedge R(z_1, z_2) \wedge \ldots \wedge R(z_k, y),$$

as well as $Q_k = \bigvee_{j \leq k} q_j$ testing if at most k stops suffice.

But what about the *reachability* query: can we get from x to y? That is, one wants to compute the transitive closure of R. The problem with this is that we do not know in advance what k is supposed to be. So the query that we need to write is

$$\bigvee_{k \in \mathbb{N}} q_k,$$

but this is not a first-order formula! Of course this is not a formal proof that reachability is not expressible in first-order logic (we shall see a proof of this fact in Chap. 3), but at least it gives a hint as to what the limitations of first-order logic are.

The inability of first-order logic to express some important queries motivated a lot of research on extensions of first-order logic that can do queries such as transitive closure or cardinality comparisons. We shall see a number of extensions of these kinds – fixed point logics, (fragments of) second-order logic, counting logics – that are important for database theory, and we shall study properties of these extensions as well.

1.2 An Example from Complexity Theory

We now turn to a different area, and to more expressive logics. Suppose that we have a graph, this time *undirected*, given to us as a pair $\langle V, E\rangle$, where V is the set of vertices, or nodes, and E is the edge relation. Assume that now we can specify graph properties in *second-order logic*; that is, we can quantify over *sets* (or relations) of nodes.

Consider a well-known property of *Hamiltonicity.* A *simple circuit* in a graph G is a sequence $(a_1, \ldots, a_n)$ of distinct nodes such that there are edges $(a_1, a_2), (a_2, a_3), \ldots, (a_{n-1}, a_n), (a_n, a_1)$. A simple circuit is Hamiltonian if $V = \{a_1, \ldots, a_n\}$. A graph is *Hamiltonian* if it has a Hamiltonian circuit.

We now consider the following formula:

$$\exists L\ \exists S \begin{pmatrix} \textit{linear order}(L) \\ \wedge\ S \textit{ is the successor relation of } L \\ \wedge\ \forall x \exists y\ (L(x,y) \vee L(y,x)) \\ \wedge\ \forall x \forall y\ (S(x,y) \rightarrow E(x,y)) \end{pmatrix} \tag{1.1}$$

The quantifiers $\exists L\ \exists S$ state the existence of two binary relations, L and S, that satisfy the formula in parentheses. That formula uses some abbreviations. The subformula *linear order*(L) in (1.1) states that the relation L is a linear ordering; it can be defined as

$$\begin{aligned} &(\forall x \neg L(x,x))\ \wedge\ (\forall x \forall y \forall z\ (L(x,y) \wedge L(y,z) \rightarrow L(x,z))) \\ \wedge\ \ &\forall x \forall y\ \Big((x \neq y) \rightarrow \big(L(x,y) \vee L(y,x)\big)\Big). \end{aligned}$$

The subformula *S is the successor relation of* L states that S is the successor relation associated with the linear ordering L; it can be defined as

$$\forall x \forall y\ \ S(x,y) \leftrightarrow \begin{pmatrix} \big(L(x,y) \wedge \neg\exists z (L(x,z) \wedge L(z,y))\big) \\ \vee\ (\neg\exists z\ L(x,z)\ \wedge\ \neg\exists z\ L(z,y)) \end{pmatrix}$$

Note that S is the circular successor relation, as it also includes the pair (x, y) where x is the maximal and y the minimal element with respect to L.

Then (1.1) says that L and S are defined on all nodes of the graph, and that S is a subset of E. Hence, S is a Hamiltonian circuit, and thus (1.1) tests if a graph is Hamiltonian.

It it well known that testing Hamiltonicity is an NP-complete problem. Is this a coincidence, or is there a natural connection between NP and second-order logic? Let us turn our attention to two other well-known NP-complete problems: *3-colorability* and *clique*.

To test if a graph is 3-colorable, we have to check that there exist three disjoint sets A, B, C covering the nodes of the graph such that for every edge $(a, b) \in E$, the nodes a and b cannot belong to the same set. The sentence below does precisely that:

$$\exists A \exists B \exists C \left(\begin{array}{l} \forall x \begin{bmatrix} (A(x) \wedge \neg B(x) \wedge \neg C(x)) \\ \vee\, (\neg A(x) \wedge B(x) \wedge \neg C(x)) \\ \vee\, (\neg A(x) \wedge \neg B(x) \wedge C(x)) \end{bmatrix} \\ \wedge \\ \forall x, y\ E(x, y) \rightarrow \neg \begin{bmatrix} (A(x) \wedge A(y)) \\ \vee\, (B(x) \wedge B(y)) \\ \vee\, (C(x) \wedge C(y)) \end{bmatrix} \end{array} \right) \tag{1.2}$$

For *clique*, typically one has a parameter k, and the problem is to check whether a clique of size k exists. Here, to stay purely within the formalism of second-order logic, we assume that the input is a graph E and a set of nodes (a unary relation) U, and we ask if E has a clique of size $|\, U\,|$. We do it by testing if there is a set C (nodes of the clique) and a binary relation F that is a one-to-one correspondence between C and U. Testing that the restriction of E to C is a clique, and that F is one-to-one, can be done in first-order logic. Thus, the test is done by the following second-order sentence:

$$\exists C \exists F \left(\begin{array}{l} \forall x \forall y\ \big(F(x, y) \rightarrow (C(x) \wedge U(y))\big) \\ \wedge\ \forall x\ \big(C(x)\ \rightarrow\ \exists! y (F(x, y) \wedge U(y))\big) \\ \wedge\ \forall y\ \big(U(y) \rightarrow \exists! x (F(x, y) \wedge C(x))\big) \\ \wedge\ \forall x \forall y\ \big(C(x) \wedge C(y) \rightarrow E(x, y)\big) \end{array} \right) \tag{1.3}$$

Here $\exists! x \varphi(x)$ means "there exists exactly one x such that $\varphi(x)$"; this is an abbreviation for $\exists x \big(\varphi(x)\ \wedge\ \forall y\ (\varphi(y) \rightarrow x = y)\big)$.

Notice that (1.1), (1.2), and (1.3) all follow the same pattern: they start with existential second-order quantifiers, followed by a first-order formula. Such formulas form what is called *existential second-order logic*, abbreviated as $\exists$SO. The connection to NP can easily be seen: existential second-order quantifiers correspond to the guessing stage of an NP algorithm, and the remaining first-order formula corresponds to the polynomial time verification stage of an NP algorithm.

It turns out that the connection between NP and ∃SO is exact, as was shown by Fagin in his celebrated 1974 theorem, stating that NP = ∃SO. This connection opened up a new area, called descriptive complexity. The goals of descriptive complexity are to describe complexity classes by means of logical formalisms, and then use tools from mathematical logic to analyze those classes. We shall prove Fagin's theorem later, and we shall also see logical characterizations of a number of other familiar complexity classes.

1.3 An Example from Formal Language Theory

Now we turn our attention to strings over a finite alphabet, say $\Sigma = \{a, b\}$. We want to represent a string as a structure, much like a graph.

Given a string $s = s_1 s_2 \ldots s_n$, we create a structure M_s as follows: the universe is $\{1, \ldots, n\}$ (corresponding to positions in the string), we have one binary relation $<$ whose meaning of course is the usual order on the natural numbers, and two unary relations A and B. Then $A(i)$ is true if $s_i = a$, and $B(i)$ is true if $s_i = b$. For example, M_{abba} has universe $\{1, 2, 3, 4\}$, with A interpreted as $\{1, 4\}$ and B as $\{2, 3\}$.

Let us look at the following second-order sentence in which quantifiers range over sets of positions in a string:

$$\Phi \equiv \exists X \exists Y \begin{pmatrix} \forall x\ (X(x) \leftrightarrow \neg Y(x)) \\ \wedge\ \forall x\ \forall y\ (X(x) \wedge Y(y) \rightarrow x < y) \\ \wedge\ \forall x\ (X(x) \rightarrow A(x) \wedge Y(x) \rightarrow B(x)) \end{pmatrix}$$

When is M_s a model of Φ? This happens iff there exists two sets of positions, X and Y, such that X and Y form a partition of the universe (this is what the first conjunct says), that all positions in X precede the positions in Y (that is what the second conjunct says), and that for each position i in X, the ith symbol of s is a, for each position j in Y, the jth symbol is b (this is stated in the third conjunct). That is, the string starts with some a's, and then switches to all b's. Using the language of regular expressions, we can say that

$$M_s \models \Phi \quad \text{iff} \quad s \in a^* b^*.$$

Is quantification over sets really necessary in this example? It turns out that the answer is *no*: one can express the fact that s is in $a^* b^*$ by saying that there are no two positions $i < j$ such that the ith symbol is b and the jth symbol is a. This, of course, can be done in first-order logic:

$$\neg \exists i \exists j \Big((i < j)\ \wedge\ B(i)\ \wedge\ A(j) \Big).$$

A natural question that arises then is the following: are second-order quantifiers of no use if one wants to describe regular languages by logical means? The answer is *no*, as we shall see later. For now, we can give an example.

First, consider the sentence $\Phi_a \equiv \forall i\ A(i)$, which is true in M_s iff $s \in a^*$. Next, define a relation $i \prec j$ saying that j is the successor of i. It can be defined by the formula $\big((i < j) \wedge \forall k\ ((k \leq i) \vee (k \geq j))\big)$. Now consider the sentence

$$\Phi_1 \equiv \exists X \exists Y \left(\begin{array}{l} \forall i\ \big(X(i) \leftrightarrow \neg Y(i)\big) \\ \wedge\ \forall i\ \big(\neg\exists j (j < i) \rightarrow X(i)\big) \\ \wedge\ \forall i\ \big(\neg\exists j (j > i) \rightarrow Y(i)\big) \\ \wedge\ \forall i \forall j\ \big((i \prec j) \wedge X(i) \rightarrow Y(j)\big) \\ \wedge\ \forall i \forall j\ \big((i \prec j) \wedge Y(i) \rightarrow X(j)\big) \end{array} \right)$$

This sentence says that the universe $\{1, \ldots, n\}$ can be partitioned into two sets X and Y such that $1 \in X$, $n \in Y$, and the successor of an element of X is in Y and vice versa; that is, the size of the universe is even.

Now what is $\Phi_1 \wedge \Phi_a$? It says that the string is of even length, and has only a's in it – hence, $M_s \models \Phi_1 \wedge \Phi_a$ iff $s \in (aa)^*$. It turns out that one cannot define $(aa)^*$ using first-order logic alone: one needs second-order quantifiers. Moreover, with second-order quantifiers ranging over sets of positions, one defines *precisely* the regular languages. We shall deal with both expressibility and inexpressibility results related to logics over strings later in this book.

There are a number of common themes in the examples presented above. In all the cases, we are talking about the expressive power of logics over finite objects: relational databases, graphs, and strings. There is a close connection between logical formalisms and familiar concepts from computer science: first-order logic corresponds to relational calculus, existential second-order logic to the complexity class NP, and second-order logic with quantifiers ranging over sets describes regular languages.

Of equal importance is the fact that in all the examples we want to show some *inexpressibility* results. In the database example, we want to show that the transitive closure is not expressible in first-order logic. In the complexity example, it would be nice to show that certain problems cannot be expressed in ∃SO – any such result would give us bounds on the class NP, and this would hopefully lead to separation results for complexity classes. In the example from formal languages, we want to show that certain regular languages (e.g., $(aa)^*$) cannot be expressed in first-order logic.

Inexpressibility results have traditionally been a core theme of finite model theory. The main explanation for that is the source of motivating examples for finite model theory. Most of them come from computer science, where one is dealing not with natural phenomena, but rather with artificial creations. Thus, we often want to know the limitations of these creations. In general, this explains the popularity of impossibility results in computer science. After all, the most famous open problem of computer science, the PTIME vs NP problem, is so fascinating because the expected answer would tell us that a large number of important problems *cannot* be solved efficiently.

Concentrating on inexpressibility results highlights another important feature of finite model theory: since we are often interested in counterexamples, many constructions and techniques of interest apply only to a "small" fraction of structures. In fact, we shall see that some techniques (e.g., locality) degenerate to trivial statements on almost all structures, and yet it is that small fraction of structures on which they behave interestingly that gives us important techniques for analyzing expressiveness of logics, query languages, etc. Towards the end of the book, we shall also see that on most typical structures, some very expressive logics collapse to rather weak ones; however, all interesting separation examples occur outside the class of "typical" structures.

1.4 An Overview of the Book

In Chap. 2, we review the background material from mathematical logic, computability theory, and complexity theory.

In Chap. 3 we introduce the fundamental tool of Ehrenfeucht-Fraïssé games, and prove their completeness for expressibility in first-order logic (FO). The game is played by two players, the spoiler and the duplicator, on two structures. The spoiler tries to show that the structures are different, while the duplicator tries to show that they are the same. If the duplicator can succeed for k rounds of such a game, it means that the structures cannot be distinguished by FO sentences whose depth of quantifier nesting does not exceed k. We also define types, which play a very important role in many aspects of finite model theory. In the same chapter, we see some bounds on the expressive power of FO, proved via Ehrenfeucht-Fraïssé games.

Finding winning strategies in Ehrenfeucht-Fraïssé games becomes quite hard for nontrivial structures. Thus, in Chap. 4, we introduce some sufficient conditions that guarantee a win for the duplicator. These conditions are based on the idea of locality. Intuitively, local formulae cannot see very far from their free variables. We show several different ways of formalizing this intuition, and explain how each of those ways gives us easy proofs of bounds on the expressiveness of FO.

In Chap. 5 we continue to study first-order logic, but this time over structures whose universe is ordered. Here we see the phenomenon that is very common for logics over finite structures. We call a property of structures order-invariant if it can be defined with a linear order, but is independent of a particular linear order used. It turns out that there are order-invariant FO-definable properties that are not definable in FO alone. We also show that such order-invariant properties continue to be local.

Chap. 6 deals with the complexity of FO. We distinguish two kinds of complexity: data complexity, meaning that a formula is fixed and the structure on which it is evaluated varies, and combined complexity, meaning that both the formula and the structure are part of the input. We show how to evaluate

FO formulae by Boolean circuits, and use this to derive drastically different bounds for the complexity of FO: AC^0 for data complexity, and PSPACE for combined complexity. We also consider the parametric complexity of FO: in this case, the formula is viewed as a parameter of the input. Finally, we study a subclass of FO queries, called conjunctive queries, which is very important in database theory, and prove complexity bounds for it.

In Chap. 7, we move away from FO, and consider its extension with monadic second-order quantifiers: such quantifiers can range over subsets of the universe. The resulting logic is called monadic second-order logic, or MSO. We also consider two restrictions of MSO: an ∃MSO formula starts with a sequence of existential second-order quantifiers, which is followed by an FO formula, and an ∀MSO formula starts with a sequence of universal second-order quantifiers, followed by an FO formula. We first study ∃MSO and ∀MSO on graphs, where they are shown to be different. We then move to strings, where MSO collapses to ∃MSO and captures precisely the regular languages. Further restricting our attention to FO over strings, we prove that it captures the star-free languages. We also cover MSO over trees, and tree automata.

In Chap. 8 we study a different extension of FO: this time, we add mechanisms for counting, such as counting terms, counting quantifiers, or certain generalized unary quantifiers. We also introduce a logic that has a lot of counting power, and prove that it remains local, much as FO. We apply these results in the database setting, considering a standard feature of many query languages – aggregate functions – and proving bounds on the expressiveness of languages with aggregation.

In Chap. 9 we present the technique of coding Turing machines as finite structures, and use it to prove two results: Trakhtenbrot's theorem, which says that the set of finitely satisfiable sentences is not recursive, and Fagin's theorem, which says that NP problems are precisely those expressible in existential second-order logic.

Chapter 10 deals with extensions of FO for expressing properties that, algorithmically, require recursion. Such extensions have fixed point operators. There are three flavors of them: least, inflationary, and partial fixed point operators. We study properties of resulting fixed point logics, and prove that in the presence of a linear order, they capture complexity classes PTIME (for least and inflationary fixed points) and PSPACE (for partial fixed points). We also deal with a well-known database query language that adds fixed points to FO: DATALOG. In the same chapter, we consider a closely related logic based on adding the transitive closure operator to FO, and prove that over order structures it captures nondeterministic logarithmic space.

Fixed point logics are not very easy to analyze. Nevertheless, they can be embedded into a logic which uses infinitary connectives, but has a restriction that every formula only mentions finitely many variables. This logic, and its fragments, are studied in Chap. 11. We introduce the logic $\mathcal{L}^{\omega}_{\infty\omega}$, define games for it, and prove that fixed point logics are embeddable into it. We study definability of types for finite variable logics, and use them to provide a purely logical counterpart of the PTIME vs. PSPACE question.

In Chap. 12 we study the asymptotic behavior of FO and prove that every FO sentence is either true in almost all structures, or false in almost all structures. This phenomenon is known as the zero-one law. We also prove that $\mathcal{L}^{\omega}_{\infty\omega}$, and hence fixed point logics, have the zero-one law. In the same chapter we define an infinite structure whose theory consists precisely of FO sentences that hold in almost all structures. We also prove that almost everywhere, fixed point logics collapse to FO.

In Chap. 13, we show how finite and infinite model theory mix: we look at finite structures that live in an infinite one, and study the power of FO over such hybrid structures. We prove that for some underlying infinite structures, like $\langle \mathbb{N}, +, \cdot \rangle$, every computable property of finite structures embedded into them can be defined, but for others, like $\langle \mathbb{R}, +, \cdot \rangle$, one can only define properties which are already expressible in FO over the finite structure alone. We also explain connections between such mixed logics and database query languages.

Finally, in Chap. 14, we outline other applications of finite model theory: in decision problems in mathematical logic, in formal verification of properties of finite state systems, and in constraint satisfaction.

1.5 Exercises

Exercise 1.1. Show how to express the following properties of graphs in first-order logic:

- A graph is complete.
- A graph has an isolated vertex.
- A graph has at least two vertices of out-degree 3.
- Every vertex is connected by an edge to a vertex of out-degree 3.

Exercise 1.2. Show how to express the following properties of graphs in existential second-order logic:

- A graph has a kernel, i.e., a set of vertices X such that there is no edge between any two vertices in X, and every vertex outside of X is connected by an edge to a vertex of X.
- A graph on n vertices has an independent set X (i.e., no two nodes in X are connected by an edge) of size at least $n/2$.
- A graph has an even number of vertices.
- A graph has an even number of edges.
- A graph with m edges has a bipartite subgraph with at least $m/2$ edges.

Exercise 1.3. (a) Show how to define the following regular languages in monadic second-order logic:

- $a^*(b+c)^*aa^*$;

- $(aaa)^*(bb)^+$;
- $\Big(\big((a+b)^*cc^*\big)^*(aa)^*\Big)^*a$.

For the first language, provide a first-order definition as well.

(b) Let Φ be a monadic second-order logic sentence over strings. Show how to construct a sentence Ψ such that $M_s \models \Psi$ iff there is a string s' such that $|s|=|s'|$ and $M_{s \cdot s'} \models \Phi$. Here $|s|$ refers to the length of s, and $s \cdot s'$ is the concatenation of s and s'.

Remark: once we prove Büchi's theorem in Chap. 7, you will see that the above statement says that if L is a regular language, then the language

$$\frac{1}{2}L \;=\; \{s \mid \text{for some } s',\; |s|=|s'| \;\text{ and } s \cdot s' \in L\}$$

is regular too (see, e.g., Exercise 3.16 in Hopcroft and Ullman [126]).

2

Preliminaries

The goal of this chapter is to provide the necessary background from mathematical logic, formal languages, and complexity theory.

2.1 Background from Mathematical Logic

We now briefly review some standard definitions from mathematical logic.

Definition 2.1. *A* vocabulary σ *is a collection of* constant *symbols (denoted* $c_1, \ldots, c_n, \ldots$*),* relation, *or* predicate, *symbols* ($P_1, \ldots, P_n, \ldots$) *and* function *symbols* ($f_1, \ldots, f_n, \ldots$)*. Each relation and function symbol has an associated arity.*

A σ-structure *(also called a* model*)*

$$\mathfrak{A} = \langle A, \{c_i^{\mathfrak{A}}\}, \{P_i^{\mathfrak{A}}\}, \{f_i^{\mathfrak{A}}\} \rangle$$

consists of a universe A together with an interpretation of

- *each constant symbol* c_i *from* σ *as an element* $c_i^{\mathfrak{A}} \in A$*;*
- *each k-ary relation symbol* P_i *from* σ *as a k-ary relation on A; that is, a set* $P_i^{\mathfrak{A}} \subseteq A^k$*; and*
- *each k-ary function symbol* f_i *from* σ *as a function* $f_i^{\mathfrak{A}} : A^k \to A$.

A structure $\mathfrak{A}$ *is called* finite *if its universe A is a finite set. The universe of a structure is typically denoted by a Roman letter corresponding to the name of the structure; that is, the universe of* $\mathfrak{A}$ *is A, the universe of* $\mathfrak{B}$ *is B, and so on. We shall also occasionally write* $x \in \mathfrak{A}$ *instead of* $x \in A$.

For example, if σ has constant symbols 0, 1, a binary relation symbol $<$, and two binary function symbols $\cdot$ and $+$, then one possible structure for σ is the real field $\mathbf{R} = \langle \mathbb{R}, 0^{\mathbf{R}}, 1^{\mathbf{R}}, <^{\mathbf{R}}, +^{\mathbf{R}}, \cdot^{\mathbf{R}} \rangle$, where $0^{\mathbf{R}}, 1^{\mathbf{R}}, <^{\mathbf{R}}, +^{\mathbf{R}}, \cdot^{\mathbf{R}}$ have the expected meaning. Quite often – in fact, typically – we shall omit the superscript with the name of the structure, using the same symbol for both a symbol in the vocabulary, and its interpretation in a structure. For example, we shall write $\mathbf{R} = \langle \mathbb{R}, 0, 1, <, +, \cdot \rangle$ for the real field.

A few notes on restrictions on vocabularies are in order. Constants can be treated as functions of arity zero; however, we often need them separately, as in the finite case, we typically restrict vocabularies to *relational* ones: such vocabularies contain only relation symbols and constants. This is not a serious restriction, as first-order logic defines, for each k-ary function f, its graph, which is a $(k+1)$-ary relation $\{(\vec{x}, f(\vec{x})) \mid \vec{x} \in A^k\}$. A vocabulary that consists exclusively of relation symbols (i.e., does not have constant and function symbols) is called *purely relational.*

Unless stated explicitly otherwise, we shall assume that:

- any vocabulary σ is at most countable;
- when we deal with finite structures, vocabularies σ are finite and relational.

If σ is a relational vocabulary, then STRUCT$[\sigma]$ denotes the class of all finite σ-structures.

Next, we define first-order (FO) formulae, free and bound variables, and the semantics of FO formulae.

Definition 2.2. *We assume a countably infinite set of variables. Variables will be typically denoted by $x, y, z, \ldots$, with subscripts and superscripts. We inductively define terms and formulae of the* first-order predicate calculus *over vocabulary σ as follows:*

- *Each variable x is a term.*
- *Each constant symbol c is a term.*
- *If $t_1, \ldots, t_k$ are terms and f is a k-ary function symbol, then $f(t_1, \ldots, t_k)$ is a term.*
- *If t_1, t_2 are terms, then $t_1 = t_2$ is an (atomic) formula.*
- *If $t_1, \ldots, t_k$ are terms and P is a k-ary relation symbol, then $P(t_1, \ldots, t_k)$ is an (atomic) formula.*
- *If φ_1, φ_2 are formulae, then $\varphi_1 \wedge \varphi_2$, $\varphi_1 \vee \varphi_2$, and $\neg\varphi_1$ are formulae.*
- *If φ is a formula, then $\exists x\varphi$ and $\forall x\varphi$ are formulae.*

A formula that does not use existential ($\exists$) and universal ($\forall$) quantifiers is called *quantifier-free.*

We shall use the standard shorthand $\varphi \rightarrow \psi$ for $\neg\varphi \vee \psi$ and $\varphi \leftrightarrow \psi$ for $(\varphi \rightarrow \psi) \wedge (\psi \rightarrow \varphi)$.

Free variables of a formula or a term are defined as follows:

- The only free variable of a term x is x; a constant term c does not have free variables.
- Free variables of $t_1 = t_2$ are the free variables of t_1 and t_2; free variables of $P(t_1, \ldots, t_k)$ or $f(t_1, \ldots, t_k)$ are the free variables of $t_1, \ldots, t_k$.
- Negation ($\neg$) does not change the list of free variables; the free variables of $\varphi_1 \vee \varphi_2$ (and of $\varphi_1 \wedge \varphi_2$) are the free variables of φ_1 and φ_2.
- Free variables of $\forall x \varphi$ and $\exists x \varphi$ are the free variables of φ except x.

Variables that are not free are called bound.

If $\vec{x}$ is the tuple of all the free variables of φ, we write $\varphi(\vec{x})$. A sentence is a formula without free variables. We often use capital Greek letters for sentences.

Given a set of formulae $\mathcal{S}$, formulae constructed from formulae in $\mathcal{S}$ using only the Boolean connectives $\vee$, $\wedge$, and $\neg$ are called *Boolean combinations* of formulae in $\mathcal{S}$.

Given a σ-structure $\mathfrak{A}$, we define inductively for each term t with free variables $(x_1, \ldots, x_n)$ the value $t^{\mathfrak{A}}(\vec{a})$, where $\vec{a} \in A^n$, and for each formula $\varphi(x_1, \ldots, x_n)$, the notion of $\mathfrak{A} \models \varphi(\vec{a})$ (i.e., $\varphi(\vec{a})$ is true in $\mathfrak{A}$).

- If t is a constant symbol c, then the value of t in $\mathfrak{A}$ is $c^{\mathfrak{A}}$.
- If t is a variable x_i, then the value of $t^{\mathfrak{A}}(\vec{a})$ is a_i.
- If $t = f(t_1, \ldots, t_k)$, then the value of $t^{\mathfrak{A}}(\vec{a})$ is $f^{\mathfrak{A}}(t_1^{\mathfrak{A}}(\vec{a}), \ldots, t_k^{\mathfrak{A}}(\vec{a}))$.
- If $\varphi \equiv (t_1 = t_2)$, then $\mathfrak{A} \models \varphi(\vec{a})$ iff $t_1^{\mathfrak{A}}(\vec{a}) = t_2^{\mathfrak{A}}(\vec{a})$.
- If $\varphi \equiv P(t_1, \ldots, t_k)$, then $\mathfrak{A} \models \varphi(\vec{a})$ iff $(t_1^{\mathfrak{A}}(\vec{a}), \ldots, t_k^{\mathfrak{A}}(\vec{a})) \in P^{\mathfrak{A}}$.
- $\mathfrak{A} \models \neg\varphi(\vec{a})$ iff $\mathfrak{A} \models \varphi(\vec{a})$ does not hold.
- $\mathfrak{A} \models \varphi_1(\vec{a}) \wedge \varphi_2(\vec{a})$ iff $\mathfrak{A} \models \varphi_1(\vec{a})$ and $\mathfrak{A} \models \varphi_2(\vec{a})$.
- $\mathfrak{A} \models \varphi_1(\vec{a}) \vee \varphi_2(\vec{a})$ iff $\mathfrak{A} \models \varphi_1(\vec{a})$ or $\mathfrak{A} \models \varphi_2(\vec{a})$.
- If $\psi(\vec{x}) \equiv \exists y \varphi(y, \vec{x})$, then $\mathfrak{A} \models \psi(\vec{a})$ iff $\mathfrak{A} \models \varphi(a', \vec{a})$ for some $a' \in A$.
- If $\psi(\vec{x}) \equiv \forall y \varphi(y, \vec{x})$, then $\mathfrak{A} \models \psi(\vec{a})$ iff $\mathfrak{A} \models \varphi(a', \vec{a})$ for all $a' \in A$.

If $\mathfrak{A} \in \text{STRUCT}[\sigma]$ and $A_0 \subseteq A$, the *substructure* of A generated by A_0 is a σ-structure $\mathfrak{B}$ whose universe is $B = A_0 \cup \{c^{\mathfrak{A}} \mid c \text{ a constant symbol in } \sigma\}$, with $c^{\mathfrak{B}} = c^{\mathfrak{A}}$ for every c, and with each k-ary relation R interpreted as the restriction of $R^{\mathfrak{A}}$ to B: that is, $R^{\mathfrak{B}} = R^{\mathfrak{A}} \cap B^k$.

Let σ' be a vocabulary disjoint from σ. Let $\mathfrak{A}$ be a σ-structure, and let $\mathfrak{A}'$ be a σ'-structure with the same universe A. We then write $(\mathfrak{A}, \mathfrak{A}')$ for a $\sigma \cup \sigma'$-structure on A in which all constant and relation symbols in σ are interpreted as in $\mathfrak{A}$, and all constant and relation symbols in σ' are interpreted as in $\mathfrak{A}'$.

One of the most common instances of such an expansion is when σ' only contains constant symbols; in this case, the expansion allows us to go back and forth between formulae and sentences, which will be very convenient when we talk about games and expressiveness of formulas as well as sentences.

From now on, we shall use the notation σ_n for the expansion of vocabulary σ with n new constant symbols $c_1, \ldots, c_n$.

Let $\varphi(x_1, \ldots, x_n)$ be a formula in vocabulary σ. Consider a σ_n sentence Φ obtained from φ by replacing each x_i with c_i, $i \leq n$. Let $(a_1, \ldots, a_n) \in A^n$. Then one can easily show (the proof is left as an exercise) the following:

Lemma 2.3. $\mathfrak{A} \models \varphi(a_1, \ldots, a_n)$ *iff* $(\mathfrak{A}, a_1, \ldots, a_n) \models \Phi$. □

This correspondence is rather convenient: we often do not need separate treatment for sentences and formulae with free variables.

Most classical theorems from model theory fail in the finite case, as will be seen later. However, two fundamental facts – compactness and the Löwenheim-Skolem theorem – will be used to prove results about finite models. To state them, we need the following definition.

Definition 2.4. *A* theory *(over σ) is a set of sentences. A σ-structure $\mathfrak{A}$ is a model of a theory T iff for every sentence Φ of T, the structure $\mathfrak{A}$ is a model of Φ; that is, $\mathfrak{A} \models \Phi$. A theory T is called* consistent *if it has a model.*

Theorem 2.5 (Compactness). *A theory T is consistent iff every finite subset of T is consistent.* □

Theorem 2.6 (Löwenheim-Skolem). *If T has an infinite model, then it has a countable model.* □

In general, Theorem 2.1 allows one to construct a model of cardinality $\max\{\omega, |\sigma|\}$, but we shall never deal with uncountable vocabularies here.

Compactness follows from the *completeness theorem*, stating that $T \models \varphi$ iff $T \vdash \varphi$, where $\vdash$ refers to a derivation in a formal proof system. We shall see some other important corollaries of this result.

We say that a sentence Φ is *satisfiable* if it has a model, and it is *valid* if it is true in every structure. These notions are closely related: Φ is not valid iff $\neg\Phi$ is satisfiable. It follows from completeness that the set of valid sentences is recursively enumerable (if you forgot the definition of recursively enumerable, it is given in the next section). This is true when one considers validity with respect to arbitrary models; we shall see later that validity over finite models in not recursively enumerable.

Given two structures $\mathfrak{A}$ and $\mathfrak{B}$ of a relational vocabulary σ, a *homomorphism* between them is a mapping $h : A \to B$ such that for each constant symbol c in σ, we have $h(c^{\mathfrak{A}}) = c^{\mathfrak{B}}$, and for each k-ary relation symbol R and a tuple $(a_1, \ldots, a_k) \in R^{\mathfrak{A}}$, the tuple $(h(a_1), \ldots, h(a_k))$ is in $R^{\mathfrak{B}}$. A bijective homomorphism h whose inverse is also a homomorphism is called an *isomorphism.* If there is an isomorphism between two structures $\mathfrak{A}$ and $\mathfrak{B}$, we say that they are isomorphic, and we write $\mathfrak{A} \cong \mathfrak{B}$.

Next, we need the following basic definition of *m-ary queries.*

Definition 2.7 (Queries). *An m-ary query, $m \geq 0$, on σ-structures, is a mapping Q that associates with each structure $\mathfrak{A}$ a subset of A^m, such that Q is closed under isomorphism: if $\mathfrak{A} \cong \mathfrak{B}$ via isomorphism $h : A \to B$, then $Q(\mathfrak{B}) = h(Q(\mathfrak{A}))$.*

We say that Q is definable in a logic $\mathcal{L}$ if there is a formula $\varphi(x_1, \ldots, x_m)$ of $\mathcal{L}$ in vocabulary σ such that for every $\mathfrak{A}$,

$$Q(\mathfrak{A}) \;=\; \{(a_1, \ldots, a_m) \in A^m \mid \mathfrak{A} \models \varphi(a_1, \ldots, a_m)\}.$$

If Q is definable by φ, we shall also write $\varphi(\mathfrak{A})$ instead of $Q(\mathfrak{A})$. Furthermore, for a formula $\varphi(\vec{x}, \vec{y})$, we write $\varphi(\mathfrak{A}, \vec{b})$ for $\{\vec{a} \in A^{|\vec{a}|} \mid \mathfrak{A} \models \varphi(\vec{a}, \vec{b})\}$.

A very important special case is that of $m = 0$. We assume that A^0 is a one-element set, and there are only two subsets of A^0. Hence, a 0-ary query is just a mapping from σ-structures to a two-element set, which can be assumed to contain *true* and *false.* Such queries will be called *Boolean.* A Boolean query can be associated with a subset $\mathcal{C} \subseteq \mathrm{STRUCT}[\sigma]$ closed under isomorphism:

$$\mathfrak{A} \in \mathcal{C} \;\text{ iff }\; Q(\mathfrak{A}) = \mathit{true}.$$

Such a query Q is definable in a logic $\mathcal{L}$ if there is an $\mathcal{L}$-sentence Φ such that $Q(\mathfrak{A}) = \mathit{true}$ iff $\mathfrak{A} \models \Phi$.

An example of a binary ($m = 2$) query is the transitive closure of a graph. An example of a unary ($m = 1$) query is the set of all isolated nodes in a graph. An example of a Boolean ($m = 0$) query on graphs is planarity.

2.2 Background from Automata and Computability Theory

In this section we briefly review some basic concepts of finite automata and computability theory.

Let Σ be a finite nonempty alphabet; that is, a finite set of symbols. The set of all finite strings over Σ will be denoted by Σ^*. We shall use $s \cdot s'$ to denote concatenation of two strings s and s'. The empty string is denoted by ϵ. One commonly refers to subsets of Σ^* as *languages.*

A *nondeterministic finite automaton* is a tuple $\mathcal{A} = (Q, \Sigma, q_0, F, \delta)$ where Q is a finite set of states, Σ is a finite alphabet, $q_0 \in Q$ is the initial state, $F \subseteq Q$ is the set of final states, and $\delta : Q \times \Sigma \to 2^Q$ is the transition function. An automaton is *deterministic* if $|\delta(q, a)| = 1$ for every q and a; that is, if δ can be viewed as a function $Q \times \Sigma \to Q$.

Let $s = a_1 a_2 \ldots a_n$ be a string in Σ^*. Define a *run* of $\mathcal{A}$ on s as a mapping $r : \{1, \ldots, n\} \to Q$ such that

- $r(1) \in \delta(q_0, a_1)$ (or $r(1) = \delta(q_0, a_1)$ if $\mathcal{A}$ is deterministic), and
- $r(i+1) \in \delta(r(i), a_{i+1})$ (or $r(i+1) = \delta(r(i), a_{i+1})$ if $\mathcal{A}$ is deterministic).

We say that a run is accepting if $r(n) \in F$, and that $\mathcal{A}$ *accepts* s if there is an accepting run (for the case of a deterministic automaton, there is exactly one run for each string). The set of all strings accepted by $\mathcal{A}$ is denoted by $L(\mathcal{A})$.

A language L is called *regular* if there is a nondeterministic finite automaton $\mathcal{A}$ such that $L = L(\mathcal{A})$. It is well known that for any regular language L, one can find a deterministic finite automaton $\mathcal{A}$ such that $L = L(\mathcal{A})$.

Turing machines are the most general computing devices. Formally, a Turing machine M is a tuple $(Q, \Sigma, \Delta, \delta, q_0, Q_a, Q_r)$, where

- Q is a finite set of states;
- Σ is a finite *input* alphabet;
- Δ is a finite *tape* alphabet; it contains Σ and a designated blank symbol '_';
- $\delta : Q \times \Delta \to 2^{Q \times \Delta \times \{\ell, r\}}$ is the transition function;
- $q_0 \in Q$ is the initial state;
- Q_a and Q_r are the sets of accepting and rejecting states respectively; we require that $Q_a \cap Q_r = \emptyset$. We refer to states in $Q_a \cup Q_r$ as the *halting* states.

A Turing machine is called *deterministic* if $|\delta(q, a)| = 1$ for every q, a; that is, if δ can be viewed as a function $Q \times \Delta \to Q \times \Delta \times \{\ell, r\}$.

We assume that Turing machines have a one-way infinite tape, and one head. A *configuration* of a Turing machine M specifies the contents of the tape, the state, and the position of the head as follows. Let the tape contain symbols $w_1, w_2, \ldots$, where $w_i \in \Delta$ is the symbol in the ith position of the tape. Assume that the head is in position j, and $n \geq j$ is such that for all $n' > n$, $w_{n'} = \_$ (the blank symbol). If M is in state q, we denote this configuration by $w_1 w_2 \ldots w_{j-1} q w_j \ldots w_n$. We define the relation $C \vdash_\delta C'$ as follows. If $C = s \cdot q \cdot a \cdot s'$, where $s, s' \in \Delta^*$, $a \in \Delta$, and $q \notin Q_a \cup Q_r$, then

- if $(q', b, \ell) \in \delta(q, a)$, then $C \vdash_\delta s_0 \cdot q' \cdot c \cdot b \cdot s'$, where $s = s_0 \cdot c$ (that is, a is replaced by b, the new state is q', and the head moves left; if $s = \epsilon$, then $C \vdash_\delta q' \cdot b \cdot s'$), and
- if $(q', b, r) \in \delta(q, a)$, then $C \vdash_\delta s \cdot b \cdot q' \cdot s'$ (that is, a is replaced by b, the new state is q', and the head moves right).

A configuration $s \cdot q \cdot s'$ is accepting if $q \in Q_a$, and rejecting if $q \in Q_r$.

Suppose we have a string $s \in \Sigma^*$. The initial configuration $C(s)$ corresponding to this string is $q_0 \cdot s$; that is, the state is q_0, the head points to the first position of s, and the tape contains s followed by blanks. We say that s is accepted by M if there is a sequence of configurations $C_0, C_1, \ldots, C_n$ such that $C_0 = C(s)$, $C_i \vdash_\delta C_{i+1}, i < n$, and C_n is an accepting configuration. The set of all strings accepted by M is denoted by $L(M)$.

We call a subset L of Σ^* *recursively enumerable*, or *r.e.* for short, if there is a Turing machine M such that $L = L(M)$.

Notice that in general, there are three possibilities for computations by a Turing machine M on input s: M accepts s, or M eventually enters a rejecting state, or M loops; that is, it never enters a halting state. We call a Turing machine *halting* if the last outcome is impossible. In other words, on every input, M eventually enters a halting state.

We call a subset L of Σ^* *recursive* if there is a halting Turing machine M such that $L = L(M)$. Halting Turing machines can be seen as deciders for some sets L: for every string s, M eventually enters either an accepting or a rejecting state, which decides whether $s \in L$. For that reason, one sometimes uses *decidable* instead of *recursive*. When we speak of decidable problems, we mean that a suitable encoding of the problem as a subset of Σ^* for some finite Σ is decidable.

A canonical example of an undecidable problem is the *halting problem*: given a Turing machine M and an input w, does M halt on w (i.e., eventually enters a halting state)? In general, any nontrivial property of recursively enumerable sets is undecidable. One result we shall use later is that it is undecidable whether a given Turing machine halts on the empty input.

2.3 Background from Complexity Theory

Let L be a language accepted by a halting Turing machine M. Assume that for some function $f : \mathbb{N} \to \mathbb{N}$, it is the case that the number of transitions M makes before accepting or rejecting a string s is at most $f(|s|)$, where $|s|$ is the length of s. If M is deterministic, then we write $L \in \text{DTIME}(f)$; if M is nondeterministic, then we write $L \in \text{NTIME}(f)$.

We define the class PTIME of polynomial-time computable problems as

$$\text{PTIME} \quad = \quad \bigcup_{k \in \mathbb{N}} \text{DTIME}(n^k),$$

and the class NP of problems computable by nondeterministic polynomial time Turing machines as

$$\text{NP} \quad = \quad \bigcup_{k \in \mathbb{N}} \text{NTIME}(n^k).$$

The class coNP is defined as the class of languages whose complements are in NP. Notice that PTIME is closed under complementation, but this is not clear in the case of NP. We have PTIME $\subseteq$ NP $\cap$ coNP, but it is not known whether the containment is proper, and whether NP equals coNP.

Now assume that $f(n) \geq n$ for all $n \in \mathbb{N}$. Define DSPACE(f) as the class of languages L that are accepted by deterministic halting Turing machines M such that for every string s, the length of the longest configuration of M that occurs during the computation on s is at most $f(|s|)$. In other words, M does not use more than $f(|s|)$ cells of the tape. Similarly, we define the class NSPACE(f) by using nondeterministic machines. We then let

$$\text{PSPACE} = \bigcup_{k \in \mathbb{N}} \text{DSPACE}(n^k).$$

In the case of space complexity, the nondeterministic case collapses to the deterministic one: by Savitch's theorem, PSPACE $= \bigcup_{k \in \mathbb{N}} \text{NSPACE}(n^k)$.

To define space complexity for sublinear functions f, we use a model of Turing machines with a work tape. In such a model, a machine M has two tapes, and two heads. The first tape is the input tape: it stores the input, and the machine cannot write on it (but can move the head). The second tape is the work tape, which operates as the normal tape of a Turing machine. We define the class NLOG as the class of languages accepted by such nondeterministic machines where the size of the work tape does not exceed $O(\log |s|)$, on the input s. Likewise, we define the class DLOG as the class of language accepted by deterministic machines with the work tape, where at most $O(\log |s|)$ cells of the work tape are used.

Finally, we define the *polynomial hierarchy* PH. Let $\Sigma_0^p = \Pi_0^p =$ PTIME. Define inductively $\Sigma_i^p = \text{NP}^{\Sigma_{i-1}^p}$, for $i \geq 1$. That is, languages in Σ_i^p are those accepted by a nondeterministic Turing machine running in polynomial time such that this machine can make "calls" to another machine that computes a language in Σ_{i-1}^p. Such a call is assumed to have unit cost. We define the class Π_i^p as the class of languages whose complements are in Σ_i^p.

Notice that $\Sigma_1^p = $ NP and $\Pi_1^p =$ coNP. We define the polynomial hierarchy as

$$\text{PH} = \bigcup_{i \in \mathbb{N}} \Sigma_i^p = \bigcup_{i \in \mathbb{N}} \Pi_i^p.$$

This will be sufficient for our purposes, but there is another interesting definition of PH in terms of alternating Turing machines.

The relationship between the complexity classes we introduced is as follows:

$$\text{DLOG} \subseteq \text{NLOG} \subseteq \text{PTIME} \subseteq \left\{ \begin{matrix} \text{NP} \\ \text{coNP} \end{matrix} \right\} \subseteq \text{PH} \subseteq \text{PSPACE}.$$

None of the containments of any two consecutive classes in this sequence is known to be proper, although it is known that NLOG $\subsetneq$ PSPACE.

We shall also refer to two classes based on exponential running time. These are

$$\text{EXPTIME} = \bigcup_{k \in \mathbb{N}} \text{DTIME}(2^{n^k}) \text{ and } \text{NEXPTIME} = \bigcup_{k \in \mathbb{N}} \text{NTIME}(2^{n^k}).$$

Both of these contain PSPACE.

Later in the book we shall see a number of other complexity classes, in particular circuit-based classes AC^0 and TC^0 (which are both contained in DLOG).

2.4 Bibliographic Notes

Standard mathematical logic texts are Enderton [66], Ebbinghaus, Flum, and Thomas [61], and van Dalen [241]; infinite model theory is the subject of Chang and Keisler [35], Hodges [125], and Poizat [201]. Good references on complexity theory are Papadimitriou [195], Johnson [139], and Du and Ko [59]. For the basics on automata and computability, see Hopcroft and Ullman [126], Khoussainov and Nerode [145], and Sipser [221].

3

Ehrenfeucht-Fraïssé Games

We start this chapter by giving a few examples of inexpressibility proofs, using the standard model-theoretic machinery (compactness, the Löwenheim-Skolem theorem). We then show that this machinery is not generally applicable in the finite model theory context, and introduce the notion of Ehrenfeucht-Fraïssé games for first-order logic. We prove the Ehrenfeucht-Fraïssé theorem, characterizing the expressive power of FO via games, and introduce the notion of types, which will be central throughout the book.

3.1 First Inexpressibility Proofs

How can one prove that a certain property is inexpressible in FO? Certainly logicians must have invented tools for proving such results, and we shall now see a few examples. The problem is that these tools are not particularly well suited to the finite context, so in the next section, we introduce a different technique that will be used for FO and other logics over finite models.

In the first example, we deal with connectivity: given a graph G, is it connected? Recall that a graph with an edge relation E is connected if for every two nodes a, b one can find a number n and nodes $c_1, \ldots, c_n \in V$ such that $(a, c_1), (c_1, c_2), \ldots, (c_n, b)$ are all edges in the graph. A standard model-theoretic argument below shows that connectivity is not FO-definable.

Proposition 3.1. *Connectivity of arbitrary graphs is not FO-definable.*

Proof. Assume that connectivity is definable by a sentence Φ, over vocabulary $\sigma = \{E\}$. Let σ_2 expand σ with two constant symbols, c_1 and c_2. For every n, let Ψ_n be the sentence

$$\neg(\exists x_1 \ldots \exists x_n \; (E(c_1, x_1) \wedge E(x_1, x_2) \wedge \ldots \wedge E(x_n, c_2))),$$

saying that there is no path of length $n + 1$ from c_1 to c_2.

Let T be the theory

$$\{\Psi_n \mid n > 0\} \ \cup \ \{\neg(c_1 = c_2), \neg E(c_1, c_2)\} \ \cup \ \{\Phi\}.$$

We claim that T is consistent. By compactness, we have to show that every finite subset $T' \subseteq T$ is consistent. Indeed, let N be such that for all $\Psi_n \in T'$, $n < N$. Then a connected graph in which the shortest path from c_1 to c_2 has length $N + 1$ is a model of T'.

Since T is consistent, it has a model. Let $\mathfrak{G}$ be a model of T. Then $\mathfrak{G}$ is connected, but there is no path from c_1 to c_2 of length n, for any n. This contradiction shows that connectivity is not FO-definable. □

Does the proof above tell us that FO, or relational calculus, cannot express the connectivity test over *finite* graphs? Unfortunately, it does not. While connectivity is not definable in FO over *arbitrary* graphs, the proof above leaves open the possibility that there is a first-order sentence that correctly tests connectivity only for *finite* graphs. But to prove the desired result for relational calculus, one has to show inexpressibility of connectivity over *finite* graphs.

Can one modify the proof above for finite models? An obvious way to do so would be to use compactness over finite graphs (i.e., if every finite subset of T has a finite model, then T has a finite model), assuming this holds. Unfortunately, this turns out not to be the case.

Proposition 3.2. *Compactness fails over finite models: there is a theory T such that*

1. *T has no finite models, and*
2. *every finite subset of T has a finite model.*

Proof. We assume that $\sigma = \emptyset$, and define λ_n as a sentence stating that the universe has at least n distinct elements:

$$\lambda_n \equiv \exists x_1 \ldots \exists x_n \bigwedge_{i \neq j} \neg(x_i = x_j). \tag{3.1}$$

Now $T = \{\lambda_n | n \geq 0\}$. Clearly, T has no finite model, but for each finite subset $\{\lambda_{n_1}, \ldots, \lambda_{n_k}\}$ of T, a set whose cardinality exceeds all the n_i's is a model. □

However, sometimes a compactness argument works nicely in the finite context. We now consider a very important property, which will be seen many times in this book. We want to test if the cardinality of the universe is even. That is, we are interested in query EVEN defined as

$$\text{EVEN}(\mathfrak{A}) = \mathit{true} \quad \text{iff} \quad |A| \bmod 2 = 0.$$

Note that this only makes sense over finite models; for infinite $\mathfrak{A}$ the value of EVEN could be arbitrary.

Proposition 3.3. *Assume that $\sigma = \emptyset$. Then* EVEN *is not* FO*-definable.*

Proof. Suppose EVEN is definable by a sentence Φ. Consider sentences λ_n (3.1) from the proof of Proposition 3.2 and two theories:

$$T_1 \;=\; \{\Phi\} \cup \{\lambda_k \mid k > 0\}, \qquad T_2 \;=\; \{\neg\Phi\} \cup \{\lambda_k \mid k > 0\}.$$

By compactness, both are consistent. These theories only have infinite models, so by the Löwenheim-Skolem theorem, both have countable models, $\mathfrak{A}_1$ and $\mathfrak{A}_2$. Since $\sigma = \emptyset$, the structures $\mathfrak{A}_1$ and $\mathfrak{A}_2$ are just countable sets, and hence isomorphic. Thus, we have two isomorphic models, $\mathfrak{A}_1$ and $\mathfrak{A}_2$, with $\mathfrak{A}_1 \models \Phi$ and $\mathfrak{A}_2 \models \neg\Phi$. This contradiction proves the result. □

This is nice, but there is a small problem: we assumed that the vocabulary is empty. But what if we have, for example, $\sigma = \{<\}$, and we want to prove that evenness of ordered sets is not definable? In this case we would expand T_1 and T_2 with axioms of ordered sets, and we would obtain, by compactness and Löwenheim-Skolem, two countable linear orderings $\mathfrak{A}_1$ and $\mathfrak{A}_2$, one a model of Φ, the other a model of $\neg\Phi$. This is a dead end, since two arbitrary countable linear orders need not be isomorphic (in fact, some can be distinguished by first-order sentences: think, for example, of a discrete order like $\langle \mathbb{N}, < \rangle$ and a dense one like $\langle \mathbb{Q}, < \rangle$).

Thus, while traditional tools from model theory may help us prove some results, they are often not sufficient for proving results about finite models. We shall examine, in subsequent chapters, tools designed for proving expressivity bounds in the finite case.

As an introduction to these tools, let us revisit the proof of Proposition 3.3. In the proof, we constructed two models, $\mathfrak{A}_1$ and $\mathfrak{A}_2$, that agree on *all* FO sentences (since they are isomorphic), and yet compactness tells us that they disagree on Φ, which was assumed to define EVEN – hence EVEN is not first-order.

Can we extend this technique to prove inexpressibility results over finite models? The most straightforward attempt to do so fails due to the following.

Lemma 3.4. *For every finite structure $\mathfrak{A}$, there is a sentence $\Phi_{\mathfrak{A}}$ such that $\mathfrak{B} \models \Phi_{\mathfrak{A}}$ iff $\mathfrak{B} \cong \mathfrak{A}$.*

Proof. Assume without loss of generality that $\mathfrak{A}$ is a graph: $\sigma = \{E\}$. Let $\mathfrak{A} = \langle \{a_1, \ldots, a_n\}, E \rangle$. Define $\Phi_{\mathfrak{A}}$ as

$$\exists x_1 \ldots \exists x_n \left(\begin{array}{l} \big(\bigwedge_{i \neq j} \neg(x_i = x_j)\big) \;\wedge\; \big(\forall y \bigvee_i y = x_i\big) \\ \wedge\; \big(\bigwedge_{(a_i,a_j) \in E} E(x_i, x_j)\big) \;\wedge\; \big(\bigwedge_{(a_i,a_j) \notin E} \neg E(x_i, x_j)\big) \end{array} \right).$$

Then $\mathfrak{B} \models \Phi_{\mathfrak{A}}$ iff $\mathfrak{B} \cong \mathfrak{A}$. □

In particular, every two finite structures that agree on all FO sentences are isomorphic, and hence agree on any Boolean query (as Boolean queries are closed under isomorphism).

The idea that is prevalent in inexpressibility proofs in finite model theory is, nevertheless, very close to the original idea of finding structures $\mathfrak{A}$ and $\mathfrak{B}$ that agree on all FO sentences but disagree on a given query. But instead of two structures, $\mathfrak{A}$ and $\mathfrak{B}$, we consider two *families* of structures, $\{\mathfrak{A}_k \mid k \in \mathbb{N}\}$ and $\{\mathfrak{B}_k \mid k \in \mathbb{N}\}$, and instead of all FO sentences, we consider a certain partition of FO sentences into infinitely many classes.

In general, the methodology is as follows. Suppose we want to prove that a property $\mathcal{P}$ is not expressible in a logic $\mathcal{L}$. We then partition the set of all sentences of $\mathcal{L}$ into countably many classes, $\mathcal{L}[0], \mathcal{L}[1], \ldots, \mathcal{L}[k], \ldots$ (we shall see in Sect. 3.3 how to do it), and find two families of structures, $\{\mathfrak{A}_k \mid k \in \mathbb{N}\}$ and $\{\mathfrak{B}_k \mid k \in \mathbb{N}\}$, such that

- $\mathfrak{A}_k \models \Phi$ iff $\mathfrak{B}_k \models \Phi$ for every $\mathcal{L}[k]$ sentence Φ; and
- $\mathfrak{A}_k$ has property $\mathcal{P}$, but $\mathfrak{B}_k$ does not.

Clearly, this would show $\mathcal{P} \notin \mathcal{L}$; it "only" remains to show what $\mathcal{L}[k]$ is, and give techniques that help us prove that two structures agree on $\mathcal{L}[k]$. We shall do precisely that in the rest of the chapter, for the case of $\mathcal{L} = \text{FO}$, and later for other logics.

3.2 Definition and Examples of Ehrenfeucht-Fraïssé Games

Ehrenfeucht-Fraïssé games give us a nice tool for describing expressiveness of logics over finite models. In general, games are applicable for both finite and infinite models (at least for FO), but we have seen that in the infinite case we have a number of more powerful tools. In fact, in some model theory texts, Ehrenfeucht-Fraïssé games are only briefly mentioned (or even appear only as exercises), but in the finite case, their applicability makes them a central notion.

The idea of the game – for FO and other logics as well – is almost invariably the same. There are two players, called the *spoiler* and the *duplicator* (or, less imaginatively, player I and player II). The board of the game consists of two structures, say $\mathfrak{A}$ and $\mathfrak{B}$. The goal of the spoiler is to show that these two structures are different; the goal of the duplicator is to show that they are the same.

In the classical Ehrenfeucht-Fraïssé game, the players play a certain number of rounds. Each round consists of the following steps:

1. The spoiler picks a structure ($\mathfrak{A}$ or $\mathfrak{B}$).

2. The spoiler makes a move by picking an element of that structure: either $a \in \mathfrak{A}$ or $b \in \mathfrak{B}$.
3. The duplicator responds by picking an element in the other structure.

An illustration is given in Fig. 3.1. The spoiler's moves are shown as filled circles, and the duplicator's moves as empty circles. In the first round, the spoiler picks $\mathfrak{B}$ and selects $b_1 \in \mathfrak{B}$; the duplicator responds by $a_1 \in \mathfrak{A}$. In the next round, the spoiler changes structures and picks $a_2 \in \mathfrak{A}$; the duplicator responds by $b_2 \in \mathfrak{B}$. In the third round the spoiler plays $b_3 \in \mathfrak{B}$; the response of the duplicator is $a_3 \in \mathfrak{A}$.

Since there is a game, someone must win it. To define the winning condition we need a crucial definition of a *partial isomorphism*. Recall that all finite structures have a relational vocabulary (no function symbols).

Definition 3.5 (Partial isomorphism). *Let $\mathfrak{A}$, $\mathfrak{B}$ be two σ-structures, where σ is relational, and $\vec{a} = (a_1, \ldots, a_n)$ and $\vec{b} = (b_1, \ldots, b_n)$ two tuples in $\mathfrak{A}$ and $\mathfrak{B}$ respectively. Then $(\vec{a}, \vec{b})$ defines a partial isomorphism between $\mathfrak{A}$ and $\mathfrak{B}$ if the following conditions hold:*

- *For every $i, j \leq n$,*
$$a_i = a_j \quad \textit{iff} \quad b_i = b_j.$$
- *For every constant symbol c from σ, and every $i \leq n$,*
$$a_i = c^{\mathfrak{A}} \quad \textit{iff} \quad b_i = c^{\mathfrak{B}}.$$
- *For every k-ary relation symbol P from σ and every sequence $(i_1, \ldots, i_k)$ of (not necessarily distinct) numbers from $[1, n]$,*
$$(a_{i_1}, \ldots, a_{i_k}) \in P^{\mathfrak{A}} \quad \textit{iff} \quad (b_{i_1}, \ldots, b_{i_k}) \in P^{\mathfrak{B}}.$$

In the absence of constant symbols, this definition says that the mapping $a_i \mapsto b_i, i \leq n$, is an isomorphism between the substructures of $\mathfrak{A}$ and $\mathfrak{B}$ generated by $\{a_1, \ldots, a_n\}$ and $\{b_1, \ldots, b_n\}$, respectively.

After n rounds of an Ehrenfeucht-Fraïssé game, we have moves $(a_1, \ldots, a_n)$ and $(b_1, \ldots, b_n)$. Let $c_1, \ldots, c_l$ be the constant symbols in σ; then $\vec{c}^{\,\mathfrak{A}}$ denotes $(c_1^{\mathfrak{A}}, \ldots, c_l^{\mathfrak{A}})$ and likewise for $\vec{c}^{\,\mathfrak{B}}$. We say that $(\vec{a}, \vec{b})$ is a *winning position for the duplicator* if
$$((\vec{a}, \vec{c}^{\,\mathfrak{A}}), \ (\vec{b}, \vec{c}^{\,\mathfrak{B}}))$$
is a partial isomorphism between $\mathfrak{A}$ and $\mathfrak{B}$. In other words, the map that sends each a_i into b_i and each $c_j^{\mathfrak{A}}$ into $c_j^{\mathfrak{B}}$ is an isomorphism between the substructures of $\mathfrak{A}$ and $\mathfrak{B}$ generated by $\{a_1, \ldots, a_n, c_1^{\mathfrak{A}}, \ldots, c_l^{\mathfrak{A}}\}$ and $\{b_1, \ldots, b_n, c_1^{\mathfrak{B}}, \ldots, c_l^{\mathfrak{B}}\}$ respectively.

We say that the duplicator has an *n-round winning strategy in the Ehrenfeucht-Fraïssé game on $\mathfrak{A}$ and $\mathfrak{B}$* if the duplicator can play in a way

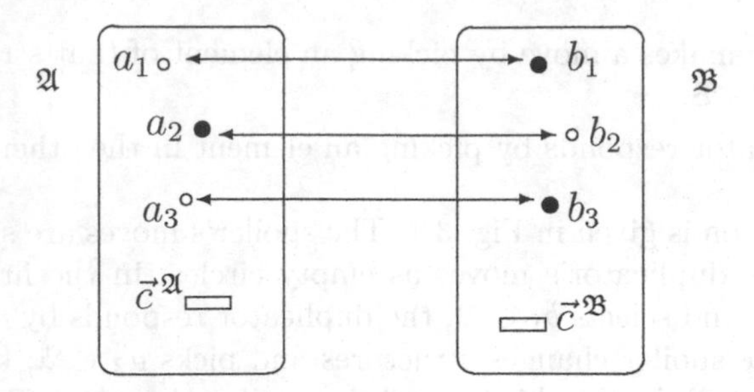

Fig. 3.1. Ehrenfeucht-Fraïssé game

that guarantees a winning position after n rounds, no matter how the spoiler plays. Otherwise, the spoiler has an n-round winning strategy. If the duplicator has an n-round winning strategy, we write $\mathfrak{A} \equiv_n \mathfrak{B}$.

Observe that $\mathfrak{A} \equiv_n \mathfrak{B}$ implies $\mathfrak{A} \equiv_k \mathfrak{B}$ for every $k \leq n$.

Before we connect Ehrenfeucht-Fraïssé games and FO-definability, we give some examples of winning strategies.

Games on Sets

In this example, the vocabulary σ is empty. That is, a structure is just a set. Let $|A|, |B| \geq n$. Then $A \equiv_n B$.

The strategy for the duplicator works as follows. Suppose i rounds have been played, and the position is $((a_1, \ldots, a_i), (b_1, \ldots, b_i))$. Assume the spoiler picks an element $a_{i+1} \in A$. If $a_{i+1} = a_j$ for $j \leq i$, then the duplicator responds with $b_{i+1} = b_j$; otherwise, the duplicator responds with any $b_{j+1} \in B - \{b_1, \ldots, b_i\}$ (which exists since $|B| \geq n$).

Games on Linear Orders

Our next example is a bit more complicated, as we add a binary relation $<$ to σ, to be interpreted as a linear order. Now suppose L_1, L_2 are two linear orders of size at least n (i.e., structures of the form $\langle\{1, \ldots, m\}, <\rangle$, $m \geq n$). Is it true that $L_1 \equiv_n L_2$?

It is very easy to see that the answer is negative even for the case of $n = 2$. Let L_1 contain three elements (say $\{1, 2, 3\}$), and L_2 two elements ($\{1, 2\}$). In the first move, the spoiler plays 2 in L_1. The duplicator has to respond with either 1 or 2 in L_2. Suppose the duplicator responds with $1 \in L_2$; then the spoiler plays $1 \in L_1$ and the duplicator is lost, since he has to respond with an element less than 1 in L_1, and there is no such element. If the duplicator selects $2 \in L_2$ as his first-round move, the spoiler plays $3 \in L_1$, and the duplicator is lost again. Hence, $L_1 \not\equiv_2 L_2$.

However, a winning strategy for the duplicator can be guaranteed if L_1, L_2 are much larger than the number of rounds.

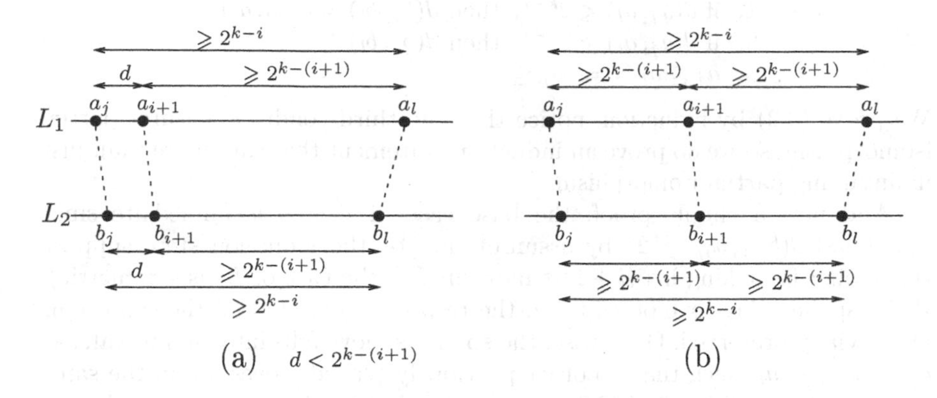

Fig. 3.2. Illustration for the proof of Theorem 3.6

Theorem 3.6. *Let $k > 0$, and let L_1, L_2 be linear orders of length at least 2^k. Then $L_1 \equiv_k L_2$.*

We shall give two different proofs of this result that illustrate two different techniques often used in game proofs.

Theorem 3.6, Proof # 1. The idea of the first proof is as follows. We use induction on the number of rounds of the game, and our induction hypothesis is ***stronger*** than just the partial isomorphism claim. The reason is that if we simply state that after i rounds we have a partial isomorphism, the induction step will not get off the ground as there are too few assumptions. Hence, we have to make additional assumptions. But if we try to impose too many conditions, there is no guarantee that a game can proceed in a way that preserves them. The main challenge in proofs of this type is to find the right induction hypothesis: the one that is strong enough to imply partial isomorphism, and that has enough conditions to make the inductive proof possible.

We now illustrate this general principle by proving Theorem 3.6. We expand the vocabulary with two new constant symbols $\underline{\min}$ and $\underline{\max}$, to be interpreted as the minimum and the maximum element of a linear ordering, and we prove a stronger fact that $L_1 \equiv_k L_2$ in the expanded vocabulary.

Let L_1 have the universe $\{1, \ldots, n\}$ and L_2 have the universe $\{1, \ldots, m\}$. Assume that the lengths of L_1 and L_2 are at least 2^k; that is, $n, m \geq 2^k+1$. The distance between two elements x, y of the universe, $d(x, y)$, is simply $|x - y|$. We claim that the duplicator can play in such a way that the following holds after each round i. Let $\vec{a} = (a_{-1}, a_0, a_1, \ldots, a_i)$ consist of $a_{-1} = \underline{\min}^{L_1}, a_0 = \underline{\max}^{L_1}$ and the i moves $a_1, \ldots, a_i$ in L_1, and likewise let $\vec{b} = (b_{-1}, b_0, b_1, \ldots, b_i)$ consist of $b_{-1} = \underline{\min}^{L_2}, b_0 = \underline{\max}^{L_2}$ and the i moves in L_2. Then, for $-1 \leq j, l \leq i$:

$$\begin{array}{ll}1. \text{ if } d(a_j, a_l) < 2^{k-i}, \text{ then } d(b_j, b_l) = d(a_j, a_l). & \\ 2. \text{ if } d(a_j, a_l) \geq 2^{k-i}, \text{ then } d(b_j, b_l) \geq 2^{k-i}. & (3.2) \\ 3. \ a_j \leq a_l \iff b_j \leq b_l. & \end{array}$$

We prove (3.2) by induction; notice that the third condition ensures partial isomorphism, so we do prove an induction statement that says more than just maintaining partial isomorphism.

And now a simple proof: the base case of $i = 0$ is immediate since $d(a_{-1}, a_0), d(b_{-1}, b_0) \geq 2^k$ by assumption. For the induction step, suppose the spoiler is making his $(i+1)$st move in L_1 (the case of L_2 is symmetric). If the spoiler plays one of $a_j, j \leq i$, the response is b_j, and all the conditions are trivially preserved. Otherwise, the spoiler's move falls into an interval, say $a_j < a_{i+1} < a_l$, such that no other previously played moves are in the same interval. By condition 3 of (3.2), this means that the interval between b_j and b_l contains no other elements of $\vec{b}$. There are two cases:

- $d(a_j, a_l) < 2^{k-i}$. Then $d(b_j, b_l) = d(a_j, a_l)$, and the intervals $[a_j, a_l]$ and $[b_j, b_l]$ are isomorphic. Then we simply find b_{i+1} so that $d(a_j, a_{i+1}) = d(b_j, b_{i+1})$ and $d(a_{i+1}, a_l) = d(b_{i+1}, b_l)$. Clearly, this ensures that all the conditions in (3.2) hold.
- $d(a_j, a_l) \geq 2^{k-i}$. In this case $d(b_j, b_l) \geq 2^{k-i}$. We have three possibilities:
  1. $d(a_j, a_{i+1}) < 2^{k-(i+1)}$. Then $d(a_{i+1}, a_l) \geq 2^{k-(i+1)}$, and we can choose b_{i+1} so that $d(b_j, b_{i+1}) = d(a_j, a_{i+1})$ and $d(b_{i+1}, b_l) \geq 2^{k-(i+1)}$. This is illustrated in Fig. 3.2 (a), where d stands for $d(a_j, a_{i+1})$.
  2. $d(a_{i+1}, a_l) < 2^{k-(i+1)}$. This case is similar to the previous one.
  3. $d(a_j, a_{i+1}) \geq 2^{k-(i+1)}$, $d(a_{i+1}, a_l) \geq 2^{k-(i+1)}$. Since $d(b_j, b_l) \geq 2^{k-i}$, by choosing b_{i+1} to be the middle of the interval $[b_j, b_l]$ we ensure that $d(b_j, b_{i+1}) \geq 2^{k-(i+1)}$ and $d(b_{i+1}, b_l) \geq 2^{k-(i+1)}$. This case is illustrated in Fig. 3.2 (b).

Thus, in all the cases, (3.2) is preserved.

This completes the inductive proof; hence we have shown that the duplicator can win a k-round Ehrenfeucht-Fraïssé game on L_1 and L_2. □

Theorem 3.6, Proof # 2. The second proof relies on the *composition method*: a way of composing simpler games into more complicated ones.

Before we proceed, we make the following observation. Suppose $L_1 \equiv_k L_2$. Then we can assume, without loss of generality, that the duplicator has a winning strategy in which he responds to the minimal element of one ordering by the minimal element of the other ordering (and likewise for the maximal elements).

Indeed, suppose the spoiler plays $\underline{\min}^{L_1}$, the minimal element of L_1. If the duplicator responds by $b > \underline{\min}^{L_2}$ and there is at least one round left, then in the next round the spoiler plays $\underline{\min}^{L_2}$ and the duplicator loses. If this is the last round of the game, then the duplicator can respond by any element that does not exceed those previously played in L_2, in particular, $\underline{\min}^{L_2}$. The proof for other cases is similar.

Let L be a linear ordering, and $a \in L$. By $L^{\leq a}$ we mean the substructure of L that consists of all the elements $b \leq a$, and by $L^{\geq a}$ the substructure of L that consists of all the elements $b \geq a$. The composition result we shall need says the following.

Lemma 3.7. *Let L_1, L_2, $a \in L_1$, and $b \in L_2$ be such that*

$$L_1^{\leq a} \equiv_k L_2^{\leq b} \quad \text{and} \quad L_1^{\geq a} \equiv_k L_2^{\geq b}.$$

Then $(L_1, a) \equiv_k (L_2, b)$.

Proof of Lemma 3.7. The strategy for the duplicator is very simple: if the spoiler plays in $L_1^{\leq a}$, the duplicator uses the winning strategy for $L_1^{\leq a} \equiv_k L_2^{\leq b}$, and if the spoiler plays in $L_1^{\geq a}$, the duplicator uses the winning strategy for $L_1^{\geq a} \equiv_k L_2^{\geq b}$ (the case when the spoiler plays in L_2 is symmetric). By the remark preceding the lemma, the duplicator always responds to a by b and to b by a, which implies that the strategy allows him to win in the k-round game on (L_1, a) and (L_2, b). □

And now we prove Theorem 3.6. The proof again is by induction on k, and the base case is easily verified. For the induction step, assume we have two linear orderings, L_1 and L_2, of length at least 2^k. Suppose the spoiler plays $a \in L_1$ (the case when the spoiler plays in L_2 is symmetric). We will show how to find $b \in L_2$ so that $(L_1, a) \equiv_{k-1} (L_2, b)$. There are three cases:

- The length of $L_1^{\leq a}$ is less than 2^{k-1}. Then let b be an element of L_2 such that $d(\underline{\min}^{L_1}, a) = d(\underline{\min}^{L_2}, b)$; in other words, $L_1^{\leq a} \cong L_2^{\leq b}$. Since the length of each of $L_1^{\geq a}$ and $L_2^{\geq b}$ is at least 2^{k-1}, by the induction hypothesis, $L_1^{\geq a} \equiv_{k-1} L_2^{\geq b}$. Hence, by Lemma 3.7, $(L_1, a) \equiv_{k-1} (L_2, b)$.
- The length of $L_1^{\geq a}$ is less than 2^{k-1}. This case is symmetric to the previous case.
- The lengths of both $L_1^{\leq a}$ and $L_1^{\geq a}$ are at least 2^{k-1}. Since the length of L_2 is at least 2^k, we can find $b \in L_2$ such that the lengths of both $L_2^{\leq b}$ and $L_2^{\geq b}$ are at least 2^{k-1}. Then, by the induction hypothesis, $L_1^{\leq a} \equiv_{k-1} L_2^{\leq b}$ and $L_1^{\geq a} \equiv_{k-1} L_2^{\geq b}$, and by Lemma 3.7, $(L_1, a) \equiv_{k-1} (L_2, b)$.

Thus, for every $a \in L_1$, we can find $b \in L_2$ such that $(L_1, a) \equiv_{k-1} (L_2, b)$ (and symmetrically with the roles of L_1 and L_2 reversed). This proves $L_1 \equiv_k L_2$, and completes the proof of the theorem. □

3.3 Games and the Expressive Power of FO

And now it is time to see why games are important. For this, we need a crucial definition of quantifier rank.

Definition 3.8 (Quantifier rank). *The* quantifier rank *of a formula* $\mathsf{qr}(\varphi)$ *is its depth of quantifier nesting. That is:*

- *If φ is atomic, then* $\mathsf{qr}(\varphi) = 0$.
- $\mathsf{qr}(\varphi_1 \vee \varphi_2) = \mathsf{qr}(\varphi_1 \wedge \varphi_2) = \max(\mathsf{qr}(\varphi_1), \mathsf{qr}(\varphi_2))$.
- $\mathsf{qr}(\neg\varphi) = \mathsf{qr}(\varphi)$.
- $\mathsf{qr}(\exists x \varphi) = \mathsf{qr}(\forall x \varphi) = \mathsf{qr}(\varphi) + 1$.

We use the notation FO[k] *for all* FO *formulae of quantifier rank up to k.*

In general, quantifier rank of a formula is different from the total of number of quantifiers used. For example, we can define a family of formulae by induction: $d_0(x, y) \equiv E(x, y)$, and $d_k \equiv \exists z\ d_{k-1}(x, z) \wedge d_{k-1}(z, y)$. The quantifier rank of d_k is k, but the total number of quantifiers used in d_k is $2^k - 1$. For formulae in the prenex form (i.e., all quantifiers are in front, followed by a quantifier-free formula), quantifier rank is the same as the total number of quantifiers.

Given a set S of FO sentences (over vocabulary σ), we say that two σ-structures $\mathfrak{A}$ and $\mathfrak{B}$ *agree on* S if for every sentence Φ of S, it is the case that $\mathfrak{A} \models \Phi \Leftrightarrow \mathfrak{B} \models \Phi$.

Theorem 3.9 (Ehrenfeucht-Fraïssé). *Let $\mathfrak{A}$ and $\mathfrak{B}$ be two structures in a relational vocabulary. Then the following are equivalent:*

1. *$\mathfrak{A}$ and $\mathfrak{B}$ agree on* FO[k].
2. $\mathfrak{A} \equiv_k \mathfrak{B}$.

We will prove this theorem shortly, but first we discuss how this is useful for proving inexpressibility results.

Characterizing the expressive power of FO via games gives rise to the following methodology for proving inexpressibility results.

Corollary 3.10. *A property $\mathcal{P}$ of finite σ-structures is* not expressible *in* FO *if for every $k \in \mathbb{N}$, there exist two finite σ-structures, $\mathfrak{A}_k$ and $\mathfrak{B}_k$, such that:*

- $\mathfrak{A}_k \equiv_k \mathfrak{B}_k$, *and*
- *$\mathfrak{A}_k$ has property $\mathcal{P}$, and $\mathfrak{B}_k$ does not.*

Proof. Assume to the contrary that $\mathcal{P}$ is definable by a sentence Φ. Let $k = \mathsf{qr}(\Phi)$, and pick $\mathfrak{A}_k$ and $\mathfrak{B}_k$ as above. Then $\mathfrak{A}_k \equiv_k \mathfrak{B}_k$, and thus if $\mathfrak{A}_k$ has property $\mathcal{P}$, then so does $\mathfrak{B}_k$, which contradicts the assumptions. □

We shall see in the next section that the *if* of Corollary 3.10 can be replaced by *iff*; that is, Ehrenfeucht-Fraïssé games are complete for first-order definability.

The methodology above extends from sentences to formulas with free variables.

Corollary 3.11. *An m-ary query Q on σ-structures is* not expressible *in* FO *iff for every $k \in \mathbb{N}$, there exist two finite σ-structures, $\mathfrak{A}_k$ and $\mathfrak{B}_k$, and two m-tuples $\vec{a}$ and $\vec{b}$ in them such that:*

- $(\mathfrak{A}_k, \vec{a}) \equiv_k (\mathfrak{B}_k, \vec{b})$, *and*
- $\vec{a} \in Q(\mathfrak{A}_k)$ *and* $\vec{b} \notin Q(\mathfrak{B}_k)$. □

We next see some simple examples of using games; more examples will be given in Sect. 3.6. An immediate application of the Ehrenfeucht-Fraïssé theorem is that EVEN is not FO-expressible when σ is empty: we take $\mathfrak{A}_k$ to contain k elements, and $\mathfrak{B}_k$ to contain $k+1$ elements. However, we have already proved this by a simple compactness argument in Sect. 3.1. But we could not prove, by the same argument, that EVEN is not expressible over finite linear orders. Now we get this for free:

Corollary 3.12. EVEN *is not* FO*-expressible over linear orders.*

Proof. Pick $\mathfrak{A}_k$ to be a linear order of length 2^k, and $\mathfrak{B}_k$ to be a linear order of length 2^k+1. By Theorem 3.6, $\mathfrak{A}_k \equiv_k \mathfrak{B}_k$. The statement now follows from Corollary 3.10. □

3.4 Rank-k Types

We now further analyze FO[k] and introduce the concept of types (more precisely, rank-k types).

First, what is FO[0]? It contains Boolean combinations of atomic formulas. If we are interested in sentences in FO[0], these are precisely *atomic* sentences: that is, sentences without quantifiers. In a relational vocabulary, such sentences are Boolean combinations of formulae of the form $c = c'$ and $R(c_1, \ldots, c_k)$, where $c, c', c_1, \ldots, c_k$ are constant symbols from σ.

Next, assume that φ is an FO[$k+1$] formula. If $\varphi = \varphi_1 \vee \varphi_2$, then both φ_1, φ_2 are FO[$k+1$] formulae, and likewise for $\wedge$; if $\varphi = \neg\varphi_1$, then $\varphi_1 \in$ FO[$k+1$]. However, if $\varphi = \exists x\psi$ or $\varphi = \forall x\psi$, then ψ is an FO[k] formula. Hence, every formula from FO[$k+1$] is equivalent to a Boolean combination of formulae of the form $\exists x\psi$, where $\psi \in$ FO[k]. Using this, we show:

Lemma 3.13. *If σ is finite, then up to logical equivalence,* FO$[k]$ *over σ contains only finitely many formulae in m free variables $x_1, \ldots, x_m$.*

Proof. The proof is by induction on k. The base case is FO$[0]$; there are only finitely many atomic formulae, and hence only finitely many Boolean combinations of those, up to logical equivalence. Going from k to $k+1$, recall that each formula $\varphi(x_1, \ldots, x_m)$ from FO$[k+1]$ is a Boolean combination of $\exists x_{m+1} \psi(x_1, \ldots, x_m, x_{m+1})$, where $\psi \in$ FO$[k]$. By the hypothesis, the number of FO$[k]$ formulae in $m+1$ free variables $x_1, \ldots, x_{m+1}$ is finite (up to logical equivalence) and hence the same can be concluded about FO$[k+1]$ formulas in m free variables. □

In model theory, a *type* (or m-type) of an m-tuple $\vec{a}$ over a σ structure $\mathfrak{A}$ is the set of all FO formulae φ in m free variables such that $\mathfrak{A} \models \varphi(\vec{a})$. This notion is too general in our setting, as the type of $\vec{a}$ over a finite $\mathfrak{A}$ describes $(\mathfrak{A}, \vec{a})$ up to isomorphism.

Definition 3.14 (Types). *Fix a relational vocabulary σ. Let $\mathfrak{A}$ be a σ-structure, and $\vec{a}$ an m-tuple over A. Then the* rank-k m-type *of $\vec{a}$ over $\mathfrak{A}$ is defined as*

$$\text{tp}_k(\mathfrak{A}, \vec{a}) = \{\varphi \in \text{FO}[k] \mid \mathfrak{A} \models \varphi(\vec{a})\}.$$

A rank-k m-type *is any set of formulae of the form $\text{tp}_k(\mathfrak{A}, \vec{a})$, where $|\vec{a}| = m$. When m is clear from the context, we speak of rank-k types.*

In the special case of $m = 0$ we deal with $\text{tp}_k(\mathfrak{A})$, defined as the set of FO$[k]$ sentences that hold in $\mathfrak{A}$. Also note that rank-k types are maximally consistent sets of formulae: that is, each rank-k type S is consistent, and for every $\varphi(x_1, \ldots, x_m) \in$ FO$[k]$, either $\varphi \in S$ or $\neg\varphi \in S$.

At this point, it seems that rank-k types are inherently infinite objects, but they are not, because of Lemma 3.13. We know that up to logical equivalence, FO$[k]$ is finite, for a fixed number m of free variables. Let $\varphi_1(\vec{x}), \ldots, \varphi_M(\vec{x})$ enumerate all the nonequivalent formulae in FO$[k]$ with free variables $\vec{x} = (x_1, \ldots, x_m)$. Then a rank-$k$ type is uniquely determined by a subset K of $\{1, \ldots, M\}$ specifying which of the φ_i's belong to it. Moreover, testing that $\vec{x}$ satisfies all the φ_i's with $i \in K$ and does not satisfy all the φ_j's with $j \notin K$ can be done by a single formula

$$\alpha_K(\vec{x}) \equiv \bigwedge_{i \in K} \varphi_i \wedge \bigwedge_{j \notin K} \neg\varphi_j. \tag{3.3}$$

Note that $\alpha_K(\vec{x})$ itself is an FO$[k]$ formula, since no new quantifiers were introduced.

Furthermore, all the α_K's are mutually exclusive: for $K \neq K'$, if $\mathfrak{A} \models \alpha_K(\vec{a})$, then $\mathfrak{A} \models \neg\alpha_{K'}(\vec{a})$. Every FO$[k]$ formula is a disjunction of some of the α_K's: indeed, every FO$[k]$ formula is equivalent to some φ_i in the above enumeration, which is the disjunction of all α_K's with $i \in K$.

Summing up, we have the following.

Theorem 3.15. *a) For a finite relational vocabulary σ, the number of different rank-k m-types is finite.*

b) Let $T_1, \ldots, T_r$ enumerate all the rank-k m-types. There exist FO$[k]$ *formulae $\alpha_1(\vec{x}), \ldots, \alpha_r(\vec{x})$ such that:*

- *for every $\mathfrak{A}$ and $\vec{a} \in A^m$, it is the case that $\mathfrak{A} \models \alpha_i(\vec{a})$ iff $\text{tp}_k(\mathfrak{A}, \vec{a}) = T_i$, and*
- *every* FO$[k]$ *formula $\varphi(\vec{x})$ in m free variables is equivalent to a disjunction of some α_i's.*

Thus, in what follows we normally associate types with their defining formulae α_i's (3.3). It is important to remember that these defining formulae for rank-k types have the same quantifier rank, k.

From the Ehrenfeucht-Fraïssé theorem and Theorem 3.15, we obtain:

Corollary 3.16. *The equivalence relation $\equiv_k$ is of finite index (that is, has finitely many equivalence classes).*

As promised in the last section, we now show that games are complete for characterizing the expressive power of FO: that is, the *if* of Corollary 3.10 can be replaced by *iff*.

Corollary 3.17. *A property $\mathcal{P}$ is expressible in* FO *iff there exists a number k such that for every two structures $\mathfrak{A}, \mathfrak{B}$, if $\mathfrak{A} \in \mathcal{P}$ and $\mathfrak{A} \equiv_k \mathfrak{B}$, then $\mathfrak{B} \in \mathcal{P}$.*

Proof. If $\mathcal{P}$ is expressible by an FO sentence Φ, let $k = \mathsf{qr}(\Phi)$. If $\mathfrak{A} \in \mathcal{P}$, then $\mathfrak{A} \models \Phi$, and hence for $\mathfrak{B}$ with $\mathfrak{A} \equiv_k \mathfrak{B}$, we have $\mathfrak{B} \models \Phi$. Thus, $\mathfrak{B} \in \mathcal{P}$.

Conversely, if $\mathfrak{A} \in \mathcal{P}$ and $\mathfrak{A} \equiv_k \mathfrak{B}$ imply $\mathfrak{B} \in \mathcal{P}$, then any two structures with the same rank-k type agree on $\mathcal{P}$, and hence $\mathcal{P}$ is a union of types, and thus definable by a disjunction of some of the α_i's defined by (3.3). $\square$

Thus, a property $\mathcal{P}$ is *not* expressible in FO *iff* for every k, one can find two structures, $\mathfrak{A}_k \equiv_k \mathfrak{B}_k$, such that $\mathfrak{A}_k$ has $\mathcal{P}$ and $\mathfrak{B}_k$ does not.

3.5 Proof of the Ehrenfeucht-Fraïssé Theorem

We shall prove the equivalence of 1 and 2 in the Ehrenfeucht-Fraïssé theorem, as well as a new important condition, the *back-and-forth* equivalence. Before stating this condition, we briefly analyze the equivalence relation $\equiv_0$.

When does the duplicator win the game without even starting? This happens iff $(\emptyset, \emptyset)$ is a partial isomorphism between two structures $\mathfrak{A}$ and $\mathfrak{B}$. That is, if $\vec{c}$ is the tuple of constant symbols, then $c_i^{\mathfrak{A}} = c_j^{\mathfrak{A}}$ iff $c_i^{\mathfrak{B}} = c_j^{\mathfrak{B}}$ for every i, j, and for each relation symbol R, the tuple $(c_{i_1}^{\mathfrak{A}}, \ldots, c_{i_k}^{\mathfrak{A}})$ is in $R^{\mathfrak{A}}$ iff the tuple $(c_{i_1}^{\mathfrak{B}}, \ldots, c_{i_k}^{\mathfrak{B}})$ is in $R^{\mathfrak{B}}$. In other words, $(\emptyset, \emptyset)$ is a partial isomorphism between $\mathfrak{A}$ and $\mathfrak{B}$ iff $\mathfrak{A}$ and $\mathfrak{B}$ satisfy the same atomic sentences.

We now use this as the basis for the inductive definition of back-and-forth relations on $\mathfrak{A}$ and $\mathfrak{B}$. More precisely, we define a family of relations $\simeq_k$ on pairs of structures of the same vocabulary as follows:

- $\mathfrak{A} \simeq_0 \mathfrak{B}$ iff $\mathfrak{A} \equiv_0 \mathfrak{B}$; that is, $\mathfrak{A}$ and $\mathfrak{B}$ satisfy the same atomic sentences.
- $\mathfrak{A} \simeq_{k+1} \mathfrak{B}$ iff the following two conditions hold:

forth: for every $a \in A$, there exists $b \in B$ such that $(\mathfrak{A}, a) \simeq_k (\mathfrak{B}, b)$;
back: for every $b \in B$, there exists $a \in A$ such that $(\mathfrak{A}, a) \simeq_k (\mathfrak{B}, b)$.

We now prove the following extension of Theorem 3.9.

Theorem 3.18. *Let $\mathfrak{A}$ and $\mathfrak{B}$ be two structures in a relational vocabulary σ. Then the following are equivalent:*

1. $\mathfrak{A}$ and $\mathfrak{B}$ agree on FO$[k]$.
2. $\mathfrak{A} \equiv_k \mathfrak{B}$.
3. $\mathfrak{A} \simeq_k \mathfrak{B}$.

Proof. By induction on k. The case of $k = 0$ is obvious. We first show the equivalence of 2 and 3. Going from k to $k + 1$, assume $\mathfrak{A} \simeq_{k+1} \mathfrak{B}$; we must show $\mathfrak{A} \equiv_{k+1} \mathfrak{B}$. Assume for the first move the spoiler plays $a \in A$; we find $b \in \mathfrak{B}$ with $(\mathfrak{A}, a) \simeq_k (\mathfrak{B}, b)$, and thus by the hypothesis $(\mathfrak{A}, a) \equiv_k (\mathfrak{B}, b)$. Hence the duplicator can continue to play for k moves, and thus wins the $k + 1$-move game. The other direction is similar.

With games replaced by the back-and-forth relation, we show the equivalence of 1 and 3. Assume $\mathfrak{A}$ and $\mathfrak{B}$ agree on all quantifier-rank $k+1$ sentences; we must show $\mathfrak{A} \simeq_{k+1} \mathfrak{B}$. We prove the *forth* case; the *back* case is identical. Pick $a \in A$, and let α_i define its rank-k 1-type. Then $\mathfrak{A} \models \exists x \alpha_i(x)$. Since $\mathsf{qr}(\alpha_i) = k$, this is a sentence of quantifier-rank $k+1$; hence $\mathfrak{B} \models \exists x \alpha_i(x)$. Let b be the witness for the existential quantifier; that is, $\text{tp}_k(\mathfrak{A}, a) = \text{tp}_k(\mathfrak{B}, b)$. Hence for every σ_1 sentence Ψ of $\mathsf{qr}(\Psi) = k$, we have $(\mathfrak{A}, a) \models \Psi$ iff $(\mathfrak{B}, b) \models \Psi$, and thus $(\mathfrak{A}, a)$ and $(\mathfrak{B}, b)$ agree on quantifier-rank k sentences. By the hypothesis, this implies $(\mathfrak{A}, a) \simeq_k (\mathfrak{B}, b)$.

For the implication $3 \rightarrow 1$, we need to prove that $\mathfrak{A} \simeq_{k+1} \mathfrak{B}$ implies that $\mathfrak{A}$ and $\mathfrak{B}$ agree on FO$[k+1]$. Every FO$[k+1]$ sentence is a Boolean combination of $\exists x \varphi(x)$, where $\varphi \in$ FO$[k]$, so it suffices to prove the result for sentences of the form $\exists x \varphi(x)$. Assume that $\mathfrak{A} \models \exists x \varphi(x)$, so $\mathfrak{A} \models \varphi(a)$ for some $a \in A$. By **forth**, find $b \in B$ such that $(\mathfrak{A}, a) \simeq_k (\mathfrak{B}, b)$; hence $(\mathfrak{A}, a)$ and $(\mathfrak{B}, b)$ agree on FO$[k]$ by the hypothesis. Hence, $\mathfrak{B} \models \varphi(b)$, and thus $\mathfrak{B} \models \exists x \varphi(x)$. The converse (that $\mathfrak{B} \models \exists x \varphi(x)$ implies $\mathfrak{A} \models \exists x \varphi(x)$) is identical, which completes the proof. $\square$

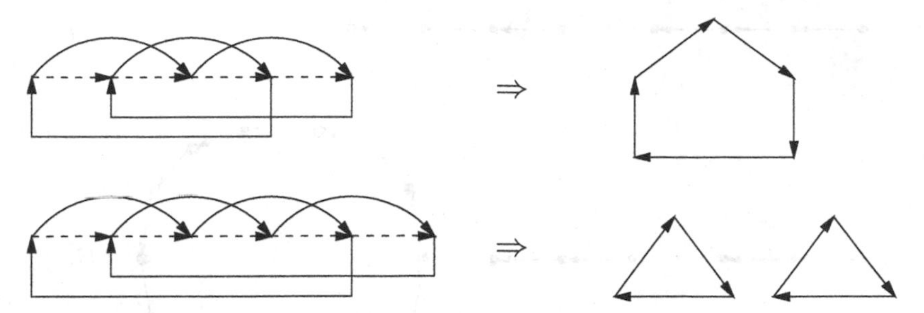

Fig. 3.3. Reduction of parity to connectivity

3.6 More Inexpressibility Results

So far we have used games to prove that EVEN is not expressible in FO, in both ordered and unordered settings. Next, we show inexpressibility of graph connectivity over finite graphs. In Sect. 3.1 we used compactness to show that connectivity of arbitrary graphs is inexpressible, leaving open the possibility that it may be FO-definable over finite graphs. We now show that this cannot happen. It turns out that no new game argument is needed, as the proof uses a reduction from EVEN over linear orders.

Assume that connectivity of finite graphs is definable by an FO sentence Φ, in the vocabulary that consists of one binary relation symbol E. Next, given a linear ordering, we define a directed graph from it as described below. First, from a linear ordering $<$ we define the successor relation

$$\text{succ}(x, y) \equiv (x < y) \wedge \forall z\big((z \leq x) \vee (z \geq y)\big).$$

Using this, we define an FO formula $\gamma(x, y)$ such that $\gamma(x, y)$ is true iff one of the following holds:

- y is the successor of the successor of x: $\exists z\ \big(\text{succ}(x, z) \wedge \text{succ}(z, y)\big)$, or
- x is the predecessor of the last element, and y is the first element: $\big(\exists z\ (\text{succ}(x, z) \wedge \forall u(u \leq z))\big) \wedge \forall u(y \leq u)$, or
- x is the last element and y is the successor of the first element (the FO formula is similar to the one above).

Thus, $\gamma(x, y)$ defines a new graph on the elements of the linear ordering; the construction is illustrated in Fig. 3.3.

Now observe that the graph defined by γ is connected iff the size of the underlying linear ordering is odd. Hence, taking $\neg\Phi$, and substituting γ for every occurrence of the predicate E, we get a sentence that tests EVEN for linear orderings. Since this is impossible, we obtain the following.

Corollary 3.19. *Connectivity of finite graphs is not* FO*-definable.*

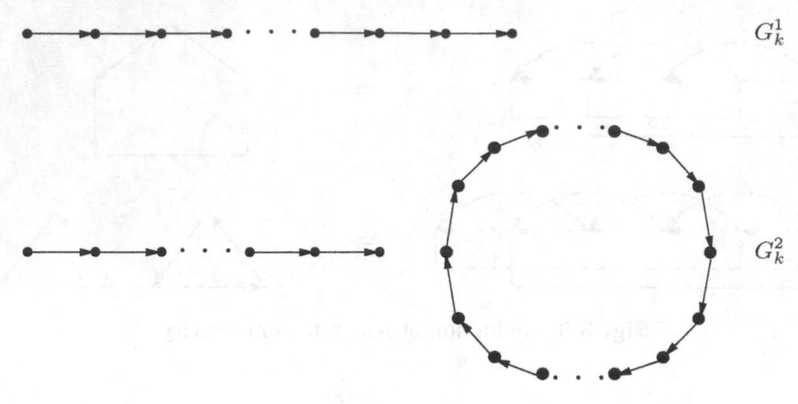

Fig. 3.4. Graphs G_k^1 and G_k^2

So far all the examples of inexpressibility results proved via Ehrenfeucht-Fraïssé games were fairly simple. Unfortunately, this is a rather unusual situation; typically game proofs are hard, and often some nontrivial combinatorial arguments are required. We now present an additional example of a game proof, as well as a few more problems that could possibly be handled by games, but are better left until we have seen more powerful techniques. These show how the difficulty of game proofs can rapidly increase as the problems become more complex.

Suppose that we want to test if a graph is a tree. By trees we mean directed rooted trees. This seems to be impossible in FO. To prove this, we follow the general methodology: that is, for each k we must find two graphs, $G_k^1 \equiv_k G_k^2$, such that one of them is a tree, and the other one is not.

We choose these graphs as follows: G_k^1 is the graph of a successor relation of length $2m$, and G_k^2 has two connected components: one is the graph of a successor relation of length m, and the other one is a cycle of length m. We did not say what m is, and it will be clear from the proof what it should be: at this point we just say that m depends only on k, and is sufficiently large.

Clearly G_k^1 is a tree (of degree 1), and G_k^2 is not, so we must show $G_k^1 \equiv_k G_k^2$. In each of these two graphs there are two special points: the start and the endpoint of the successor relation. Clearly these must be preserved in the game, so we may just assume that the game starts in a position where these points were played. That is, we let a_{-1}, a_0 be the start and the endpoint of G_k^1, and b_{-1}, b_0 be the start and the endpoint of the successor part of G_k^2. We let a_i's stand for the points played in G_k^1, and b_i's for the points played in G_k^2.

What do we put in the inductive hypothesis? The approach we take is very similar to the first proof of Theorem 3.6. We define the distance between two elements as the length of the shortest path between them. Notice that in the case of G_k^2, the distance could be infinity, as the graph has two connected components. We then show that the duplicator can play in a way that ensures the following conditions after each round i:

$$\begin{array}{l} 1. \text{ if } d(a_j, a_l) \le 2^{k-i}, \text{ then } d(b_j, b_l) = d(a_j, a_l). \\ 2. \text{ if } d(a_j, a_l) > 2^{k-i}, \text{ then } d(b_j, b_l) > 2^{k-i}. \end{array} \tag{3.4}$$

These are very similar to conditions (3.2) used in the proof of Theorem 3.6.

How do we prove that the duplicator can maintain these conditions? Suppose i rounds have been played, and the spoiler makes his move in round $i+1$. If the spoiler plays close (at a distance at most $2^{k-(i+1)}$) to a previously played point, we can apply the proof of Theorem 3.6 to show that the duplicator has a response.

But what if the spoiler plays at a distance greater than $2^{k-(i+1)}$ from all the previously played points? In the proof of Theorem 3.6 we were able to place that move into some interval on a linear ordering and use some knowledge of that interval to find the response – but this does not work any more, since our graphs now have a different structure. Nevertheless, there is a way to ensure that the duplicator can maintain the winning conditions: simply by choosing m "very large", we can always be sure that if fewer than k rounds of the game have been played, there is a point at a distance greater than $2^{k-(i+1)}$ from all the previously played points in the graph. We leave it to the reader to calculate m for a given k (it is not that much different from the bound we had in Theorem 3.6).

Thus, the duplicator can maintain all the conditions (3.4). In the proof of Theorem 3.6, one of the conditions of (3.2) stated that the moves in the game define a partial isomorphism. Here, we do not have this property, but we can still derive that after k rounds, the duplicator achieves a partial isomorphism. Indeed, suppose all k rounds have been played, and we have two elements a_i, a_j such that there is an edge between a_i and a_j. This means that $d(a_i, a_j) = 1$, and, by (3.4), $d(b_i, b_j) = 1$. Therefore, there is an edge between b_i and b_j. Conversely, let there be an edge between b_i and b_j. If there is no edge between a_i and a_j, then $d(a_i, a_j) > 1$, and, by (3.4), $d(b_i, b_j) > 1$, which contradicts our assumption that there is an edge between them.

Thus, we have shown that $G_k^1 \equiv_k G_k^2$, which proves the following.

Proposition 3.20. *It is impossible to test, by an* FO *sentence, if a finite graph is a tree.* □

This proof is combinatorially slightly more involved than other game proofs we have seen, and yet it uses trees with only unary branching. So it does not tell us whether testing the property of being an n-ary tree, for $n > 1$, is expressible. Moreover, one can easily imagine that the combinatorics in a game argument even for binary trees will be much harder. And what if we are interested in more complex properties? For example, testing if a graph is:

- a balanced binary tree (the branching factor is 2, and all the maximal branches are of the same length);
- a binary tree with all the maximal branches of different length;
- or even a bit different: assuming that we know that the input is a binary tree, can we check, in FO, if it is balanced?

It would thus be nice to have some easily verifiable criteria that guarantee a winning strategy for the duplicator, and that is exactly what we shall do in the next chapter.

3.7 Bibliographic Notes

Examples of using compactness for proving some very easy inexpressibility results over finite models are taken from Väänänen [239] and Gaifman and Vardi [89].

Characterization of the expressive power of FO in terms of the back-and-forth equivalence is due to Fraïssé [84]; the game description of the back-and-forth equivalence is due to Ehrenfeucht [62].

Theorem 3.6 is a classical application of Ehrenfeucht-Fraïssé games, and was rediscovered many times, cf. Gurevich [117] and Rosenstein [209]. The composition method, used in the second proof of Theorem 3.6, will be discussed elsewhere in the book (e.g., exercise 3.15 in this chapter, as well as Chap. 7). For a recent survey, see Makowsky [177].

The proof of inexpressibility of connectivity is standard, see, e.g., [60, 133].

Types are a central concept of model theory, see [35, 125, 201]. The proof of the Ehrenfeucht-Fraïssé theorem given here is slightly different from the proof one finds in most texts (e.g., [60, 125]); an alternative proof using what is called *Hintikka formulae* is presented in Exercise 3.11.

Some of the exercises for this chapter show that several classical theorems in model theory (not only compactness) fail over finite models. For this line of work, see Gurevich [116], Rosen [207], Rosen and Weinstein [208], Feder and Vardi [78].

Sources for exercises:

Exercise 3.11:	Ebbinghaus and Flum [60]
Exercises 3.12 and 3.13:	Gurevich [116]
Exercise 3.14:	Ebbinghaus and Flum [60]
Exercise 3.17:	Cook and Liu [41]
Exercise 3.18:	Pezzoli [199]

3.8 Exercises

Exercise 3.1. Use compactness to show that the following is not FO-expressible over finite structures in the vocabulary of one unary relation symbol U: for a structure $\mathfrak{A}$, both $|U^{\mathfrak{A}}|$ and $|A - U^{\mathfrak{A}}|$ are even.

Exercise 3.2. Prove Lemma 3.4 for an arbitrary vocabulary.

Exercise 3.3. Prove Corollary 3.11.

Exercise 3.4. Using Ehrenfeucht-Fraïssé games, show that acyclicity of finite graphs is not FO-definable.

Exercise 3.5. Same as in the previous exercise, for the following properties of finite graphs:

1. Planarity.
2. Hamiltonicity.
3. 2-colorability.
4. k-colorability for any $k > 2$.
5. Existence of a clique of size at least $n/2$, where n is the number of nodes.

Exercise 3.6. We now consider a query closely related to EVEN. Let σ be a vocabulary that includes a unary relation symbol U. We then define a Boolean query PARITY$_U$ as follows: a finite σ-structure $\mathfrak{A}$ satisfies PARITY$_U$ iff

$$|U^{\mathfrak{A}}| = 0 \pmod 2).$$

Prove that if $\sigma = \{<, U\}$, where $<$ is interpreted as a linear ordering on the universe, then PARITY$_U$ is not FO-definable.

Exercise 3.7. Theorem 3.6 tells us that $L_1 \equiv_k L_2$ for two linear orders of length at least 2^k. Is the bound 2^k tight? If it is not, what is the tight bound?

Exercise 3.8. Just as for linear orders, the following can be proved for $\mathfrak{G}_n$, the graph of successor relation on $\{1, \ldots, n\}$. There is a function $f : \mathbb{N} \to \mathbb{N}$ such that $\mathfrak{G}_n \equiv_k \mathfrak{G}_m$ whenever $n, m \geq f(k)$. Calculate $f(k)$.

Exercise 3.9. Consider sets of the form $X_\Phi = \{n \in \mathbb{N} \mid L_n \models \Phi\}$, where Φ is an FO sentence, and L_n is a linear order with n elements. Describe these sets.

Exercise 3.10. Find an upper bound, in terms of k, on the number of rank-k types.

Exercise 3.11. The goal of this exercise is to give another proof of the Ehrenfeucht-Fraïssé theorem. In this proof, one constructs formulae defining rank-k types explicitly, by specifying inductively a winning condition for the duplicator.

Assume that σ is relational. For any σ-structure $\mathfrak{A}$ and $\vec{a} \in A^m$, we define inductively formulae $\alpha^k_{\mathfrak{A},\vec{a}}(x_1, \ldots, x_m)$ as follows:

- $\alpha^0_{\mathfrak{A},\vec{a}}(\vec{x}) = \bigwedge \chi(\vec{x})$ where the conjunction is taken over all atomic or negated atomic χ such that $\mathfrak{A} \models \chi(\vec{a})$. Note that the conjunction is finite.

- Assuming α^k's are defined, we define

$$\alpha^{k+1}_{\mathfrak{A},\vec{a}}(\vec{x}) = \Big(\bigwedge_{c\in A} \exists z\ \alpha^k_{\mathfrak{A},\vec{a}c}(\vec{x},z)\Big) \wedge \Big(\forall z \bigvee_{c\in A} \alpha^k_{\mathfrak{A},\vec{a}c}(\vec{x},z)\Big).$$

Prove that the following are equivalent:
1. $(\mathfrak{A},\vec{a}) \equiv_k (\mathfrak{B},\vec{b})$;
2. $(\mathfrak{A},\vec{a}) \simeq_k (\mathfrak{B},\vec{b})$;
3. for every $\varphi(\vec{x})$ with $\mathsf{qr}(\varphi) \le k$, we have $\mathfrak{A} \models \varphi(\vec{a})$ iff $\mathfrak{B} \models \varphi(\vec{b})$;
4. $\mathfrak{B} \models \alpha^k_{\mathfrak{A},\vec{a}}(\vec{b})$.

Using this, prove the following statement. Let Q be a query definable in FO by a formula of quantifier rank k. Then Q is definable by the following formula:

$$\bigvee_{\vec{a}\in Q(\mathfrak{A})} \alpha^k_{\mathfrak{A},\vec{a}}(\vec{x}).$$

Note that the disjunction is finite, by Lemma 3.13.

Exercise 3.12. Beth's definability theorem is a classical result in mathematical logic: it says that a property is definable implicitly iff it is definable explicitly. Explicit definability of a k-ary query Q on σ-structures means that there is a formula $\varphi(x_1,\ldots,x_k)$ such that $\varphi(\mathfrak{A}) = Q(\mathfrak{A})$. Implicit definability means that there is a sentence Φ in the language of σ expanded with a single k-ary relation P such that for every σ-structure $\mathfrak{A}$, there exists a unique set $P \subseteq A^k$ such that $(\mathfrak{A}, P) \models \Phi$ and $P = Q(\mathfrak{A})$.

Prove that Beth's theorem fails over finite models.

Hint: P is a unary query that returns the set of even elements in a linear order.

Exercise 3.13. Craig's interpolation is another classical result from mathematical logic. Let σ^1, σ^2 be two vocabularies, and $\sigma = \sigma^1 \cap \sigma^2$. Let Φ^i be a sentence over σ^i, $i = 1, 2$. Assume that $\Phi^1 \vdash \Phi^2$. Craig's theorem says that there exists a sentence Φ over σ such that $\Phi^1 \vdash \Phi$ and $\Phi \vdash \Phi^2$.

Using techniques similar to those in the previous exercise, prove that Craig's interpolation fails over finite models.

Exercise 3.14. This exercise demonstrates another example of a result from mathematical logic that fails over finite models. The Łos-Tarski theorem says that a sentence which is preserved under extensions (that is, $\mathfrak{A} \subseteq \mathfrak{B}$ and $\mathfrak{A} \models \Phi$ implies $\mathfrak{B} \models \Phi$) is equivalent to an existential sentence: a sentence built from atomic and negated atomic formulae by using $\vee, \wedge$, and $\exists$.

Prove that the Łos-Tarski theorem fails over finite models.

Exercise 3.15. Winning strategies for complex structures can be composed from winning strategies for simpler structures. Two commonly used examples of such compositions are the subject of this exercise.

Given two structures $\mathfrak{A}, \mathfrak{B}$ of the same vocabulary σ, their Cartesian product $\mathfrak{A}\times\mathfrak{B}$ is defined as a σ-structure whose universe is $A\times B$, each constant c is interpreted as a pair $(c^{\mathfrak{A}}, c^{\mathfrak{B}})$, and each m-ary relation P is interpreted as $\{((a_1,b_1),\ldots,(a_m,b_m)) \mid (a_1,\ldots,a_m) \in P^{\mathfrak{A}}, (b_1,\ldots,b_m) \in P^{\mathfrak{B}}\}$.

If the vocabulary contains only relation symbols, the disjoint union $\mathfrak{A} \coprod \mathfrak{B}$ for two structures with $A \cap B = \emptyset$ has the universe $A \cup B$, and each relation P is interpreted as $P^{\mathfrak{A}} \cup P^{\mathfrak{B}}$.

Assume $\mathfrak{A}_1 \equiv_k \mathfrak{A}_2$ and $\mathfrak{B}_1 \equiv_k \mathfrak{B}_2$. Show that:

- $\mathfrak{A}_1 \times \mathfrak{B}_1 \equiv_k \mathfrak{A}_2 \times \mathfrak{B}_2$;
- $\mathfrak{A}_1 \coprod \mathfrak{B}_1 \equiv_k \mathfrak{A}_2 \coprod \mathfrak{B}_2$.

Exercise 3.16. The $n \times m$ *grid* is a graph whose set of nodes is $\{(i,j) \mid i \leq n, j \leq m\}$ for some $n, m \in \mathbb{N}$, and whose edges go from (i,j) to $(i+1,j)$ and to $(i,j+1)$. Use composition of Ehrenfeucht-Fraïssé games to show that there are no FO sentences testing if $n = m$ ($n > m$) for the $n \times m$ grid.

Exercise 3.17. Consider finite structures which are disjoint unions of finite linear orderings. Such structures occur in AI applications under the name of *blocks world.* Use Ehrenfeucht-Fraïssé games to show that the theory of such structures is decidable, and finitely axiomatizable.

Exercise 3.18. Fix a relational vocabulary σ that has at least one unary and one ternary relation. Prove that the following is PSPACE-complete. Given k, and two σ-structures $\mathfrak{A}$ and $\mathfrak{B}$, is $\mathfrak{A} \equiv_k \mathfrak{B}$?

What happens if k is fixed?

Exercise 3.19.* A sentence Φ of vocabulary σ is called positive if no symbol from σ occurs under the scope of an odd number of negations in Φ. We say that a sentence Φ is preserved under surjective homomorphisms if $\mathfrak{A} \models \Phi$ and $h(\mathfrak{A}) = \mathfrak{B}$ implies $\mathfrak{B} \models \Phi$, where $h : A \to B$ is a homomorphism such that $h(A) = B$. Lyndon's theorem says that if Φ is preserved under surjective homomorphisms (where $\mathfrak{A}, \mathfrak{B}$ could be arbitrary structures), then Φ is equivalent to a positive sentence.

Does Lyndon's theorem hold in the finite? That is, if Φ is preserved under surjective homomorphisms over finite structures, is it the case that, over finite structures, Φ is equivalent to a positive sentence?

4

Locality and Winning Games

Winning games becomes nontrivial even for fairly simple examples. But often we can avoid complicated combinatorial arguments, by using rather simple sufficient conditions that guarantee a winning strategy for the duplicator. For first-order logic, most such conditions are based on the idea of *locality*, best illustrated by the example in Fig. 4.1.

Suppose we want to show that the transitive closure query is not expressible in FO. We assume, to the contrary, that it is definable by a formula $\varphi(x, y)$, and then use the *locality of* FO to conclude that such a formula can only see up to some distance r from its free variables, where r is determined by φ. Then we take a successor relation $\mathfrak{A}$ long enough so that the distance from a and b to each other and the endpoints is bigger than $2r$ – in that case, φ cannot see the difference between (a, b) and (b, a), but our assumption implies that $\mathfrak{A} \models \varphi(a, b) \wedge \neg\varphi(b, a)$ since a precedes b.

The goal of this chapter is to formalize this type of reasoning, and use it to provide winning strategies for the duplicator. Such strategies will help us find easy criteria for FO-definability.

Throughout the chapter, we assume that the vocabulary σ is purely relational; that is, contains only relation symbols. All the results extend easily to the case of vocabularies that have constant symbols (see Exercise 4.1), but restricting to purely relational vocabularies often makes notations simpler.

4.1 Neighborhoods, Hanf-locality, and Gaifman-locality

We start by defining *neighborhoods* that formalize the concept of "seeing up to distance r from the free variables".

Definition 4.1. *Given a σ-structure $\mathfrak{A}$, its* Gaifman graph, *denoted by $\mathcal{G}(\mathfrak{A})$, is defined as follows. The set of nodes of $G(\mathfrak{A})$ is A, the universe of $\mathfrak{A}$. There is an edge (a_1, a_2) in $\mathcal{G}(\mathfrak{A})$ iff $a_1 = a_2$, or there is a relation R in σ such that for some tuple $t \in R^{\mathfrak{A}}$, both a_1, a_2 occur in t.*

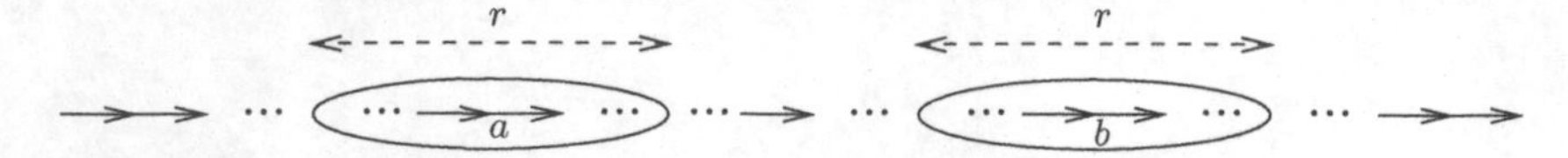

Fig. 4.1. A local formula cannot distinguish (a,b) from (b,a)

Note that $\mathcal{G}(\mathfrak{A})$ is an undirected graph. If $\mathfrak{A}$ is an undirected graph to start with, then $\mathcal{G}(\mathfrak{A})$ is simply $\mathfrak{A}$ together with the diagonal $\{(a,a) \mid a \in A\}$. If $\mathfrak{A}$ is a directed graph, then $\mathcal{G}(\mathfrak{A})$ simply forgets about the orientation (and adds the diagonal as well).

By the *distance* $d_{\mathfrak{A}}(x,y)$ we mean the distance in the Gaifman graph: that is, the length of the shortest path from x to y in $\mathcal{G}(\mathfrak{A})$. If there is no such path, then $d_{\mathfrak{A}}(x,a) = \infty$. It is easy to verify that the distance satisfies all the usual properties of a metric: $d_{\mathfrak{A}}(x,y) = 0$ iff $x = y$, $d_{\mathfrak{A}}(x,y) = d_{\mathfrak{A}}(y,x)$, and $d_{\mathfrak{A}}(x,z) \leq d_{\mathfrak{A}}(x,y) + d_{\mathfrak{A}}(y,z)$, for all x,y,z.

If we are given two tuples, $\vec{a} = (a_1,\ldots,a_n)$ and $\vec{b} = (b_1,\ldots,b_m)$, and an element c, then

$$d_{\mathfrak{A}}(\vec{a},c) \;=\; \min_{1\leq i\leq n} d_{\mathfrak{A}}(a_i,c),$$

$$d_{\mathfrak{A}}(\vec{a},\vec{b}) \;=\; \min_{1\leq i\leq n,\,1\leq j\leq m} d_{\mathfrak{A}}(a_i,b_j).$$

Furthermore, $\vec{a}c$ stands for the $n+1$-tuple $(a_1,\ldots,a_n,c)$, and $\vec{a}\vec{b}$ stands for the $n+m$-tuple $(a_1,\ldots,a_n,b_1,\ldots,b_m)$.

Recall that we use the notation σ_n for σ expanded with n constant symbols.

Definition 4.2. *Let σ contain only relation symbols, and let $\mathfrak{A}$ be a σ-structure, and $\vec{a} = (a_1,\ldots,a_n) \in A^n$. The* radius r ball *around $\vec{a}$ is the set*

$$B_r^{\mathfrak{A}}(\vec{a}) \;=\; \{b \in A \mid d_{\mathfrak{A}}(\vec{a},b) \leq r\}.$$

The r-neighborhood *of $\vec{a}$ in $\mathfrak{A}$ is the σ_n-structure $N_r^{\mathfrak{A}}(\vec{a})$, where:*

- *the universe is $B_r^{\mathfrak{A}}(\vec{a})$;*
- *each k-ary relation R is interpreted as $R^{\mathfrak{A}}$ restricted to $B_r^{\mathfrak{A}}(\vec{a})$; that is, $R^{\mathfrak{A}} \cap (B_r^{\mathfrak{A}}(\vec{a}))^k$;*
- *n additional constants are interpreted as $a_1,\ldots,a_n$.*

Note that since we define a neighborhood around an n-tuple as a σ_n-structure, for any isomorphism h between two isomorphic neighborhoods $N_r^{\mathfrak{A}}(a_1,\ldots,a_n)$ and $N_r^{\mathfrak{B}}(b_1,\ldots,b_n)$, it must be the case that $h(a_i) = b_i, 1 \leq i \leq n$.

Definition 4.3. *Let $\mathfrak{A}, \mathfrak{B}$ be σ-structures, where σ only contains relation symbols. Let $\vec{a} \in A^n$ and $\vec{b} \in B^n$. We write*

$$(\mathfrak{A}, \vec{a}) \leftrightarrows_d (\mathfrak{B}, \vec{b})$$

if there exists a bijection $f : A \to B$ such that for every $c \in A$,

$$N_d^{\mathfrak{A}}(\vec{a}c) \cong N_d^{\mathfrak{B}}(\vec{b}f(c)).$$

We shall often deal with the case of $n = 0$; then $\mathfrak{A} \leftrightarrows_d \mathfrak{B}$ means that for some bijection $f : A \to B$,

$$N_d^{\mathfrak{A}}(c) \cong N_d^{\mathfrak{B}}(f(c)) \text{ for all } c \in A.$$

The $\leftrightarrows_d$ relation says, in a sense, that locally two structures look the same, with respect to a certain bijection f; that is, f sends each element c into $f(c)$ that has the same neighborhood. The lemma below summarizes some properties of this relation:

Lemma 4.4. *1. $(\mathfrak{A}, \vec{a}) \leftrightarrows_d (\mathfrak{B}, \vec{b}) \Rightarrow |A| = |B|$.*
2. $(\mathfrak{A}, \vec{a}) \leftrightarrows_d (\mathfrak{B}, \vec{b}) \Rightarrow (\mathfrak{A}, \vec{a}) \leftrightarrows_{d'} (\mathfrak{B}, \vec{b})$, for $d' \leq d$.
3. $(\mathfrak{A}, \vec{a}) \leftrightarrows_d (\mathfrak{B}, \vec{b}) \Rightarrow N_d^{\mathfrak{A}}(\vec{a}) \cong N_d^{\mathfrak{B}}(\vec{b})$.

Recall that a neighborhood of an n-tuple is a σ_n-structure. By an *isomorphism type* of such structures we mean an equivalence class of $\cong$ on STRUCT$[\sigma_n]$. We shall use the letter τ (with sub- and superscripts) to denote isomorphism types. Instead of saying that a structure belongs to τ, we shall say that it is of the isomorphism type τ.

If τ is an isomorphism type of σ_n-structures, and $\vec{a} \in A^n$, we say that $\vec{a}$ *d-realizes* τ in $\mathfrak{A}$ if $N_d^{\mathfrak{A}}(\vec{a})$ is of type τ. If d is understood from the context, we say that $\vec{a}$ realizes τ.

The following is now easily proved from the definition of the $\leftrightarrows_d$ relation.

Lemma 4.5. *Let $\mathfrak{A}, \mathfrak{B} \in$ STRUCT$[\sigma]$. Then $\mathfrak{A} \leftrightarrows_d \mathfrak{B}$ iff for each isomorphism type τ of σ_1-structures, the number of elements of $\mathfrak{A}$ and $\mathfrak{B}$ that d-realize τ is the same.* □

We now formulate the first locality criterion.

Definition 4.6 (Hanf-locality). *An m-ary query Q on σ-structures is Hanf-local if there exists a number $d \geq 0$ such that for every $\mathfrak{A}, \mathfrak{B} \in$ STRUCT$[\sigma]$, $\vec{a} \in A^m, \vec{b} \in B^m$,*

$$(\mathfrak{A}, \vec{a}) \leftrightarrows_d (\mathfrak{B}, \vec{b}) \;\; implies \;\; \big(\vec{a} \in Q(\mathfrak{A}) \;\Leftrightarrow\; \vec{b} \in Q(\mathfrak{B}) \big).$$

The smallest d for which the above condition holds is called the Hanf-locality rank *of Q and is denoted by* hlr(Q).

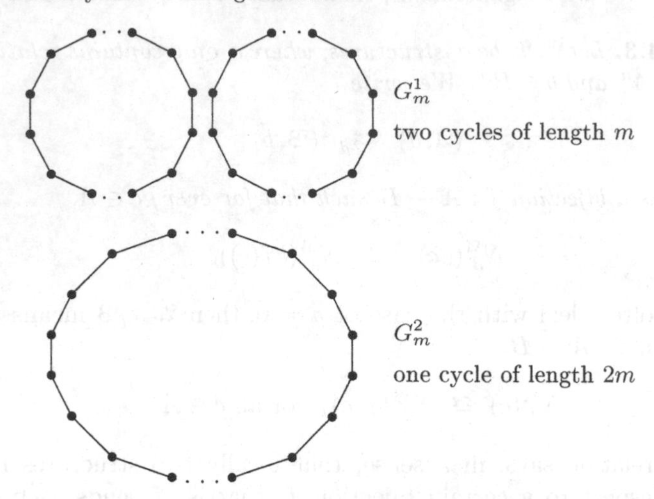

Fig. 4.2. Connectivity is not Hanf-local

Most commonly Hanf-locality is used for Boolean queries; then the definition says that for some $d \geq 0$, for every $\mathfrak{A}, \mathfrak{B} \in \text{STRUCT}[\sigma]$, the condition $\mathfrak{A} \leftrightarrows_d \mathfrak{B}$ implies that $\mathfrak{A}$ and $\mathfrak{B}$ agree on Q.

Using Hanf-locality for proving that a query Q is not definable in a logic $\mathcal{L}$ then amounts to showing:

- that every $\mathcal{L}$-definable query is Hanf-local, and
- that Q is not Hanf-local.

We now give the canonical example of using Hanf-locality. We show, by a very simple argument, that graph connectivity is not Hanf-local; it will then follow that graph connectivity is not expressible in any logic that only defines Hanf-local Boolean queries.

Assume to the contrary that the graph connectivity query Q is Hanf-local, and $\text{hlr}(Q) = d$. Let $m > 2d + 1$, and choose two graphs G_m^1 and G_m^2 as shown in Fig. 4.2. Their sets of nodes have the same cardinality. Let f be an arbitrary bijection between the nodes of G_m^1 and G_m^2. Since each cycle is of length $> 2d + 1$, the d-neighborhood of *any* node a is the same: it is a chain of length $2d$ with a in the middle. Hence, $G_m^1 \leftrightarrows_d G_m^2$, and they must agree on Q, but G_m^2 is connected, and G_m^1 is not. Thus, graph connectivity is not Hanf-local.

While Hanf-locality works well for Boolean queries, a different notion is often helpful for m-ary queries, $m > 0$.

Definition 4.7 (Gaifman-locality). *An m-ary query Q, $m > 0$, on σ-structures, is called* Gaifman-local *if there exists a number $d \geq 0$ such that for every σ-structure $\mathfrak{A}$ and every $\vec{a}_1, \vec{a}_2 \in A^m$,*

$$N_d^{\mathfrak{A}}(\vec{a}_1) \cong N_d^{\mathfrak{A}}(\vec{a}_2) \quad \textit{implies} \quad (\vec{a}_1 \in Q(\mathfrak{A}) \Leftrightarrow \vec{a}_2 \in Q(\mathfrak{A})).$$

The minimum d for which the above condition holds is called the locality rank *of Q, and is denoted by* $\mathsf{lr}(Q)$.

Note the difference between Hanf- and Gaifman-locality: the former relates two different structures, while the latter is talking about definability in one structure.

The methodology for proving inexpressibility of queries using Gaifman-locality is then as follows:

- first we show that all m-ary queries, $m > 0$, definable in a logic $\mathcal{L}$ are Gaifman-local,
- then we show that a given query Q is not Gaifman-local.

We shall see many examples of logics that define only Gaifman-local queries. At this point, we give a typical example of a query that is *not* Gaifman-local. The query is transitive closure, and we already saw that it is not Gaifman-local. Recall Fig. 4.1. Assume that the transitive closure query Q is Gaifman-local, and let $\mathsf{lr}(Q) = r$. If a, b are at a distance $> 2r+1$ from each other and the start and the endpoints, then the r-neighborhoods of (a, b) and (b, a) are isomorphic, since each is a disjoint union of two chains of length $2r$. We know that (a, b) belongs to the output of Q; hence by Gaifman-locality, (b, a) is in the output as well, which contradicts the assumption that Q defines transitive closure.

These examples demonstrate that locality tools are rather easy to use to obtain inexpressibility results. Our goal now is to show that FO-definable queries are both Hanf-local and Gaifman-local.

4.2 Combinatorics of Neighborhoods

The main technical tool for proving locality is combinatorial reasoning about neighborhoods. We start by presenting simple properties of neighborhoods; proofs are left as an exercise for the reader.

Lemma 4.8.
- *Assume that* $\mathfrak{A}, \mathfrak{B} \in \mathrm{STRUCT}[\sigma]$ *and* $h : N_r^{\mathfrak{A}}(\vec{a}) \to N_r^{\mathfrak{B}}(\vec{b})$ *is an isomorphism. Let* $d \leq r$. *Then h restricted to* $B_d^{\mathfrak{A}}(\vec{a})$ *is an isomorphism between* $N_d^{\mathfrak{A}}(\vec{a})$ *and* $N_d^{\mathfrak{B}}(\vec{b})$.
- *Assume that* $\mathfrak{A}, \mathfrak{B} \in \mathrm{STRUCT}[\sigma]$ *and* $h : N_r^{\mathfrak{A}}(\vec{a}) \to N_r^{\mathfrak{B}}(\vec{b})$ *is an isomorphism. Let* $d + l \leq r$ *and* $\vec{x}$ *be a tuple from* $B_l^{\mathfrak{A}}(\vec{a})$. *Then* $h(B_d^{\mathfrak{A}}(\vec{x})) = B_d^{\mathfrak{B}}(h(\vec{x}))$, *and* $N_d^{\mathfrak{A}}(\vec{x})$ *and* $N_d^{\mathfrak{B}}(h(\vec{x}))$ *are isomorphic.*
- *Let* $\mathfrak{A}, \mathfrak{B} \in \mathrm{STRUCT}[\sigma]$ *and let* $\vec{a}_1 \in A^n, \vec{b}_1 \in B^n$ *for* $n \geq 1$, *and* $\vec{a}_2 \in A^m, \vec{b}_2 \in B^m$ *for* $m \geq 1$. *Assume that* $N_r^{\mathfrak{A}}(\vec{a}_1) \cong N_r^{\mathfrak{B}}(\vec{b}_1)$, $N_r^{\mathfrak{A}}(\vec{a}_2) \cong N_r^{\mathfrak{B}}(\vec{b}_2)$, *and* $d_{\mathfrak{A}}(\vec{a}_1, \vec{a}_2)$, $d_{\mathfrak{B}}(\vec{b}_1, \vec{b}_2) > 2r+1$. *Then* $N_r^{\mathfrak{A}}(\vec{a}_1\vec{a}_2) \cong N_r^{\mathfrak{B}}(\vec{b}_1\vec{b}_2)$.

From now on, we shall use the notation

$$\vec{a} \approx_r^{\mathfrak{A},\mathfrak{B}} \vec{b}$$

for $N_r^{\mathfrak{A}}(\vec{a}) \cong N_r^{\mathfrak{B}}(\vec{b})$, omitting $\mathfrak{A}$ and $\mathfrak{B}$ when they are understood. We shall also write $d(\cdot,\cdot)$ instead of $d_{\mathfrak{A}}(\cdot,\cdot)$ when $\mathfrak{A}$ is understood.

The main technical result of this section is the lemma below.

Lemma 4.9. *If* $\mathfrak{A} \leftrightarrows_d \mathfrak{B}$ *and* $\vec{a} \approx_{3d+1}^{\mathfrak{A},\mathfrak{B}} \vec{b}$, *then* $(\mathfrak{A}, \vec{a}) \leftrightarrows_d (\mathfrak{B}, \vec{b})$.

Proof. We need to define a bijection $f : A \to B$ such that $\vec{a}c \approx_d^{\mathfrak{A},\mathfrak{B}} \vec{b}f(c)$ for every $c \in A$. Since $\vec{a} \approx_{3d+1}^{\mathfrak{A},\mathfrak{B}} \vec{b}$, there is an isomorphism $h : N_{3d+1}^{\mathfrak{A}}(\vec{a}) \to N_{3d+1}^{\mathfrak{B}}(\vec{b})$. Then the restriction of h to $B_{2d+1}^{\mathfrak{A}}(\vec{a})$ is an isomorphism between $N_{2d+1}^{\mathfrak{A}}(\vec{a})$ and $N_{2d+1}^{\mathfrak{B}}(\vec{b})$. Since $|A|=|B|$, we obtain

$$|A - B_{2d+1}^{\mathfrak{A}}(\vec{a})| \;=\; |B - B_{2d+1}^{\mathfrak{B}}(\vec{b})|\,.$$

Now consider an arbitrary isomorphism type τ of a d-neighborhood of a single point. Assume that $c \in B_{2d+1}^{\mathfrak{A}}(\vec{a})$ realizes τ in $\mathfrak{A}$. Since h is an isomorphism of $3d+1$-neighborhoods, $B_d^{\mathfrak{A}}(c) \subseteq B_{3d+1}^{\mathfrak{A}}(\vec{a})$ and thus $h(c) \in B_{2d+1}^{\mathfrak{B}}(\vec{b})$ realizes τ. Similarly, if $c \in B_{2d+1}^{\mathfrak{B}}(\vec{b})$ realizes τ, then so does $h^{-1}(c) \in B_{2d+1}^{\mathfrak{A}}(\vec{a})$. Hence, the number of elements in $B_{2d+1}^{\mathfrak{A}}(\vec{a})$ and $B_{2d+1}^{\mathfrak{B}}(\vec{b})$ that realize τ is the same.

Since $\mathfrak{A} \leftrightarrows_d \mathfrak{B}$, the number of elements of A and of B that realize τ is the same. Therefore,

$$\begin{aligned} &|\{a \in A - B_{2d+1}^{\mathfrak{A}}(\vec{a}) \mid a \text{ } d\text{-realizes } \tau\}| \\ &= |\{b \in B - B_{2d+1}^{\mathfrak{B}}(\vec{b}) \mid b \text{ } d\text{-realizes } \tau\}| \end{aligned} \tag{4.1}$$

for every τ. Using (4.1), we can find a bijection $g : A - B_{2d+1}^{\mathfrak{A}}(\vec{a}) \to B - B_{2d+1}^{\mathfrak{B}}(\vec{b})$ such that $c \approx_d g(c)$ for every $c \in A - B_{2d+1}^{\mathfrak{A}}(\vec{a})$.

We now define f by

$$f(c) \;=\; \begin{cases} h(c) & \text{if } c \in B_{2d+1}^{\mathfrak{A}}(\vec{a}) \\ g(c) & \text{if } c \notin B_{2d+1}^{\mathfrak{A}}(\vec{a}). \end{cases}$$

It is clear that f is a bijection $A \to B$.

We claim that $\vec{a}c \approx_d \vec{b}f(c)$ for every $c \in A$. This is illustrated in Fig. 4.3. If $c \in B_{2d+1}^{\mathfrak{A}}(\vec{a})$, then $B_d^{\mathfrak{A}}(c) \subseteq B_{3d+1}^{\mathfrak{A}}(\vec{a})$, and $\vec{a}c \approx_d \vec{b}h(c)$ because h is an isomorphism. If $c \notin B_{2d+1}^{\mathfrak{A}}(\vec{a})$, then $f(c) = g(c) \notin B_{2d+1}^{\mathfrak{B}}(\vec{b})$, and $c \approx_d g(c)$. Since $d(c,\vec{a}), d(g(c),\vec{b}) > 2d+1$, by Lemma 4.8, $\vec{a}c \approx_d \vec{b}g(c)$. □

The following corollary is very useful in establishing locality of logics.

Corollary 4.10. *If* $(\mathfrak{A},\vec{a}) \leftrightarrows_{3d+1} (\mathfrak{B},\vec{b})$, *then there exists a bijection* $f : A \to B$ *such that*

$$\forall c \in A \quad (\mathfrak{A}, \vec{a}c) \leftrightarrows_d (\mathfrak{B}, \vec{b}f(c)).$$

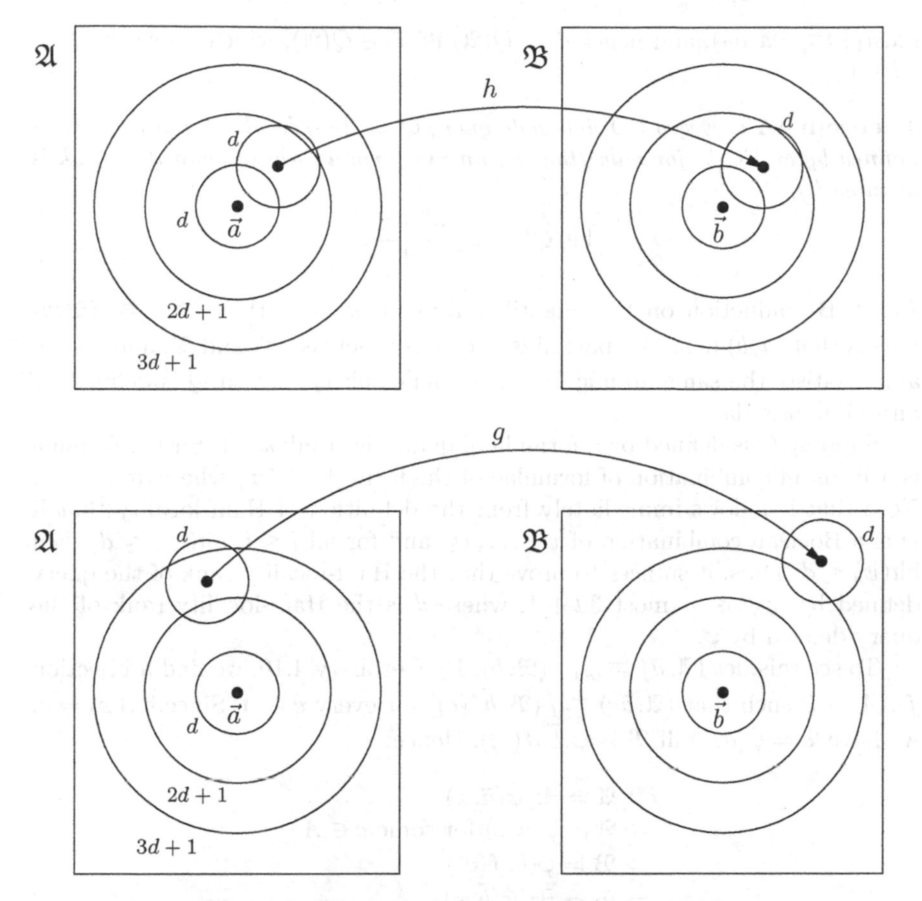

Fig. 4.3. Illustration of the proof of Lemma 4.9

Proof. By the definition of the $\leftrightarrows$ relation, there exists a bijection $f : A \to B$, such that for any $c \in A$, $\vec{a}c \approx^{\mathfrak{A},\mathfrak{B}}_{3d+1} \vec{b}f(c)$. Since $\mathfrak{A} \leftrightarrows_{3d+1} \mathfrak{B}$, we have $\mathfrak{A} \leftrightarrows_d \mathfrak{B}$. By Lemma 4.9, $(\mathfrak{A}, \vec{a}c) \leftrightarrows_d (\mathfrak{B}, \vec{b}f(c))$. ☐

4.3 Locality of FO

We now show that FO-definable queries are both Hanf-local and Gaifman-local. In fact, it suffices to prove the former, due to the following result.

Theorem 4.11. *If Q is a Hanf-local non-Boolean query, then Q is Gaifman-local, and* $\mathsf{lr}(Q) \leq 3 \cdot \mathsf{hlr}(Q) + 1$.

Proof. Suppose Q is an m-ary query on STRUCT$[\sigma]$, $m > 0$, and $\mathsf{hlr}(Q) = d$. Let $\mathfrak{A}$ be a σ-structure, and let $\vec{a}_1 \approx^{\mathfrak{A}}_{3d+1} \vec{a}_2$. Since $\mathfrak{A} \leftrightarrows_d \mathfrak{A}$, by Lemma 4.9,

$(\mathfrak{A}, \vec{a}_1) \leftrightarrows_d (\mathfrak{A}, \vec{a}_2)$, and hence $\vec{a}_1 \in Q(\mathfrak{A})$ iff $\vec{a}_2 \in Q(\mathfrak{A})$, which proves $\mathsf{lr}(Q) \leq 3d + 1$. □

Theorem 4.12. *Every* FO*-definable query Q is Hanf-local. Moreover, if Q is defined by an* FO$[k]$ *formula (that is, an* FO *formula whose quantifier rank is at most k), then*

$$\mathsf{hlr}(Q) \quad \leq \quad \frac{3^k - 1}{2}.$$

Proof. By induction on the quantifier rank. If $k = 0$, then $(\mathfrak{A}, \vec{a}) \leftrightarrows_0 (\mathfrak{B}, \vec{b})$ means that $(\vec{a}, \vec{b})$ defines a partial isomorphism between $\mathfrak{A}$ and $\mathfrak{B}$, and thus $\vec{a}$ and $\vec{b}$ satisfy the same atomic formulas. Hence $\mathsf{hlr}(Q) = 0$, if Q is defined by an FO[0] formula.

Suppose Q is defined by a formula of quantifier rank $k+1$. Such a formula is a Boolean combination of formulae of the form $\exists z \varphi(\vec{x}, z)$ where $\mathsf{qr}(\varphi) \leq k$. Note that it follows immediately from the definition of Hanf-locality that if ψ is a Boolean combination of $\psi_1, \ldots, \psi_l$, and for all $i \leq l$, $\mathsf{hlr}(\psi_i) \leq d$, then $\mathsf{hlr}(\psi) \leq d$. Thus, it suffices to prove that the Hanf-locality rank of the query defined by $\exists z \varphi$ is at most $3d + 1$, where d is the Hanf-locality rank of the query defined by φ.

To see this, let $(\mathfrak{A}, \vec{a}) \leftrightarrows_{3d+1} (\mathfrak{B}, \vec{b})$. By Corollary 4.10, we find a bijection $f : A \to B$ such that $(\mathfrak{A}, \vec{a}c) \leftrightarrows_d (\mathfrak{B}, \vec{b}f(c))$ for every $c \in A$. Since $\mathsf{hlr}(\varphi) = d$, we have $\mathfrak{A} \models \varphi(\vec{a}, c)$ iff $\mathfrak{B} \models \varphi(\vec{b}, f(c))$. Hence,

$$\begin{aligned} & \mathfrak{A} \models \exists z \ \varphi(\vec{a}, z) \\ \Rightarrow\ & \mathfrak{A} \models \varphi(\vec{a}, c) \text{ for some } c \in A \\ \Rightarrow\ & \mathfrak{B} \models \varphi(\vec{b}, f(c)) \\ \Rightarrow\ & \mathfrak{B} \models \exists z \ \varphi(\vec{b}, z). \end{aligned}$$

The same proof shows $\mathfrak{B} \models \exists z \ \varphi(\vec{b}, z)$ implies $\mathfrak{A} \models \exists z \ \varphi(\vec{a}, z)$. Thus, $\vec{a}$ and $\vec{b}$ agree on the query defined by $\exists z \varphi(\vec{x}, z)$, which completes the proof. □

Combining Theorems 4.11 and 4.12, we obtain:

Corollary 4.13. *Every* FO*-definable m-ary query Q, $m > 0$, is Gaifman-local. Moreover, if Q is definable by an* FO$[k]$ *formula, then*

$$\mathsf{lr}(Q) \quad \leq \quad \frac{3^{k+1} - 1}{2}.$$

Since we know that graph connectivity is not Hanf-local and transitive closure is not Gaifman-local, we immediately obtain, without using games, that these queries are not FO-definable.

We can give rather easy inexpressibility proofs for many queries. Below, we provide two examples.

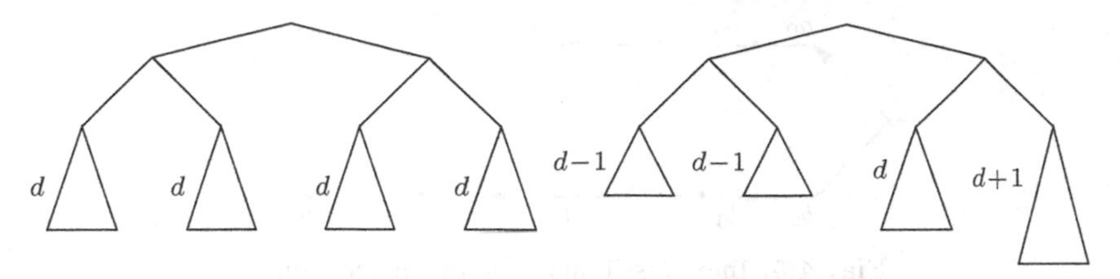

Fig. 4.4. Balanced binary trees are not FO-definable

Balanced Binary Trees

This example was mentioned at the end of Chap. 3. Suppose we are given a graph, and we want to test if it is a balanced binary tree. We now sketch the proof of inexpressibility of this query in FO; details are left as an exercise for the reader.

Suppose a test for being a balanced binary tree is definable in FO, say by a sentence Φ of quantifier rank k. Then we know that it is a Hanf-local query, with Hanf-locality rank at most $r = (3^k - 1)/2$. Choose d to be much larger than r, and consider two trees shown in Fig. 4.4.

In the first tree, denoted by T_1, the subtrees hanging at all four nodes on the second level are balanced binary trees of depth d; in the second tree, denoted by T_2, they are balanced binary trees of depths $d-1$, $d-1$, d, and $d+1$. We claim that $T_1 \leftrightarrows_r T_2$ holds.

First, notice that the number of nodes and the number of leaves in T_1 and T_2 is the same. If d is sufficiently large, these trees realize the following isomorphism types of neighborhoods:

- isomorphism types of r-neighborhoods of nodes a at a distance m from the root, $m \leq r$;
- isomorphism types of r-neighborhoods of nodes a at a distance m from a leaf, $m \leq r$;
- the isomorphism type of the r-neighborhood of a node a at a distance $> r$ from both the root and all the leaves.

Since the number of leaves and the number of nodes are the same, it is easy to see that each type of an r-neighborhood has the same number of nodes realizing it in both T_1 and T_2, and hence $T_1 \leftrightarrows_r T_2$. But this contradicts Hanf-locality of the balanced binary tree test, since T_1 is balanced, and T_2 is not.

Same Generation

The query we consider now is *same generation*: given a graph, two nodes a and b are in the same generation if there is a node c (common ancestor) such

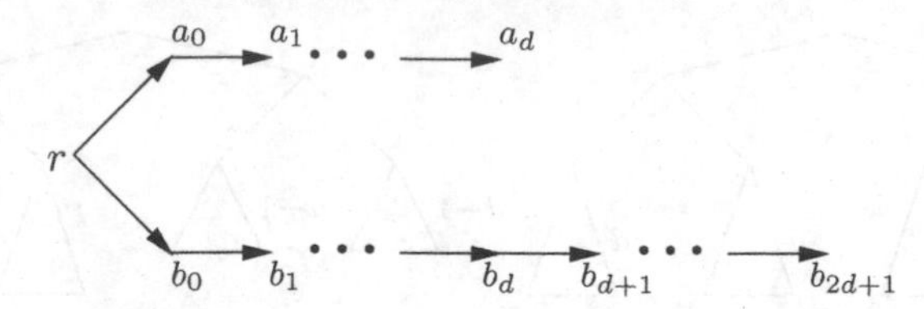

Fig. 4.5. Inexpressibility of same generation

that the shortest paths from c to a and from c to b have the same length. This query is most commonly computed on trees; in this case a, b are in the same generation if they are at the same distance from the root.

We now give a very simple proof that the same-generation query Q_{sg} is not FO-definable. Assume to the contrary that it is FO-definable, and $\mathsf{lr}(Q_{sg}) = d$. Consider a tree T with root r and two branches, one with nodes $a_0, a_1, \ldots, a_d$ (where a_{i+1} is the successor of a_i) and the other one with nodes $b_0, b_1, \ldots, b_d, \ldots, b_{2d+1}$, see Fig. 4.5.

It is clear that $(a_d, b_d) \approx_d^T (a_d, b_{d+1})$, while a_d, b_d are in the same generation, and a_d, b_{d+1} are not.

In most examples seen so far, locality ranks (for either Hanf- or Gaifman-locality) were exponential in the quantifier rank. We now show a simple exponential lower bound for the locality rank; precise bounds will be given in Exercise 4.11.

Suppose that σ is the vocabulary of undirected graphs: that is, $\sigma = \{E\}$ where E is binary. Define the following formulae:

- $d_0(x, y) \;\equiv\; E(x, y)$,
- $d_1(x, y) \equiv \exists z\, (d_0(x, z) \wedge d_0(y, z))$, ...,
- $d_{k+1}(x, y) \;\equiv\; \exists z (d_k(x, z) \wedge d_k(y, z))$.

For an undirected graph, $d_k(a, b)$ holds iff there is a path of length 2^k between a and b; that is, if the distance between a and b is at most 2^k. Hence, $\mathsf{lr}(d_k) \geq 2^{k-1}$. However, $\mathsf{qr}(d_k) = k$, which shows that locality rank can be exponential in the quantifier rank.

4.4 Structures of Small Degree

In this section, we shall see a large class of structures for which very simple criteria for FO-definability can be obtained. These are structures in which all the degrees are bounded by a constant. If we deal with undirected graphs, degrees are the usual degrees of nodes; if we deal with directed graphs, they are in- and out-degrees. In general, we use the following definition.

Definition 4.14. *Let σ be a relational vocabulary, R an m-ary symbol in σ, and $\mathfrak{A} \in \text{STRUCT}[\sigma]$. For $a \in A$ and $i \leq m$, define $degree^{\mathfrak{A}}_{R,i}(a)$ as the cardinality of the set*

$$\{(a_1, \ldots, a_{i-1}, a, a_{i+1}, \ldots, a_m) \in R^{\mathfrak{A}} \mid a_1, \ldots, a_{i-1}, a_{i+1}, \ldots, a_m \in A\}.$$

That is, $degree^{\mathfrak{A}}_{R,i}(a)$ is the number of tuples in $R^{\mathfrak{A}}$ that have a in the ith position.

Define $deg\_set(\mathfrak{A})$ to be the set of all the numbers of the form $degree^{\mathfrak{A}}_{R,i}(a)$, where $a \in A$, $R \in \sigma$, and i is at most the arity of R. That is,

$$deg\_set(\mathfrak{A}) \;=\; \{degree^{\mathfrak{A}}_{R,i}(a) \mid a \in A,\; R \in \sigma,\; i \leq arity(R)\}.$$

Finally, $\text{STRUCT}_l[\sigma]$ stands for

$$\{\mathfrak{A} \in \text{STRUCT}[\sigma] \mid deg\_set(\mathfrak{A}) \subseteq \{0, \ldots, l\}\}.$$

In other words, $\text{STRUCT}_l[\sigma]$ consists of σ-structures in which all degrees do not exceed l.

We shall also be applying *deg_set* to outputs of queries: by $deg\_set(Q(\mathfrak{A}))$, for an m-ary query Q, we mean the set of all degrees realized in the structure whose only m-ary relation is $Q(\mathfrak{A})$; that is, $deg\_set(\langle A, Q(\mathfrak{A})\rangle)$.

When we talk of structures of *small degree*, we mean $\text{STRUCT}_l[\sigma]$ for some fixed $l \in \mathbb{N}$.

There is another way of defining structures of small degree, essentially equivalent to the way we use here. Instead of defining degrees for m-ary relations, one can use only the definition of degrees for nodes of an undirected graph, and define structures of small degrees as structures $\mathfrak{A}$ where $deg\_set(\mathcal{G}(\mathfrak{A})) \subseteq \{0, \ldots, l\}$ for some $l \in \mathbb{N}$. Recall that $\mathcal{G}(\mathfrak{A})$ is the Gaifman graph of $\mathfrak{A}$, so in this case we are talking about the usual degrees in a graph. However, this is essentially the same as the definition of $\text{STRUCT}_l[\sigma]$.

Lemma 4.15. *For every relational vocabulary σ, there exist two functions $f_\sigma, g_\sigma : \mathbb{N} \to \mathbb{N}$ such that*

1. *$deg\_set(\mathcal{G}(\mathfrak{A})) \subseteq \{0, \ldots, f_\sigma(l)\}$ for every $\mathfrak{A} \in \text{STRUCT}_l[\sigma]$, and*
2. *$\mathfrak{A} \in \text{STRUCT}_{g_\sigma(l)}[\sigma]$ for every $\mathfrak{A}$ with $deg\_set(\mathcal{G}(\mathfrak{A})) \subseteq \{0, \ldots, l\}$.*

One reason to study structures of small degrees is that many queries behave particularly nicely on them. We capture this notion of nice behavior by the following definition.

Definition 4.16. *Let σ be relational. An m-ary query Q on σ-structures, $m > 0$, has the* bounded number of degrees property (BNDP) *if there exists a function $f_Q : \mathbb{N} \to \mathbb{N}$ such that for every $l \geq 0$ and every $\mathfrak{A} \in \text{STRUCT}_l[\sigma]$,*

$$|\,deg\_set(Q(\mathfrak{A}))| \;\leq\; f_Q(l).$$

Notice a certain asymmetry of this definition: our assumption is that all the numbers in *deg_set*($\mathfrak{A}$) are small, but the conclusion is that the *cardinality* of *deg_set*($Q(\mathfrak{A})$) is small. We cannot possibly ask for all the numbers in *deg_set*($Q(\mathfrak{A})$) to be small and still say anything interesting about FO-definable queries: consider, for example, the query defined by $\varphi(y, z) \equiv \exists x(x = x)$. On every structure $\mathfrak{A}$ with $| A |= n > 0$, it defines the complete graph on n nodes, where every node has the same degree n. Hence, some degrees in *deg_set*($Q(\mathfrak{A})$) do depend on $\mathfrak{A}$, but the *number of different degrees* is determined by *deg_set*($\mathfrak{A}$) and the query.

It is usually very easy to show that a query does not have the BNDP. Consider, for example, the transitive closure query. Assume that its input is a successor relation G_n on n nodes. Then *deg_set*(G_n) $= \{0, 1\}$. The transitive closure of G_n is a linear order L_n on n nodes, and *deg_set*(L_n) $= \{0, \ldots, n-1\}$, showing that the transitive closure query does not have the BNDP.

We next show that the BNDP is closely related to locality concepts.

Theorem 4.17. *Let Q be a Gaifman-local m-ary query, $m > 0$. Then Q has the BNDP.*

Proof. Let Q be Gaifman-local with $\mathsf{lr}(Q) = d$. We assume, without loss of generality, that $m \geq 2$, since unary queries clearly have the BNDP.

Next, we need the following claim. Let $n_d(k)$ be defined inductively by $n_d(0) = d, n_d(k+1) = 3 \cdot n_d(k) + 1$. That is, $n_d(k) = 3^k \cdot d + (3^k - 1)/2$ for $k \geq 0$.

Claim 4.18. *Let $\vec{a} \approx^{\mathfrak{A}}_{n_d(k)} \vec{b}$. Then there is a bijection $f : A^k \to A^k$ such that $\vec{a}\vec{c} \approx^{\mathfrak{A}}_d \vec{b}f(\vec{c})$ for every $\vec{c} \in A^k$.*

The proof of Claim 4.18 is by induction on k. For $k = 0$ there is nothing to prove. Assume that it holds for k, and prove it for $k + 1$. Let $r = n_d(k)$; then $n_d(k+1) = 3r+1$. Let $\vec{a} \approx^{\mathfrak{A}}_{3r+1} \vec{b}$. Then, by Lemma 4.9, $(\mathfrak{A}, \vec{a}) \leftrightarrows_r (\mathfrak{A}, \vec{b})$. That is, there exists a bijection $g : A \to A$ such that for every $c \in A$, $\vec{a}c \approx^{\mathfrak{A}}_r \vec{b}g(c)$. By the induction hypothesis, we then know that for each $c \in A$, there exists a bijection $g_c : A^k \to A^k$ such that for every $\vec{e} \in A^k$,

$$\vec{a}c\vec{e} \;\approx^{\mathfrak{A}}_d\; \vec{b}g(c)g_c(\vec{e}).$$

We thus define a bijection $f : A^{k+1} \to A^{k+1}$ as follows: if $\vec{c} = c\vec{e}$, where $\vec{e} \in A^k$, then $f(\vec{c}) = g(c)g_c(\vec{e})$. Clearly, $\vec{a}\vec{c} \approx^{\mathfrak{A}}_d \vec{b}f(\vec{c})$. This proves the claim.

Now we prove the BNDP. First, note that for every vocabulary σ, there exists a function $G_\sigma : \mathbb{N} \times \mathbb{N} \to \mathbb{N}$ such that for every $\mathfrak{A} \in \mathrm{STRUCT}_l[\sigma]$, the size of $B^{\mathfrak{A}}_d(a)$ is at most $G_\sigma(l, d)$. Thus, there exists a function $F_\sigma : \mathbb{N} \times \mathbb{N} \to \mathbb{N}$ such that every structure $\mathfrak{A}$ in $\mathrm{STRUCT}_l[\sigma]$ can realize at most $F_\sigma(l, d)$ isomorphism types of d-neighborhoods of a point.

Now consider $Q(\mathfrak{A})$, for $\mathfrak{A} \in \mathrm{STRUCT}_l[\sigma]$, and note that for any two $a, b \in A$ with $a \approx^{\mathfrak{A}}_{n_d(m-1)} b$,

$$|\{\vec{c} \in A^{m-1} \mid a\vec{c} \in Q(\mathfrak{A})\}| \;=\; |\{\vec{c} \in A^{m-1} \mid b\vec{c} \in Q(\mathfrak{A})\}|, \qquad (4.2)$$

by Claim 4.18. In particular, (4.2) implies that the degrees of a and b in $Q(\mathfrak{A})$ (in the first position of an m-tuple) are the same. This is because $degree_1^{Q(\mathfrak{A})}(c)$, the degree of an element c, corresponding to the first position of the m-ary relation $Q(\mathfrak{A})$, is precisely the cardinality of the set $\{\vec{c} \in A^{m-1} \mid c\vec{c} \in Q(\mathfrak{A})\}$. Thus, the number of different degrees in $Q(\mathfrak{A})$ corresponding to the first position in the m-tuple is at most $F_\sigma(l, n_d(m-1))$, and hence

$$|\,deg\_set(Q(\mathfrak{A}))\,| \;\leq\; m \cdot F_\sigma(l, n_d(m-1)). \qquad (4.3)$$

Since the upper bound in (4.3) depends on l, m, d, and σ only, this proves the BNDP. □

Corollary 4.19. *Every* FO*-definable query has the BNDP.* □

Balanced Binary Trees Revisited

We now revisit the balanced binary tree test, and give a simple proof of its inexpressibility in FO. In fact, we show that this test is inexpressible even if it is restricted to binary trees. That is, there is no FO-definable Boolean query Q_{bbt} such that, for a binary tree T, the output $Q_{bbt}(T)$ is true iff T is balanced.

Assume, to the contrary, that such a query is FO-definable. We now construct a binary FO-definable query Q which fails the BNDP – this would contradict Corollary 4.19.

The new query Q works as follows. It takes as an input a binary tree T, and for every two nonleaf nodes a, b finds their successors a', a'' and b', b''. It then constructs a new tree $T_{a,b}$ by removing the edges from a to a', a'' and from b to b', b'', and instead by adding the edges from a to b', b'' and from b to a', a''. It then puts (a, b) in the output if $Q_{bbt}(T_{a,b})$ is true (see Fig. 4.6). Clearly, Q is FO-definable, if Q_{bbt} is.

Assume that T itself is a balanced binary tree; that is a structure in $\text{STRUCT}_2[\sigma]$. Then for two nonleaf nodes a, b, the pair (a, b) is in $Q(T)$ iff a, b are at the same distance from the root. Hence, for a balanced binary tree T of depth n, the graph $Q(T)$ is a disjoint union of $n-1$ cliques of different sizes, and thus $|\; deg\_set(Q(T)) \;| = n-1$. Hence, Q fails the BNDP, which proves that Q_{bbt} is not FO-definable. □

4.5 Locality of FO Revisited

In this section, we start by analyzing the proof of Hanf-locality of FO, and discover that it establishes a stronger statement than that of Theorem 4.12. We characterize a new notion of expressibility via a stronger version of Ehrenfeucht-Fraïssé games, which will later be used to prove bounds on logics

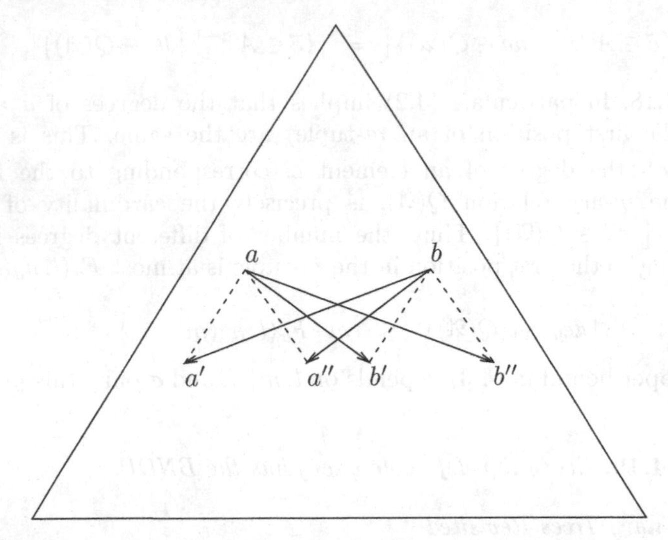

Fig. 4.6. Changing successors of nodes in a balanced binary tree

with counting quantifiers. The question that we ask then is: are there more precise and restrictive locality criteria that can be stated for FO? The answer to this is positive, and we shall present two such results: Gaifman's theorem, and the threshold equivalence criterion.

First, we show how to avoid the restriction that no constant symbols occur in σ; that is, we extend the notions of the r-ball and r-neighborhood to the case of arbitrary relational vocabularies σ (vocabularies without function symbols). Let $\vec{c} = (c_1, \ldots, c_n)$ list all the constant symbols of σ. Then

$$B_r^{\mathfrak{A}}(\vec{a}) \quad = \quad \{b \in A \mid d_{\mathfrak{A}}(b, \vec{a}) \leq r \text{ or } d_{\mathfrak{A}}(b, \vec{c}^{\mathfrak{A}}) \leq r\}.$$

The r-neighborhood of $\vec{a}$, with $|\vec{a}| = m$, is defined as the structure $N_r^{\mathfrak{A}}(\vec{a})$ in the vocabulary σ_m (σ extended with m constants), whose universe is $B_r^{\mathfrak{A}}(\vec{a})$, the interpretations of σ-relations and constants are inherited from $\mathfrak{A}$, and the m extra constants are interpreted as $\vec{a}$.

One can check that all the results proved so far extend to the setting that allows constants (see Exercise 4.1). From now on, we apply all the locality concepts to relational vocabularies.

We can also use the notion of locality to state when $\mathfrak{A} \equiv_0 \mathfrak{B}$; that is, when the duplicator wins the Ehrenfeucht-Fraïssé game on $\mathfrak{A}$ and $\mathfrak{B}$ without even starting. This happens if and only if $(\emptyset, \emptyset)$ is a partial isomorphism, or, equivalently, $N_0^{\mathfrak{A}}(\emptyset) \cong N_0^{\mathfrak{B}}(\emptyset)$.

We now define a new equivalence relation $\leftrightarrows_k^{bij}$ as follows.

- $\mathfrak{A} \leftrightarrows_0^{bij} \mathfrak{B}$ if $\mathfrak{A} \equiv_0 \mathfrak{B}$;

- $\mathfrak{A} \simeq^{bij}_{k+1} \mathfrak{B}$ if there is a bijection $f : A \to B$ such that

  forth: for each $a \in A$, we have $(\mathfrak{A}, a) \simeq^{bij}_{k} (\mathfrak{B}, f(a))$;

  back: for each $b \in B$, we have $(\mathfrak{A}, f^{-1}(b)) \simeq^{bij}_{k} (\mathfrak{B}, b)$.

One can easily see that just one of **forth** and **back** suffices: that is, **forth** and **back** are equivalent, since f is a bijection.

The notion of the back-and-forth described in Sect. 3.5 was equivalent to the Ehrenfeucht-Fraïssé game. We can also describe the new notion of back-and-forth as a game, called a *bijective Ehrenfeucht-Fraïssé game* (or just bijective game). Let $\mathfrak{A}$ and $\mathfrak{B}$ be two structures in a relational vocabulary. The k-round bijective game is played by the same two players, the spoiler and the duplicator. If $|A| \neq |B|$, then the duplicator loses before the game even starts. In the ith round, the duplicator first selects a bijection $f_i : A \to B$. Then the spoiler moves in exactly the same way as in the Ehrenfeucht-Fraïssé game: that is, he plays either $a_i \in A$ or $b_i \in B$. The duplicator responds by either $f(a_i)$ or $f^{-1}(b_i)$. As in the Ehrenfeucht-Fraïssé game, the duplicator wins if, after k rounds, the moves $(\vec{a}, \vec{b})$ form a winning position: that is, $(\vec{a}, \vec{c}^{\mathfrak{A}})$ and $(\vec{b}, \vec{c}^{\mathfrak{B}})$ are a partial isomorphism between $\mathfrak{A}$ and $\mathfrak{B}$.

If the duplicator has a winning strategy in the k-round bijective game on $\mathfrak{A}$ and $\mathfrak{B}$, we write $\mathfrak{A} \equiv^{bij}_{k} \mathfrak{B}$. Clearly, it is harder for the duplicator to win the bijective game; that is, $\mathfrak{A} \equiv^{bij}_{k} \mathfrak{B}$ implies $\mathfrak{A} \equiv_k \mathfrak{B}$. In the bijective game, the duplicator does not simply come up with responses to all the possible moves by the spoiler, but he has to establish a one-to-one correspondence between the spoiler's moves and his responses.

The following is immediate from the definitions.

Lemma 4.20. $\mathfrak{A} \simeq^{bij}_{k} \mathfrak{B}$ *iff* $\mathfrak{A} \equiv^{bij}_{k} \mathfrak{B}$. □

By Corollary 4.10, $(\mathfrak{A}, \vec{u}) \leftrightarrows_{3d+1} (\mathfrak{B}, \vec{v})$ implies the existence of a bijection $f : A \to B$ such that $(\mathfrak{A}, \vec{u}c) \leftrightarrows_d (\mathfrak{B}, \vec{v}f(c))$ for all $c \in A$. Since the winning condition in the bijective game is that $N^{\mathfrak{A}}_0(\vec{a}) \cong N^{\mathfrak{B}}_0(\vec{b})$, where $\vec{a}$ and $\vec{b}$ are the moves of the game on $\mathfrak{A}$ and $\mathfrak{B}$, by induction on k we conclude:

Corollary 4.21. *If* $(\mathfrak{A}, \vec{a}) \leftrightarrows_{(3^k-1)/2} (\mathfrak{B}, \vec{b})$, *then* $(\mathfrak{A}, \vec{a}) \equiv^{bij}_{k} (\mathfrak{B}, \vec{b})$. □

Bijective games, as will be seen, characterize the expressive power of a certain logic. Since the bijective game is harder to win for the duplicator than the ordinary Ehrenfeucht-Fraïssé game, such a logic must be more expressive than FO. Hence, the tool of Hanf-locality will be applicable to a certain extension of FO. We shall see how it works when we discuss logics with *counting* in Chap. 8.

Since the most general locality-based bounds apply to more restricted games than the ordinary Ehrenfeucht-Fraïssé games, and hence to more expressive logics, it is natural to ask whether more specific locality criteria can be stated for FO. We now present two such criteria.

We start with Gaifman's theorem. First, a few observations are needed. If σ is a relational vocabulary, and m is the maximum arity of a relation symbol in it, $m \geq 2$, then the Gaifman graph $\mathcal{G}(\mathfrak{A})$ is definable by a formula of quantifier rank $m-2$. (Note that for the case of unary relations, the Gaifman graph is simply $\{(a,a) \mid a \in A\}$ and hence is definable by the formula $x = y$.)

We show this for the case of a single ternary relation R; a general proof should be obvious. The Gaifman graph is then defined by the formula

$$(x = y) \vee \exists z \begin{pmatrix} R(x,y,z) \vee R(x,z,y) \vee R(y,x,z) \\ \vee\, R(y,z,x) \vee R(z,x,y) \vee R(z,y,x) \end{pmatrix}.$$

Since the Gaifman graph is FO-definable, so is the r-ball of any tuple $\vec{x}$. That is, for any fixed r, there is a formula $d^{\leq r}(y,\vec{x})$ such that $\mathfrak{A} \models d^{\leq r}(b,\vec{a})$ iff $d_{\mathfrak{A}}(b,\vec{a}) \leq r$. Similarly, there are formulae $d^{=r}$ and $d^{>r}$. We can next define *local* quantification

$$\exists y \in B_r(\vec{x})\ \varphi \qquad \forall y \in B_r(\vec{x})\ \varphi$$

simply as abbreviations: $\exists y \in B_r(\vec{x})\ \varphi$ stands for $\exists y\ \left(d^{\leq r}(y,\vec{x}) \wedge \varphi\right)$, and $\forall y \in B_r(\vec{x})\ \varphi$ stands for $\forall y\ \left(d^{\leq r}(y,\vec{x}) \rightarrow \varphi\right)$.

For a fixed r, we say that a formula $\psi(\vec{x})$ is *r-local around $\vec{x}$*, and write this as $\psi^{(r)}(\vec{x})$, if all quantification in ψ is of the form $\exists y \in B_r(\vec{x})$ or $\forall y \in B_r(\vec{x})$.

Theorem 4.22 (Gaifman). *Let σ be relational. Then every* FO *formula $\varphi(\vec{x})$ over σ is equivalent to a Boolean combination of the following:*

- *local formulae $\psi^{(r)}(\vec{x})$ around $\vec{x}$;*
- *sentences of the form*

$$\exists x_1,\ldots,x_s \left(\bigwedge_{i=1}^{s} \alpha^{(r)}(x_i) \ \wedge \bigwedge_{1\leq i<j\leq s} d^{>2r}(x_i,x_j) \right).$$

Furthermore,

- *the transformation from φ to such a Boolean combination is effective;*
- *if φ itself is a sentence, then only sentences of the above form appear in the Boolean combination;*
- *if $\mathsf{qr}(\varphi) = k$, and n is the length of $\vec{x}$, then the bounds on r and s are $r \leq 7^k$, $s \leq k+n$.* □

Notice that Gaifman-locality of FO is an immediate corollary of Gaifman's theorem (hence the name). However, the proof we presented earlier is much simpler than the proof of Gaifman's theorem (Exercise 4.9), and the bounds obtained are better.

Thus, Gaifman-locality can be strengthened for the case of FO formulae. Then what about Hanf-locality? The answer, as it turns out, is positive, if one's attention is restricted to structures in which degrees are bounded. We start with the following definition.

Definition 4.23 (Threshold equivalence). *Given two structures $\mathfrak{A}, \mathfrak{B}$ in a relational vocabulary, we write $\mathfrak{A} \leftrightarrows_{d,m}^{thr} \mathfrak{B}$ if for every isomorphism type τ of a d-neighborhood of a point either*

- *both $\mathfrak{A}$ and $\mathfrak{B}$ have the same number of points that d-realize τ, or*
- *both $\mathfrak{A}$ and $\mathfrak{B}$ have at least m points that d-realize τ.*

Thus, if m were allowed to be infinity, $\mathfrak{A} \leftrightarrows_{d,\infty}^{thr} \mathfrak{B}$ would be the usual definition of $\mathfrak{A} \leftrightarrows_d \mathfrak{B}$. In the new definition, however, we are only interested in the number of elements that d-realize a type of neighborhood up to a threshold: below the threshold, the numbers must be the same, but above it, they do not have to be.

Theorem 4.24. *For each $k, l > 0$, there exist $d, m > 0$ such that for $\mathfrak{A}, \mathfrak{B} \in$ STRUCT$_l[\sigma]$,*

$$\mathfrak{A} \leftrightarrows_{d,m}^{thr} \mathfrak{B} \quad \textit{implies} \quad \mathfrak{A} \equiv_k \mathfrak{B}.$$

Proof. The proof is very similar to the proof of Hanf-locality of FO. We define inductively $r_0 = 0, r_{i+1} = 3r_i+1$, take $d = r_{k-1}$, and prove that the duplicator can play the Ehrenfeucht-Fraïssé game on $\mathfrak{A}$ and $\mathfrak{B}$ in such a way that after i rounds (or: with $k - i$ rounds remaining),

$$N_{r_{k-i}}^{\mathfrak{A}}(\vec{a}_i) \quad \cong \quad N_{r_{k-i}}^{\mathfrak{B}}(\vec{b}_i), \tag{4.4}$$

where $\vec{a}_i, \vec{b}_i$ are points played in the first i rounds of the game.

It only remains to specify m. Recall from the proof of Theorem 4.17 that there is a function $G_\sigma : \mathbb{N} \times \mathbb{N}$ such that the maximum size of a radius d neighborhood of a point in a structure in STRUCT$_l[\sigma]$ is $G_\sigma(d, l)$. We take m to be $k \cdot G_\sigma(r_k, l)$.

The rest is by induction on i. For the first move, suppose the spoiler plays $a \in A$. By $\mathfrak{A} \leftrightarrows_{r_k,m}^{thr} \mathfrak{B}$, the duplicator can find $b \in B$ with $N_{r_k}^{\mathfrak{A}}(a) \cong N_{r_k}^{\mathfrak{B}}(b)$.

Now assume (4.4) holds after i rounds. That is, $N_{3r+1}^{\mathfrak{A}}(\vec{a}_i) \cong N_{3r+1}^{\mathfrak{B}}(\vec{b}_i)$, where $r = r_{k-(i+1)}$. We have to show that (4.4) holds after $i+1$ rounds (i.e., with $k-(i+1)$ rounds remaining). Suppose in round $i+1$ the spoiler plays $a \in A$ (the case of a move in B is identical). If $a \in B_{2r+1}^{\mathfrak{A}}(\vec{a}_i)$, the response is by the isomorphism between $N_{3r+1}^{\mathfrak{A}}(\vec{a}_i)$ and $N_{3r+1}^{\mathfrak{B}}(\vec{b}_i)$, which guarantees (4.4). If $a \notin B_{2r+1}^{\mathfrak{A}}(\vec{a}_i)$, let τ be the isomorphism type of the r-neighborhood of a. To ensure (4.4), all we need is to find $b \in B$ such that b r-realizes τ in $\mathfrak{B}$, and $d_{\mathfrak{B}}(b, \vec{b}_i) > 2r+1$ – then such an element b would be the response of the duplicator.

Assume that there is no such element b. Since there is an element $a \in A$ that r-realizes τ in $\mathfrak{A}$, there must be an element $b' \in B$ that r-realizes τ in $\mathfrak{B}$. Then all such elements b' must be in $N_{2r+1}^{\mathfrak{B}}(\vec{b}_i)$. Let there be s of them.

Notice that the cardinality of $N_{2r+1}^{\mathfrak{B}}(\vec{b}_i)$ does not exceed $m = k \cdot G_\sigma(r_k, l)$. This is because the length of $\vec{b}_i$ is at most k, the size of each r_k neighborhood is at most $G_\sigma(r_k, l)$, and $2r + 1 \leq r_k$.

Therefore, $s \leq m$, and from $\mathfrak{A} \leftrightarrows_{d,m}^{thr} \mathfrak{B}$ we see that there are exactly s elements $a' \in A$ that r-realize τ in $\mathfrak{A}$. But by the isomorphism between $N_{3r+1}^{\mathfrak{A}}(\vec{a}_i)$ and $N_{3r+1}^{\mathfrak{B}}(\vec{b}_i)$ we know that $N_{2r+1}^{\mathfrak{A}}(\vec{a}_i)$ alone contains s such elements, and hence there are at least $s + 1$ of them in A. This contradiction shows that we can find b that r-realizes τ in $\mathfrak{B}$ outside of $N_{2r+1}^{\mathfrak{B}}(\vec{b}_i)$, which completes the proof of (4.4) and the theorem. □

The threshold equivalence is a useful tool when in the course of proving inexpressibility of a certain property, one constructs pairs of structures $\mathfrak{A}_k, \mathfrak{B}_k$ whose universes have different cardinalities: then Hanf-locality is inapplicable.

For example, consider the following query over graphs. Suppose the input graph is a simple cycle with loops on some nodes (i.e., it has edges $(a_1, a_2), (a_2, a_3), \ldots, (a_{n-1}, a_n), (a_n, a_1)$, with all a_is distinct, as well as some edges of the form (a_i, a_i)). The question is whether the number of loops is even. An attempt to prove that it is not FO-definable using Hanf-locality does not succeed: for any $d > 0$, and any two structures $\mathfrak{A}, \mathfrak{B}$ with $\mathfrak{A} \leftrightarrows_d \mathfrak{B}$, the numbers of nodes with loops in $\mathfrak{A}$ and $\mathfrak{B}$ are equal.

However, the threshold equivalence helps us. Assume that the above query Q is expressible by a sentence of quantifier rank k. Then apply Theorem 4.24 to k and 2 (the maximum degree in graphs described above), and find $d, m > 0$. We now construct a graph $G_{d,n}$ for any $n > 0$, as a cycle on which the distance between any two consecutive nodes with loops is $2d+2$, and the number of such nodes with loops is n. One can then easily check that $G_{d,m+1} \leftrightarrows_{d,m}^{thr} G_{d,m+2}$ and hence the two must agree on Q. This is certainly impossible, showing that Q is not FO-definable.

Note that in this example, $G_{d,m+1} \not\leftrightarrows_r G_{d,m+2}$ for any $r > 0$, since the cardinalities of $G_{d,m+1}$ and $G_{d,m+2}$ are different, and hence Hanf-locality is not applicable.

4.6 Bibliographic Notes

The first locality result for FO was Hanf's theorem, formulated in 1965 by Hanf [120] for infinite models. The version for the finite case was presented by Fagin, Stockmeyer, and Vardi in [76]. In fact, [76] proves what we call the threshold equivalence for FO, and what we call Hanf-locality is stated as a corollary.

Gaifman's theorem is from [88]; Gaifman-locality, inspired by it, was introduced by Hella, Libkin, and Nurmonen [123], who also proved Theorem 4.11. The proof of Hanf-locality for FO follows Libkin [167].

The bounded number of degrees property (BNDP) is from Libkin and Wong [169] (where it was called BDP, and proved only for FO-definable queries over graphs). Dong, Libkin and Wong [57] showed that every Gaifman-local query has the BNDP, and a simpler proof was given by Libkin [166].

Bijective games were introduced by Hella [121], and the connection between them and Hanf-locality is due to Nurmonen [188]; the presentation here follows [123].

Sources for exercises:

Exercise 4.9:	Gaifman [88]
Exercises 4.10, 4.11, and 4.12:	Libkin [166]
Exercise 4.13:	Dong, Libkin, and Wong [57]
Exercise 4.14:	Schwentick and Barthelmann [217]
Exercise 4.15:	Schwentick [215]

4.7 Exercises

Exercise 4.1. Verify that all the results in Sects. 4.1–4.4 extend to vocabularies with constant symbols.

Exercise 4.2. Prove Lemma 4.4.

Exercise 4.3. Prove Lemma 4.5.

Exercise 4.4. Prove Lemma 4.8.

Exercise 4.5. Prove Lemma 4.15.

Exercise 4.6. Use Hanf-locality to give a simple proof that graph acyclicity and testing if a graph is a tree are not FO-definable.

Exercise 4.7. Consider *colored* graphs: that is, structures of vocabulary $\{E, U_1, \ldots, U_k\}$ where E is binary and $U_1, \ldots, U_k$ are unary (i.e., U_i defines the set of nodes of color i). Prove that neither connectivity nor transitive closure are FO-definable over colored graphs.

Exercise 4.8. Provide a complete proof that testing if a binary tree is balanced is not FO-definable.

Exercise 4.9. Prove Theorem 4.22.

Exercise 4.10. In all the proofs in this chapter we obtained bounds on locality ranks of the order $O(3^k)$, where k is the quantifier rank. And yet the exponential lower bound was $O(2^k)$. The goal of this exercise is to reduce the upper bound from $O(3^k)$ to $O(2^k)$, at the expense of a slightly more complicated proof.

Let $\vec{x} = (x_1, \ldots, x_n)$, and let $\mathcal{I} = \{I_1, \ldots, I_m\}$ be a partition of $\{1, \ldots, n\}$. The subtuple of $\vec{x}$ that consists of the components whose indices are in I_j is denoted by $\vec{x}_j^{\mathcal{I}}$.

Let $r > 0$. Given two structures, $\mathfrak{A}$ and $\mathfrak{B}$, and $\vec{a} \in A^n, \vec{b} \in B^n$, we say that $\vec{a}$ and $\vec{b}$ are $(\mathcal{I}, r)$-*similar* if the following hold:

- $N_r^{\mathfrak{A}}(\vec{a}_j^{\mathcal{I}}) \cong N_r^{\mathfrak{B}}(\vec{b}_j^{\mathcal{I}})$ for all $j = 1, \ldots, m$;
- $d(\vec{a}_j^{\mathcal{I}}, \vec{a}_l^{\mathcal{I}}) > r$ for all $l \neq j$;
- $d(\vec{b}_j^{\mathcal{I}}, \vec{b}_l^{\mathcal{I}}) > r$ for all $l \neq j$.

We call $\vec{a}$ and $\vec{b}$ r-*similar* if there exists a partition $\mathcal{I}$ such that $\vec{a}$ and $\vec{b}$ are $(\mathcal{I}, r)$-similar. A formula φ has the r-*separation property* if $\mathfrak{A} \models \varphi(\vec{a}) \leftrightarrow \varphi(\vec{b})$ whenever $\vec{a}$ and $\vec{b}$ are r-similar.

Your first task is to prove that a formula has the separation property iff it is Gaifman-local.

Next, prove the following. If $r > 0$, $\mathfrak{A} \leftrightarrows_r \mathfrak{B}$, and $\vec{a}, \vec{b}$ are $2r$-similar, then there exists a bijection $f : A \to B$ such that, for every $c \in A$, the tuples $\vec{a}x$ and $\vec{b}f(c)$ are r-similar.

Use this result to show that $\mathsf{lr}(\varphi) \leq 2^k$ for every FO formula φ of quantifier rank k.

Exercise 4.11. Define functions $\text{Hanf\_rank}_{\text{FO}}, \text{Gaifman\_rank}_{\text{FO}} : \mathbb{N} \to \mathbb{N}$ as follows:

$$\text{Hanf\_rank}_{\text{FO}}(n) = \max\{\mathsf{hlr}(\varphi) \mid \varphi \in \text{FO}, \ \mathsf{qr}(\varphi) = n\},$$

$$\text{Gaifman\_rank}_{\text{FO}}(n) = \max\{\mathsf{lr}(\varphi) \mid \varphi \in \text{FO}, \ \mathsf{qr}(\varphi) = n\}.$$

Assume that the vocabulary is purely relational. Prove that for every $n > 1$, $\text{Hanf\_rank}_{\text{FO}}(n) = 2^{n-1} - 1$ and $\text{Gaifman\_rank}_{\text{FO}}(n) = 2^n - 1$.

Exercise 4.12. Exponential lower bounds for locality rank were achieved on formulae of quantifier rank n with the total number of quantifiers exponential in n. Could it be that locality rank is polynomial in the number of quantifiers?

Your goal is to show that the answer is negative. More precisely, show that there exist FO formulae with n quantifiers and locality rank $O(\sqrt{2}^n)$.

Exercise 4.13. The BNDP was formulated in a rather asymmetric way: the assumption was that $\forall i \in \mathit{deg\_set}(\mathfrak{A})\ (i \leq l)$, and the conclusion that $|\,\mathit{deg\_set}(Q(\mathfrak{A}))\,| \leq f_Q(l)$. A natural way to make it more symmetric is to introduce the following property of a query Q: there exists a function $f'_Q : \mathbb{N} \to \mathbb{N}$ such that

$$|\,\mathit{deg\_set}(Q(\mathfrak{A}))\,| \ \leq \ f'_Q(|\,\mathit{deg\_set}(\mathfrak{A})\,|)$$

for ever structure $\mathfrak{A}$.

Prove that there are FO-definable queries on finite graphs that violate the above property.

Exercise 4.14. Recall that a formula $\varphi(\vec{x})$ is r-local around $\vec{x}$ if all the quantification is of the form $\exists y \in B_r(\vec{x})$ and $\forall y \in B_r(\vec{x})$. We now say that $\varphi(\vec{x})$ is *basic r-local around* $\vec{x}$ if it is a Boolean combination of formulae of the form $\alpha(x_i)$, where x_i is a component of $\vec{x}$, and $\alpha(x_i)$ is r-local around x_i. A formula is local (or basic local) around $\vec{x}$ if it is r-local (or basic r-local) around $\vec{x}$ for some r.

Prove that every FO formula $\varphi(\vec{x})$ that is local around $\vec{x}$ is logically equivalent to a formula that is basic local around $\vec{x}$.

Use this result to prove that any FO sentence is logically equivalent to a sentence of the form

$$\exists x_1 \ldots \exists x_n \forall y \ \varphi(x_1, \ldots, x_n, y),$$

where $\varphi(x_1, \ldots, x_n, y)$ is local around $(x_1, \ldots, x_n, y)$.

Exercise 4.15. This exercise presents a sufficient condition that guarantees a winning strategy by the duplicator. It shows that if two structures look similar (meaning that the duplicator has a winning strategy), and are extended to bigger structures in a "similar way", then the duplicator has a winning strategy on the bigger structures as well.

Let $\mathfrak{A}$, $\mathfrak{B}$ be two structures of the same vocabulary that contains only relation symbols. Let $\mathfrak{A}_0, \mathfrak{B}_0$ be their substructures, with universes A_0 and B_0, respectively, and let $\mathfrak{A}_1$ and $\mathfrak{B}_1$ be substructures of $\mathfrak{A}$ and $\mathfrak{B}$ whose universes are $A - A_0$ and $B - B_0$.

For every $a \in A$, $d_{\mathfrak{A}}(a, \mathfrak{A}_0)$ is, as usual, $\min\{d_{\mathfrak{A}}(a, a_0) \mid a_0 \in A_0\}$, and $d_{\mathfrak{B}}(b, \mathfrak{B}_0)$ is defined similarly. Let $\mathfrak{A}_{(r)}$ ($\mathfrak{B}_{(r)}$) be the substructure of $\mathfrak{A}$ (respectively, $\mathfrak{B}$) whose universe is $\{a \mid d_{\mathfrak{A}}(a, \mathfrak{A}_0) \leq r\}$ (respectively, $\{b \mid d_{\mathfrak{B}}(b, \mathfrak{B}_0) \leq r\}$). We write

$$\mathfrak{A}_{(r)} \equiv_k^{\text{dist}} \mathfrak{B}_{(r)}$$

if $\mathfrak{A}_{(r)} \equiv_k \mathfrak{B}_{(r)}$ and, whenever a_i, b_i are moves in the ith round, $d_{\mathfrak{A}}(a_i, \mathfrak{A}_0) = d_{\mathfrak{B}}(b_i, \mathfrak{B}_0)$. We also write

$$\mathfrak{A}_1 \cong^{\text{dist}} \mathfrak{B}_1$$

if there is an isomorphism $h : \mathfrak{A}_1 \to \mathfrak{B}_1$ such that $d_{\mathfrak{A}}(a, \mathfrak{A}_0) = d_{\mathfrak{B}}(h(a), \mathfrak{B}_0)$ for every $a \in A - A_0$.

Now assume that the following two conditions hold:

1. $\mathfrak{A}_{(2^k)} \equiv_k^{\text{dist}} \mathfrak{B}_{(2^k)}$, and
2. $\mathfrak{A}_1 \cong^{\text{dist}} \mathfrak{B}_1$.

Prove that $\mathfrak{A} \equiv_k \mathfrak{B}$.

Exercise 4.16. Let σ consist of one binary relation E, and let Φ be a σ-sentence. Prove that it is decidable whether Φ has a model in $\text{STRUCT}_1[\sigma]$; that is, one can decide if there is a finite graph G in which all in- and out-degrees are 0 and 1 such that $G \models \Phi$.

5

Ordered Structures

We know how to prove basic results about FO; so now we start adding things to FO. One way to make FO more expressive is to include additional operations on the universe. For example, in database applications, data items stored in a database are numbers, strings, etc. Both numbers and strings could be ordered; on numbers we have arithmetic operations, on strings we have concatenation, substring tests, and so on. As query languages routinely use those operations, one may want to study them in the context of FO.

In this chapter, we describe a general framework of adding new operations on the domain of a finite model. The main concept is that of invariant queries, which do not depend on a particular interpretation of the new operations. We show that such an addition could increase the expressiveness of a logic, even for properties that do not mention those new operations. We then concentrate on one operation of special importance: a linear order on the finite universe. We study FO($<$) – that is, FO with an additional linear order $<$ on the universe, and study its expressive power.

Adding ordering will be of importance for almost all logics that we study (the only exception is fragments of second-order logic, where linear orderings are definable). We shall observe the following general phenomenon: for any logic that cannot define a linear ordering, adding one increases the expressive power, even for invariant queries.

5.1 Invariant Queries

We start with an example. Suppose we have a vocabulary σ, and an additional vocabulary $\sigma_{<,+} = \{<, +\}$, where $<$ is a binary relation symbol, and $+$ is a ternary relation symbol. The intended interpretation is as follows. Given a set A, the relation $<$ is interpreted as a linear ordering on it, say $a_1 < \ldots < a_n$, if $A = \{a_1, \ldots, a_n\}$. Then $+$ is interpreted as

$$\{(a_i, a_j, a_k) \mid a_i, a_j, a_k \in A \text{ and } i + j = k\}.$$

Recall that the query EVEN($\mathfrak{A}$) testing if $|A| = 0 \pmod 2$ is not expressible over σ-structures: we proved this by using Ehrenfeucht-Fraïssé games. Now assume that we are allowed to use $\sigma_{<,+}$ symbols in the query. Then we can write:

$$\Phi \;=\; (\neg\exists x\, (x = x)) \;\vee\; \exists x \exists y\, \Big((x + x = y) \;\wedge\; \neg\exists z\, (y < z)\Big).$$

That is, either the universe is empty, or y is the largest element of the universe and $y = x + x$ for some x. Then Φ tests if $|A| = 0 \pmod 2$.

However, one has to be careful with this statement. We cannot write $\mathfrak{A} \models \Phi$ iff EVEN($\mathfrak{A}$) for a σ-structure $\mathfrak{A}$, simply because Φ is not a sentence of vocabulary σ. The structure in which Φ is checked is an *expansion* of $\mathfrak{A}$ with an interpretation of predicate symbols in $\sigma_{<,+}$. That is, if $\mathfrak{A}_{<,+}$ is a structure with universe A in which $<, +$ are interpreted as was shown above, then

$$(\mathfrak{A}, \mathfrak{A}_{<,+}) \models \Phi \quad \text{iff} \quad \text{EVEN}(\mathfrak{A}).$$

Here by $(\mathfrak{A}, \mathfrak{A}_{<,+})$ we mean the structure whose universe is A, the symbols from σ are interpreted as in $\mathfrak{A}$, and $<, +$ are interpreted as in $\mathfrak{A}_{<,+}$.

Before giving a general definition, we make another important observation. If we find any other interpretation for symbols $<$ and $+$, as long as $<$ is a linear ordering on A and $+$ is the addition corresponding to $<$, the result of the query defined by Φ will be the same. This is the idea of invariance: no matter how the extra relations are interpreted, the result of the query is the same.

We now formalize this concept. Recall that if σ and σ' are two disjoint vocabularies, $\mathfrak{A} \in \text{STRUCT}[\sigma]$, $\mathfrak{A}' \in \text{STRUCT}[\sigma']$, and $\mathfrak{A}, \mathfrak{A}'$ have the same universe A, then $(\mathfrak{A}, \mathfrak{A}')$ stands for a structure of vocabulary $\sigma \cup \sigma'$, in which the universe is A, and the interpretation of σ (respectively, σ') is inherited from $\mathfrak{A}$ ($\mathfrak{A}'$).

Definition 5.1. *Let σ and σ' be two disjoint vocabularies, and let $\mathcal{C}$ be a class of σ'-structures. Let $\mathfrak{A} \in \text{STRUCT}[\sigma]$. A formula $\varphi(\vec{x})$ in the language of $\sigma \cup \sigma'$ is called* $\mathcal{C}$-invariant on $\mathfrak{A}$ *if for any two $\mathcal{C}$ structures $\mathfrak{A}'$ and $\mathfrak{A}''$ on A we have*

$$\varphi[(\mathfrak{A}, \mathfrak{A}')] \;= \varphi[(\mathfrak{A}, \mathfrak{A}'')].$$

A formula φ is $\mathcal{C}$-invariant *if it is $\mathcal{C}$-invariant on every σ-structure.*

If $\varphi(\vec{x})$ is $\mathcal{C}$-invariant, we associate with it an m-ary query Q_φ, where $m = |\vec{x}|$. It is given by

$$\vec{a} \in Q_\varphi(\mathfrak{A}) \;\text{ iff }\; (\mathfrak{A}, \mathfrak{A}') \models \varphi(\vec{a}),$$

where $\mathfrak{A}'$ is some σ'-structure in $\mathcal{C}$ whose universe is A. By invariance, it does not matter which $\mathcal{C}$-structure $\mathfrak{A}'$ is used.

We shall write FO $+\, \mathcal{C}$ for a class of all queries on $\sigma \cup \sigma'$-structures, and

$$(\text{FO} + \mathcal{C})_{\text{inv}}$$

for the class of queries Q_φ, where φ is a $\mathcal{C}$-invariant formula over $\sigma \cup \sigma'$.

The most important case for us is when $\mathcal{C}$ is the class of finite linear orderings. In that case, we write $<$ instead of $\mathcal{C}$ and use the notation

$$(\text{FO}+<)_{\text{inv}}.$$

We refer to queries in this class as *order-invariant* queries.

Notice that $(\text{FO}+<)_{\text{inv}}$ refers to a class of queries, rather than a logic. In fact, we shall see in Chap. 9 (Exercise 9.3) that it is undecidable whether an FO sentence is $<$-invariant.

Coming back to our example of expressing EVEN with $<$ and $+$, the sentence Φ is a $\mathcal{C}_{<,+}$-invariant sentence, where $\mathcal{C}_{<,+}$ is the class of finite structures $\langle A, <, + \rangle$, with $a_1 < \ldots < a_n$ being a linear order on A, and $+$ defined as $\{(a_i, a_j, a_k) \mid i + j = k\}$. The Boolean query Q_Φ defined by this invariant sentence is precisely EVEN.

In some cases, establishing bounds on $\text{FO} + \mathcal{C}$ and $(\text{FO} + \mathcal{C})_{\text{inv}}$ is easy. For example, the proof that the bounded number of degrees property (BNDP) holds for FO shows that adding any structure of bounded degree would not violate the BNDP. Thus, we have the following result.

Proposition 5.2. *Let* $\mathcal{C} \subseteq \text{STRUCT}_l[\sigma']$ *for a fixed* $l \geq 0$. *Then* $(\text{FO} + \mathcal{C})$ *queries have the BNDP. In particular,* $(\text{FO} + \mathcal{C})$ *cannot express the transitive closure query.* □

The situation becomes much more interesting when degrees are not bounded; for example, when $\mathcal{C}$ is the class of linear orderings. We study it in the next section.

5.2 The Power of Order-invariant FO

While queries in $(\text{FO}+\mathcal{C})_{\text{inv}}$ are independent of any particular structure from $\mathcal{C}$, the mere presence of such a structure can have an impact on the expressive power.

In fact, this can be demonstrated for the class of $(\text{FO}+<)_{\text{inv}}$ queries. The main result we prove here is the following.

Theorem 5.3 (Gurevich). *There are* $(\text{FO}+<)_{\text{inv}}$ *queries that are not* FO-*definable. That is,*

$$\text{FO} \subsetneq (\text{FO}+<)_{\text{inv}}.$$

In the rest of the section we present the proof of this theorem. The proof is constructive: we explicitly generate the separating query, show that it belongs to $(\text{FO}+<)_{\text{inv}}$, and then prove that it is not FO-definable.

We consider structures in the vocabulary $\sigma = \{\subseteq\}$ where $\subseteq$ is a binary relation symbol. The intended interpretation of σ-structures of interest to us is finite Boolean algebras: that is, $\langle 2^X, \subseteq \rangle$, where X is a finite set.

We first show that there is a sentence Φ_{BA} such that $\mathfrak{A} \models \Phi_{\mathrm{BA}}$ iff $\mathfrak{A}$ is of the form $\langle 2^X, \subseteq \rangle$ for a finite X. For that, we shall need the following abbreviations:

- $\bot(x) \equiv \forall z\ (x \subseteq z)$ (intended interpretation of x then is the empty set);
- $\top(x) \equiv \forall z\ (z \subseteq x)$ (x is the maximal element with respect to $\subseteq$);
- $x \cup y = z \equiv (x \subseteq z) \wedge (y \subseteq z) \wedge \forall u\ \big((x \subseteq u) \wedge (y \subseteq u) \rightarrow (z \subseteq u)\big)$;
- $x \cap y = z \equiv (z \subseteq x) \wedge (z \subseteq y) \wedge \forall u\ \big((u \subseteq x) \wedge (u \subseteq y) \rightarrow (u \subseteq z)\big)$;
- $\mathrm{atom}(x) \equiv \neg\bot(x) \wedge \forall z\ \big(z \subseteq x \rightarrow (z = x \vee \bot(z))\big)$ (i.e., x is an atom, or a singleton set);
- $x = \bar{y} \equiv \forall z\ (x \cup y = z \rightarrow \top(z)) \wedge \forall z\ (x \cap y = z \rightarrow \bot(z))$ (x is the complement of y).

The sentence Φ_{BA} is now the usual axiomatization for atomic Boolean algebras; that is, it is a conjunction of sentences that assert that $\subseteq$ is a partial ordering, $\cup$ and $\cap$ exist, are unique, and satisfy the distributivity law and the absorption law ($x \cap (x \cup y) = x$); that the least and the greatest elements $\bot$ and $\top$ are unique; and that complements are unique and satisfy De Morgan's laws. Clearly, this can be stated as an FO sentence.

We now formulate the separating query $Q^{\mathrm{even}}_{\mathrm{atom}}$:

$$Q^{\mathrm{even}}_{\mathrm{atom}}(\mathfrak{A}) = \mathrm{true} \quad \Leftrightarrow \quad \mathfrak{A} \models \Phi_{\mathrm{BA}} \ \text{ and } \ |\{a \mid \mathfrak{A} \models \mathrm{atom}(a)\}| = 0 \pmod 2.$$

That is, it checks if the number of atoms in the finite Boolean algebra $\mathfrak{A}$ is even.

Lemma 5.4. $Q^{\mathrm{even}}_{\mathrm{atom}} \in (\mathrm{FO}{+}{<})_{\mathrm{inv}}$.

Proof. Let $<$ be an ordering on the universe of $\mathfrak{A}$. It orders the atoms of the Boolean algebra: $a_0 < \ldots < a_{n-1}$. To check if the number of atoms is even, we check if there is a set that contains all the atoms in even positions (i.e., $a_0, a_2, a_4, \ldots$) and does not contain a_{n-1}. For that, we define the following formulae:

- $\mathrm{firstatom}(x) \equiv \mathrm{atom}(x) \wedge \forall y\ (\mathrm{atom}(y) \rightarrow x \leq y)$.
- $\mathrm{lastatom}(x) \equiv \mathrm{atom}(x) \wedge \forall y\ (\mathrm{atom}(y) \rightarrow y \leq x)$.
- $\mathrm{nextatom}(x, y) \equiv \left(\begin{array}{c} \mathrm{atom}(x) \wedge \mathrm{atom}(y) \wedge (x < y) \\ \wedge\ \ \neg\exists z\ (\mathrm{atom}(z) \wedge (x < z) \wedge (z < y)) \end{array} \right)$.

That is, firstatom(x) is true of a_0, lastatom(x) is true of a_{n-1}, and nextatom(x, y) is true of any pair (a_{i-1}, a_i), $0 < i \leq n-1$.

Based on these, we express $Q_{\text{atom}}^{\text{even}}$ by the sentence below:

$$\exists z \left(\begin{array}{l} \forall x \ (\text{firstatom}(x) \rightarrow x \subseteq z) \\ \wedge \forall x \ (\text{lastatom}(x) \rightarrow \neg(x \subseteq z)) \\ \wedge \forall x, y \ (\text{nextatom}(x, y) \rightarrow ((x \subseteq z) \leftrightarrow \neg(y \subseteq z))) \end{array} \right).$$

That is, the above sentence is true iff the set containing the even atoms $a_0, a_2, \ldots$ does not contain a_{n-1}. Note that the set z may be different for each different interpretation of the linear ordering $<$, but the sentence still tests if the number of atoms is even, which is a property independent of a particular ordering. □

Lemma 5.5. $Q_{\text{atom}}^{\text{even}}$ *is not* FO*-definable (in the vocabulary* $\{\subseteq\}$*).*

Proof. We shall use a game argument. Notice that locality does not help us here: in $\langle 2^X, \subseteq \rangle$, for any two sets $C, D \subseteq X$, the distance between them is at most 2, since $\emptyset \subseteq C, D$.

The proof illustrates the idea of composing a larger Ehrenfeucht-Fraïssé game from smaller and simpler games, already seen in Chap. 3.

In the proof, we shall be using games on Boolean algebras. We first observe that if $\langle 2^X, \subseteq \rangle \equiv_k \langle 2^Y, \subseteq \rangle$, then we can assume, without any loss of generality, that the duplicator has a winning strategy in which he responds to the empty set by the empty set, to X by Y, and to Y by X. Indeed, suppose the spoiler plays $\emptyset$ in 2^X, and the duplicator responds with $Y' \neq \emptyset$ in 2^Y. If there is one more round left in the game, the spoiler would play the empty set in 2^Y, and the duplicator has no response in 2^X, contradicting the assumption that he has a winning strategy. Thus, in every round but the last, the duplicator must respond to $\emptyset$ by $\emptyset$. If the spoiler plays $\emptyset$ in 2^X in the last round, it is contained in all the other moves played in 2^X, and the duplicator can respond by $\emptyset$ in 2^Y to maintain partial isomorphism. The proof for the other cases is similar.

Next, we need the following composition result.

Claim 5.6. *Let* $\langle 2^{X_1}, \subseteq \rangle \equiv_k \langle 2^{Y_1}, \subseteq \rangle$ *and* $\langle 2^{X_2}, \subseteq \rangle \equiv_k \langle 2^{Y_2}, \subseteq \rangle$*. Assume that* $X_1 \cap X_2 = Y_1 \cap Y_2 = \emptyset$*. Then*

$$\langle 2^{X_1 \cup X_2}, \subseteq, X_1, X_2 \rangle \ \equiv_k \ \langle 2^{Y_1 \cup Y_2}, \subseteq, Y_1, Y_2 \rangle. \tag{5.1}$$

Proof of Claim 5.6. Let $A_i, B_i, i \leq k$, be the moves by the spoiler and the duplicator in the game (5.1). Let $A_i^1 = A_i \cap X_1, A_i^2 = A_i \cap X_2$, and likewise $B_i^1 = B_i \cap Y_1, B_i^2 = B_i \cap Y_2$. The winning strategy for the duplicator is as follows. Suppose $i-1$ rounds have been played, and in the ith round the spoiler plays $A_i \subseteq X_1 \cup X_2$ (the case of the spoiler playing in $Y_1 \cup Y_2$ is symmetric). The duplicator considers the position

$$((A_1^1, \ldots, A_{i-1}^1), \ \ (B_1^1, \ldots, B_{i-1}^1))$$

in the game on $\langle 2^{X_1}, \subseteq\rangle$ and $\langle 2^{Y_1}, \subseteq\rangle$, and finds his response $B_i^1 \subseteq Y_1$ to A_i^1. Similarly, he finds $B_i^2 \subseteq Y_2$ as the response to A_i^2 in the position $((A_1^2, \ldots, A_{i-1}^2), (B_1^2, \ldots, B_{i-1}^2))$ in the game on $\langle 2^{X_2}, \subseteq\rangle$ and $\langle 2^{Y_2}, \subseteq\rangle$. His response to A_i is then $B_i = B_i^1 \cup B_i^2$. Clearly, playing in such a way, the duplicator preserves the $\subseteq$ relation. Furthermore, it follows from the observation made before the claim that this strategy also preserves the constants: that is, if the spoiler plays X_1, then the duplicator responds by Y_1, etc. Hence, the duplicator has a winning strategy for (5.1). This proves the claim.

The lemma now follows from the claim below.

Claim 5.7. *Let* $|X|, |Y| \geq 2^k$. *Then*

$$\langle 2^X, \subseteq\rangle \equiv_k \langle 2^Y, \subseteq\rangle.$$

Indeed, assume $Q_{\text{atom}}^{\text{even}}$ is definable by an FO-sentence of quantifier rank k. Take any X of odd cardinality and any Y of even cardinality, greater than 2^k. By Claim 5.7, $\langle 2^X, \subseteq\rangle \equiv_k \langle 2^Y, \subseteq\rangle$, and hence they must agree on $Q_{\text{atom}}^{\text{even}}$ which is clearly false.

Proof of Claim 5.7. It will be proved by induction on k. The cases of $k = 0, 1$ are obvious. Going from k to $k+1$, suppose we have X, Y with $|X|, |Y| \geq 2^{k+1}$. Assume, without loss of generality, that the spoiler plays $A \subseteq X$ in $\langle 2^X, \subseteq\rangle$. There are three possibilities.

1. $|A| < 2^k$. Pick an arbitrary $B \subseteq Y$ with $|B| = |A|$. Then both $|X - A|$ and $|Y - B|$ exceed 2^k. Thus, by the induction hypothesis, $\langle 2^{X-A}, \subseteq\rangle \equiv_k \langle 2^{Y-B}, \subseteq\rangle$. Furthermore, $\langle 2^A, \subseteq\rangle \cong \langle 2^B, \subseteq\rangle$, which implies a weaker fact that $\langle 2^A, \subseteq\rangle \equiv_k \langle 2^B, \subseteq\rangle$. By Claim 5.6,

$$\langle 2^X, \subseteq, A\rangle \equiv_k \langle 2^Y, \subseteq, B\rangle,$$

   meaning that after the duplicator responds to A with B, he can continue playing for k more rounds. This ensures a winning position, for the duplicator, after $k+1$ rounds.
2. $|X - A| < 2^k$. Pick an arbitrary $B \subseteq Y$ with $|Y - B| = |X - A|$. Then the proof follows case 1.
3. $|A| \geq 2^k$ and $|X - A| \geq 2^k$. Since $|Y| \geq 2^{k+1}$, we can find $B \subseteq Y$ with $|B| \geq 2^k$ and $|Y - B| \geq 2^k$. Then, by the induction hypothesis,

$$\langle 2^A, \subseteq\rangle \equiv_k \langle 2^B, \subseteq\rangle,$$
$$\langle 2^{X-A}, \subseteq\rangle \equiv_k \langle 2^{Y-B}, \subseteq\rangle,$$

   and we again conclude $\langle 2^X, \subseteq, A\rangle \equiv_k \langle 2^Y, \subseteq, B\rangle$, thus proving the winning strategy for the duplicator in $k+1$ moves.

This completes the proof of the claim, and of Theorem 5.3. □

Gurevich's theorem is one of many instances of the proper containment $\mathcal{L} \subsetneq (\mathcal{L}+<)_{\text{inv}}$, which holds for many logics of interest in finite model theory. We shall see similar results for logics with counting, fixed point logics, several infinitary logics, and some restrictions of second-order logic.

5.3 Locality of Order-invariant FO

We know how to establish some expressivity bounds on invariant queries: for example, if extra relations are of bounded degree, then invariant queries have the BNDP. There are important classes of auxiliary relations that are of bounded degree. For example, the class SUCC of successor relations: that is, graphs of the form $\{(a_0, a_1), (a_1, a_2), \ldots, (a_{n-1}, a_n)\}$ where all a_i's are distinct. Then the BNDP applies to FO + SUCC, because for any $\mathfrak{A} \in$ SUCC, $\mathit{deg\_set}(\mathfrak{A}) = \{0, 1\}$.

Adding order instead of successor destroys the BNDP, because for an ordering L on n elements, $\mathit{deg\_set}(L) = \{0, \ldots, n-1\}$. Moreover, while FO+$<$ is local, locality does not tell us anything interesting. With a linear ordering, the distance between any two distinct elements is 1. Therefore, if a structure $\mathfrak{A}$ is ordered by $<$, then $N_1^{(\mathfrak{A},<)}(\vec{a}) = (\mathfrak{A}, <, \vec{a})$. Hence, every query is trivially Gaifman-local with locality rank 1.

Gaifman-locality is a useful concept when applied to "sparse" structures, and structures with a linear order are not such. However, *invariant* queries do not talk about the order: they simply use it, but they are defined on σ-structures for σ that does not need to include an ordering. Hence, if we could establish locality of order-invariant FO-definable queries, it would give us very useful bounds on the expressive power of $(\text{FO}+<)_{\text{inv}}$. All the locality proofs we presented earlier would not work in this case, since FO formulae defining invariant queries do use the ordering. Nevertheless, the following is true.

Theorem 5.8 (Grohe-Schwentick). *Every m-ary query in* $(\text{FO}+<)_{\text{inv}}$, *$m \geq 1$, is Gaifman-local.*

This theorem gives us easy bounds for FO+$<$. For example, to show that the transitive closure query is not definable in FO+$<$, one notices that it is an invariant query. Hence, if it were expressible in FO+$<$, it would have been an $(\text{FO}+<)_{\text{inv}}$ query, and thus Gaifman-local. We know, however, that transitive closure is not Gaifman-local.

The proof of the theorem is quite involved, and we shall prove a slightly easier result (that is still sufficient for most inexpressibility proofs). We say that an m-ary query Q, $m > 0$, is *weakly local* if there exists a number $d \geq 0$ such that for any structure $\mathfrak{A}$ and any $\vec{a}_1, \vec{a}_2 \in A^m$ with

$$\vec{a}_1 \approx_d^{\mathfrak{A}} \vec{a}_2 \quad \text{and} \quad B_d^{\mathfrak{A}}(\vec{a}_1) \cap B_d^{\mathfrak{A}}(\vec{a}_2) = \emptyset$$

it is the case that

$$\vec{a}_1 \in Q(\mathfrak{A}) \quad \text{iff} \quad \vec{a}_2 \in Q(\mathfrak{A}).$$

That is, the only difference between weak locality and the usual Gaifman-locality is that for the former, the neighborhoods are required to be disjoint.

The result that we prove is the following.

Proposition 5.9. *Every unary query in* $(\text{FO}+<)_{\text{inv}}$ *is weakly local.*

The proof will demonstrate all the main ideas required to prove Theorem 5.8; completing the proof of the theorem is the subject of Exercises 5.8 and 5.9.

The statement of Proposition 5.9 is also very powerful, and suffices for many bounds on the expressive power of FO+<. Suppose, for example, that we want to show that the same-generation query over colored trees is not in FO+<. Since same generation is order-invariant, it suffices to show that it is not weakly local, and thus not in $(\text{FO}+<)_{\text{inv}}$.

We consider colored trees as structures of the vocabulary (E, C), where E is binary and C is unary, and assume, towards a contradiction, that a binary query Q_{sg} (same generation) is definable in FO+< by a formula $\varphi(x, y)$. Let

$$\psi(x) \quad \equiv \quad \exists y \; (C(y) \wedge \varphi(x, y)).$$

Then ψ defines a unary order-invariant query, testing if there is a node y in the set C such that (x, y) is in the output of Q_{sg}. To show that it is not weakly local, assume to the contrary that it is, and construct a tree T as follows. Let d witness the weak locality of the query defined by ψ. Then T has three branches coming from the root, two of length $d+1$ and one of length $d+2$. Let the leaves be a, b, c, with c being the leaf of the branch of length $d+2$. The set C is then $\{a\}$. Note that $b \approx_d^T c$ and their balls of radius d are disjoint, and yet $(T, <) \models \psi(b) \wedge \neg\psi(c)$ for any ordering $<$. Hence, ψ is not weakly local, and thus Q_{sg} is not definable in FO+<.

We now move to the proof of Proposition 5.9. First, we present the main idea of the proof. For that, we define the *radius* r *sphere*, $r > 0$, of a tuple $\vec{a}$ in a structure $\mathfrak{A}$ as

$$S_r^{\mathfrak{A}}(\vec{a}) \quad = \quad B_r^{\mathfrak{A}}(\vec{a}) \; - \; B_{r-1}^{\mathfrak{A}}(\vec{a}).$$

That is, $S_r^{\mathfrak{A}}(\vec{a})$ is the set of elements at distance exactly r from $\vec{a}$. As usual, the superscript $^{\mathfrak{A}}$ will be omitted when irrelevant or understood. We fix, for the proof, the vocabulary of the structure to be that of graphs; that is, $\sigma = (E)$, where E is binary. This will simplify notation without any loss of generality.

Given a structure $\mathfrak{A}$ and $a \in A$, its d-ball can be thought of as a sequence of r-spheres, $r \leq d$, where E-edges could go between $S_i(a)$ and $S_{i+1}(a)$, or between two elements of the same sphere.

Let Q be a unary $(\text{FO}+<)_{\text{inv}}$ query on STRUCT$[\sigma]$, defined by a formula $\varphi(x)$ of quantifier rank k. Fix a sufficiently large d (exact bounds will be clear from the proof), and consider $a \approx_d^{\mathfrak{A}} b$, with $B_d(a)$ and $B_d(b)$ disjoint. Let h be an isomorphism $h : N_d(a) \to N_d(b)$.

We now fix a linear ordering $\prec_a$ on $B_d(a)$ such that $d_{\mathfrak{A}}(a, x) < d_{\mathfrak{A}}(a, y)$ implies $x \prec_a y$. In particular, a is the smallest element with respect to $\prec_a$. We let $\prec_b$ be the image of $\prec_a$ under h. Let $\prec_0$ be a fixed linear ordering on $A - B_d(a, b)$. We now define a preorder $\prec$ as follows:

$$
\begin{array}{rl}
x \prec y \text{ iff } & x \prec_a y,\ x, y \in B_d(a) \\
& \text{or } x \prec_b y,\ x, y \in B_d(b) \\
& \text{or } h(x) \prec_b y,\ x \in B_d(a), y \in B_d(b) \\
& \text{or } x \prec_b h(y),\ x \in B_d(b), y \in B_d(a) \\
& \text{or } x \prec_0 y,\ x, y \notin B_d(a, b) \\
& \text{or } x \in B_d(a, b), y \notin B_d(a, b).
\end{array}
$$

In other words, $\prec$ is a preorder that does not distinguish elements x and $h(x)$, but it makes both x and $h(x)$ less than y and $h(y)$ whenever $x \prec_a y$ holds. Furthermore, each element of $B_d(a, b)$ is less than each element of the complement, $A - B_d(a, b)$, which in turn is ordered by $\prec_0$.

Our goal is to find two linear orderings, $\leq_a$ and $\leq_b$ on $\mathfrak{A}$, such that

$$
(\mathfrak{A}, a, \leq_a) \equiv_k (\mathfrak{A}, b, \leq_b). \tag{5.2}
$$

This would imply

$$
a \in Q(\mathfrak{A}) \text{ iff } (\mathfrak{A}, \leq_a) \models \varphi(a) \text{ iff } (\mathfrak{A}, \leq_b) \models \varphi(b) \text{ iff } b \in Q(\mathfrak{A}). \tag{5.3}
$$

These orderings will be refinements of $\prec$, and will be defined sphere-by-sphere. For the $\leq_a$ ordering, a is the smallest element, and for the $\leq_b$ ordering, b is the smallest. On $S_d(a) \cup S_d(b)$, the orderings $\leq_a$ and $\leq_b$ must coincide (otherwise the spoiler will win easily).

Note that $\prec$ is a preorder: the only pairs it does not order are pairs of the form $(x, h(x))$. To define ordering on them, we select two "sparse" sets of integers $J = \{j_1, \ldots, j_m\}$ and $L = \{l_1, \ldots, l_{m+1}\}$ with $0 < j_1 < \ldots < j_m < d$ and $0 < l_1 < l_2 < \ldots < l_{m+1} < d$. "Sparse" here means that the difference between two consecutive integers is at least $2^k + 1$ (other conditions will be explained in the detailed proof). Assume that $x \in S_r(a)$, $y \in S_r(b)$, and $y = h(x)$, for $r \leq d$. Then

$$
\begin{array}{l}
x \leq_a y \Leftrightarrow |\{j \in J \mid j < r\}| \text{ is even,} \\
y \leq_a x \Leftrightarrow |\{j \in J \mid j < r\}| \text{ is odd,}
\end{array} \tag{5.4}
$$

and

$$
\begin{array}{l}
x \leq_b y \Leftrightarrow |\{l \in L \mid l < r\}| \text{ is odd,} \\
y \leq_b x \Leftrightarrow |\{l \in L \mid l < r\}| \text{ is even.}
\end{array} \tag{5.5}
$$

Thus, the parity of the number of j_i's or l_i's below r tells us whether the order on pairs $(x, h(x))$ prefers the element from $B_d(a)$ or $B_d(b)$. Note that a is the least element with respect to $\leq_a$ (in particular, $a \leq_a b$), and b is the least element for $\leq_b$, but since the number of switches of preferences differs by one for $\leq_a$ and $\leq_b$, on $S_d(a,b)$ both orderings are the same.

Of course a switch can be detected by a first-order formula, but we have many of them, and they happen at spheres that are well separated. The key idea of the proof is to use the sparseness of J and L to show that the difference between them cannot be detected by the spoiler in k moves. This will ensure $(\mathfrak{A}, a, \leq_a) \equiv_k (\mathfrak{A}, b, \leq_b)$.

We now present the complete proof; that is, we show how to construct two orderings, $\leq_a$ and $\leq_b$, such that (5.2) holds. First, we may assume, without loss of generality, that no sphere $S_r(a,b)$, $r \leq d$, is empty. If any $S_r(a,b)$ were empty, $\mathfrak{A}$ would have been a disjoint union of $B_d(a)$, $B_d(b)$, and $A - B_d(a,b)$, with no E-edges between these sets. Then, using $N_d^{\mathfrak{A}}(a) \cong N_d^{\mathfrak{A}}(b)$, it is easy to find orderings $\leq_a$ and $\leq_b$ such that $(\mathfrak{A}, a, \leq_a)$ and $(\mathfrak{A}, b, \leq_b)$ are isomorphic, and hence $(\mathfrak{A}, a, \leq_a) \equiv_k (\mathfrak{A}, b, \leq_b)$ holds.

To define the radius d for a given k (the quantifier rank of a formula defining Q), we need some additional notation. Let $\sigma_{(r)}$ be the vocabulary $(E, <, U_{-r}, U_{-r+1}, \ldots, U_{-1}, U_0, U_1, \ldots, U_{r-1}, U_r)$, where all the U_i's are unary. Let t be the number of rank-$(k+1)$ types of $\sigma_{(r)}$ structures, where $r = 2^k$ (this number, as we know, depends only on k).

Let Σ be a finite alphabet of cardinality t. Recall that a string s of length n over Σ is represented as a structure M_s of the vocabulary $(<, A_1, \ldots, A_t)$ with the universe $\{1, \ldots, n\}$ ordered by $<$, and each unary A_i interpreted as the set of positions between 1 and n where the symbol is the ith symbol of Σ.

We call a subset $X = \{x_1, \ldots, x_p\}$ of $\{1, \ldots, n\}$ r-sparse if

$$\min_{i \neq j} |x_i - x_j| > r, \quad x_i > r, \quad n - x_i > r, \quad \text{for all } i \leq p.$$

Next, we need the following lemma.

Lemma 5.10. *For every $t, k \geq 0$, there exists a number $d > 0$ such that, given any string $s \in \Sigma^*$ of length $n \geq d$, where $|\Sigma| = t$, there exist two subsets $J, L \subseteq \{1, \ldots, n\}$ such that*

- $|L| = |J| + 1 > 2^k$;
- *J and L are $2^k + 1$-sparse; and*
- $(M_s, J) \equiv_{k+2} (M_s, L)$.

The proof is a standard Ehrenfeucht-Fraïssé game argument, and is left to the reader as an exercise (Exercise 5.6).

We now let d be given by Lemma 5.10, for k the quantifier rank of a formula defining Q, and t the number of rank-$(k+1)$ types of $\sigma_{(2^k)}$-structures.

Fix $a \approx_d^{\mathfrak{A}} b$, with $B_d(a)$ and $B_d(b)$ disjoint, and let h be an isomorphism $N_d(a) \to N_d(b)$. For $i, r \leq d$, let $R_r^i(a)$ be a $\sigma_{(r)}$-structure whose universe is the union

$$\bigcup_{j=i-r}^{i+r} S_j(a)$$

(if $j < 0$ or $j > d$, we take the corresponding sphere to be empty), and each U_p is interpreted as $S_{i+p}(a)$, and the ordering is $\prec_a$, the fixed linear ordering on $B_d(a)$ such that $d_{\mathfrak{A}}(x, a) < d_{\mathfrak{A}}(y, a)$ implies $x \prec_a y$ (restricted to the universe of the structure). Structures $R_r^i(b)$ are defined similarly, with the ordering being $\prec_b$, the image of $\prec_a$ under the isomorphism h. Note that $R_r^i(b) \cong R_r^i(a)$.

Let Σ be the set of rank-$(k+1)$ types of $\sigma_{(2^k)}$-structures. Define a string s of length $d+1$ which, in position $i = 1, \ldots, d+1$, has the rank-$(k+1)$ type of $R_{2^k}^{i-1}(a)$. Applying Lemma 5.10, we get two 2^k+1-sparse sets J, L such that $(M_s, J) \equiv_k (M_s, L)$. Let $J = \{j_1, \ldots, j_m\}$ with $j_0 = 0 < j_1 < \ldots < j_m < d$ and $L = \{l_1, \ldots, l_{m+1}\}$ with $l_0 = 0 < l_1 < l_2 < \ldots < l_{m+1} < d$. Using these J and L, define $\leq_a$ and $\leq_b$ as in (5.4) and (5.5).

Let $N_{d,J}(a)$ and $N_{d,L}(a)$ be two structures in the vocabulary $(E, <, U, c)$ with the universe $B_d(a)$. In both, the binary predicate E is inherited from $\mathfrak{A}$, the ordering $<$ is $\prec_a$, and the constant c is a. The only difference is the unary predicate U: it is interpreted as $\bigcup_{j \in J} S_j(a)$ in $N_{d,J}(a)$, and as $\bigcup_{l \in L} S_l(a)$ in $N_{d,L}(a)$.

Let $\mathfrak{A}_a$ stand for $(\mathfrak{A}, \leq_a, a)$ and $\mathfrak{A}_b$ for $(\mathfrak{A}, \leq_b, b)$. The winning strategy for the duplicator on $\mathfrak{A}_a$ and $\mathfrak{A}_b$ is based on the following lemma.

Lemma 5.11. *The duplicator has a winning strategy in the k-round game on $N_{d,J}(a)$ and $N_{d,L}(a)$. Moreover, if $p_1, \ldots, p_k$ are the moves on $N_{d,J}(a)$, and $q_1, \ldots, q_k$ are the moves on $N_{d,L}(a)$, then the following conditions can be guaranteed by the winning strategy:*

1. *If $p_i \in S_r(a)$ and $d - r \leq 2^{k-i}$, then $q_i = p_i$.*
2. *If $(r_1, \ldots, r_k)$ and $(r'_1, \ldots, r'_k)$ are such that each p_i is in the sphere $S_{r_i}(a)$ and q_i is in the sphere $S_{r'_i}(a)$, then $((r_1, \ldots, r_k), (r'_1, \ldots, r'_k))$ define a partial isomorphism between (M_s, J) and (M_s, L).*

The idea of Lemma 5.11 is illustrated in Fig. 5.1. We have two structures, (M_s, J) and (M_s, L), which are linear orders with extra unary predicates, and two additional unary predicates, J and L of different parity, which are shown as short horizontal segments. Using the fact that $(M_s, J) \equiv_{k+2} (M_s, L)$, we prove that $N_{d,J}(a) \equiv_k N_{d,L}(a)$. These are shown in Fig. 5.1 as two big circles, with concentric circles inside representing spheres S_r with r being in J or L, respectively. These spheres form the interpretation for an extra unary predicate in the vocabulary of structures $N_{d,J}(a)$ and $N_{d,L}(a)$.

Next, we show that Proposition 5.9 follows from Lemma 5.11; after that, we prove Lemma 5.11.

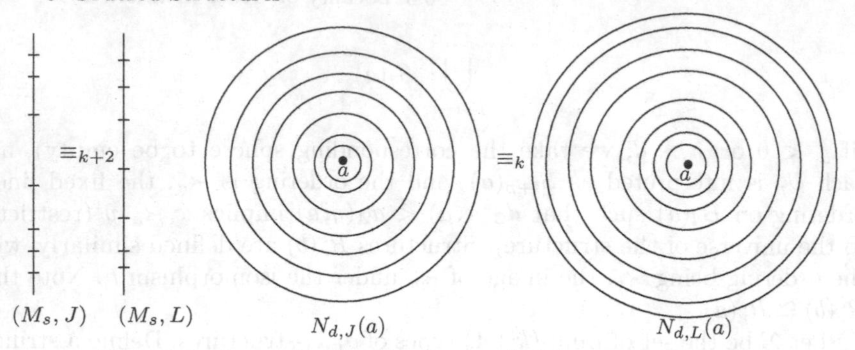

Fig. 5.1. Games involved in the proof of Proposition 5.9

From $N_{d,J}(a) \equiv_k N_{d,L}(a)$ *to* $\mathfrak{A}_a \equiv_k \mathfrak{A}_b$. We now show how Lemma 5.11 implies Proposition 5.9; that is, $\mathfrak{A}_a \equiv_k \mathfrak{A}_b$. The idea for the winning strategy on $\mathfrak{A}_a$ and $\mathfrak{A}_b$ is that it almost mimics the one in $N_{d,J}(a) \equiv_k N_{d,L}(a)$.

We shall denote moves in $\mathfrak{A}_a$ by $a_1, \ldots$, and moves in $\mathfrak{A}_b$ by $b_1, \ldots$. Suppose the spoiler plays $a_i \in \mathfrak{A}_a$ (the case of a move in $\mathfrak{A}_b$ is symmetric). If $a_i \notin B_d(a,b)$, then $b_i = a_i$, and we also set $p_i = q_i = a$.

If $a_i \in B_d(a,b)$, we define $p_i \in B_d(a)$ to be a_i if $a_i \in B_d(a)$, and $h^{-1}(a_i)$ if $a_i \in B_d(b)$. The duplicator then determines the response q_i to p_i, according to the $N_{d,J}(a) \equiv_k N_{d,L}(a)$ winning strategy. The response b_i is going to be either q_i itself, or $h(q_i)$, and we use sets J and L to determine if b_i lives in $B_d(a)$ or $B_d(b)$.

We define two mappings

$$v_J : B_d(a,b) \to \{0,1\} \quad \text{and} \quad v_L : B_d(a,b) \to \{0,1\}$$

such that for every $x \in B_d(a)$,

$$v_J(x) + v_J(h(x)) \;=\; v_L(x) + v_L(h(x)) \;=\; 1.$$

For $x \in B_d(a)$, find $r \leq d$ such that $x \in S_r(a)$. Then

$$v_J(x) \;=\; \begin{cases} 0 & \text{if } |\{j \in J \mid j < r\}| \text{ is even,} \\ 1 & \text{otherwise,} \end{cases}$$

and $v_J(h(x)) = 1 - v_J(x)$. Similarly, for $x \in B_d(b)$, we find r such that $x \in S_r(b)$ and set

$$v_L(x) \;=\; \begin{cases} 0 & \text{if } |\{l \in L \mid l < r\}| \text{ is even,} \\ 1 & \text{otherwise,} \end{cases}$$

and define $v_L(x) = 1 - v_L(h(x))$ for $x \in B_d(a)$.

We now look at q_i and $h(q_i)$; we know that $v_L(q_i) + v_L(h(q_i)) = 1$. We choose b_i to be one of q_i or $h(q_i)$ such that $v_L(b_i) = v_J(a_i)$.

This describes the strategy; now we prove that it works. Dealing with the constant is easy: if the spoiler plays a in $\mathfrak{A}_a$, then the duplicator has to respond with b in $\mathfrak{A}_b$ and vice versa.

We now move to the E-relation. Since the parity of $|J|$ and $|L|$ is different, condition 1 of Lemma 5.11 implies that for any move in $B_d(a,b) - B_{d-2^m}(a,b)$ with m moves to go, the response is the identity. Hence, if $E(a_i, a_j)$ holds, and one or both of a_i, a_j are outside of $B_d(a,b)$, then $E(b_i, b_j)$ holds (and vice versa). Therefore, it suffices to consider the case when $E(a_i, a_j)$ holds, and $a_i, a_j \in B_d(a,b)$.

Assume, without loss of generality, that $a_i, a_j \in B_d(a)$. Then $E(p_i, p_j)$ holds, and hence $E(q_i, q_j)$ holds. Given the duplicator's strategy, to conclude that $E(b_i, b_j)$ holds, we must show that both b_i and b_j belong to the same ball – $B_d(a)$ or $B_d(b)$.

The elements a_i and a_j could come either from the same sphere $S_r(a)$, or from two consecutive spheres $S_r(a)$ and $S_{r+1}(a)$. In the first case, if they come from the same sphere, $v_J(a_i) = v_J(a_j)$ and thus $v_L(b_i) = v_L(b_j)$. Furthermore, since a_i and a_j are in the same sphere, we conclude that p_i and p_j are in the same sphere, and hence, by the winning strategy of Lemma 5.11, q_i and q_j are in the same sphere. This, together with $v_L(b_i) = v_L(b_j)$, means that b_i and b_j are in the same ball.

Assume now that $a_i \in S_r(a)$ and $a_j \in S_{r+1}(a)$. From condition 2 of Lemma 5.11, for some $r' \leq d$ we have $q_i \in S_{r'}(a)$ and $q_j \in S_{r'+1}(a)$. Now there are two cases. In the first case, $v_J(a_i) = v_J(a_j)$. Then there are two possibilities. If $r, r+1 \notin J$, then $r', r'+1 \notin L$ (by condition 2 of Lemma 5.11), and hence $v_L(b_i) = v_J(a_i) = v_J(a_j) = v_L(b_j)$ implies that b_i, b_j are in the same ball, and $E(b_i, b_j)$ holds. The other possibility is that $r+1 \in J, r \notin J$. Then $r'+1 \in L, r' \notin L$, and again we conclude $E(b_i, b_j)$.

The second case is when $v_J(a_i) \neq v_J(a_j)$. This could only happen if r is in J (and thus $r+1 \notin J$). Then again by condition 2 of Lemma 5.11, $r' \in L, r'+1 \notin L$. Suppose $v_J(a_i) = 0$. Then $v_L(b_i) = 0$, and $v_L(b_j) = v_J(a_j) = 1$. Since $b_i \in S_{r'}(a,b)$ and $b_j \in S_{r'+1}(a,b)$, and $r' \in L$, both b_i and b_j must belong to the same ball ($B_d(a)$ or $B_d(b)$), and hence $E(b_i, b_j)$ holds.

Thus, $E(a_i, a_j)$ implies $E(b_i, b_j)$; the proof of the converse – that $E(b_i, b_j)$ implies $E(a_i, a_j)$ – is identical.

Finally, assume that $a_i \leq_a a_j$. If $a_i \in S_r(a,b)$, $a_j \in S_{r'}(a,b)$ and $r < r'$, then, by condition 2 of Lemma 5.11, $b_i \in S_{r_0}(a,b), b_j \in S_{r'_0}(a,b)$ for some $r_0 < r'_0$, and hence $b_i \leq_b b_j$.

Thus, it remains to consider the case of a_i, a_j being in the same sphere; that is, $a_i, a_j \in S_r(a,b)$. If $p_i \neq p_j$, then $p_i \prec_a p_j$ and hence $q_i \prec_a p_j$, which in turn implies $b_i \leq_b b_j$. The final possibility is that of $p_i = p_j$; then either (1) $a_i \in S_r(a)$ and $a_j = h(a_i)$, or (2) $a_j \in S_r(a)$ and $a_i = h(a_j)$. We prove case (1); the proof of case (2) is identical.

Note that the orderings $\leq_a$ and $\leq_b$ are defined in such a way that whenever $x = h(y)$, then $\leq_a$ orders them according to v_J; that is, if $v_J(x) < v_J(y)$, then $x \leq_a y$, and if $v_J(y) < v_J(x)$, then $y \leq_a x$. The ordering $\leq_b$ behaves likewise with respect to the function v_L. Hence, if $a_j = h(a_i)$ and $a_i \leq_a a_j$, then $v_J(a_i) = 0$ and $v_J(a_j) = 1$. From Lemma 5.11, $q_i = q_j$, and thus b_i and b_j are related by the isomorphism h. Since $v_L(b_i) = 0$ and $v_L(b_j) = 1$, we know that $b_i \leq_b b_j$.

This concludes the proof that $a_i \leq_a a_j$ implies $b_i \leq_b b_j$; the proof of the converse is identical. Thus, we have proved, using Lemma 5.11, that $\mathfrak{A}_a \equiv_k \mathfrak{A}_b$, which is precisely what is needed to conclude (weak) locality of Q. It thus remains to prove Lemma 5.11.

Proof of Lemma 5.11. We shall refer to moves in the game on $N_{d,J}(a)$ and $N_{d,L}(a)$ as p_i (in $N_{d,J}(a)$) and q_i (in $N_{d,L}(a)$), and to moves in the game on (M_s, J) and (M_s, L), provided by Lemma 5.10, as e_i for (M_s, J) and f_i for (M_s, L).

For two elements x, y in the universe of M_s (which is $\{1, \ldots, d+1\}$), the distance between them is $|x - y|$. The next claim shows that after i rounds, distances up to 2^{k-i} between played elements, and elements of the sets J and L, are preserved.

Claim 5.12. *Let $e_1, \ldots, e_i$ and $f_1, \ldots, f_i$ be elements played in the first i rounds of the game on (M_s, J) and (M_s, L). Then:*

- *if $|e_{j_1} - e_{j_2}| \leq 2^{k-i}$, then $|f_{j_1} - f_{j_2}| = |e_{j_1} - e_{j_2}|$;*
- *if $|e_{j_1} - e_{j_2}| > 2^{k-i}$, then $|f_{j_1} - f_{j_2}| > 2^{k-i}$;*
- *if $\min_{x \in J, j \leq i} |x - e_j| \leq 2^{k-i}$, then $\min_{x \in J, j \leq i} |x - e_j| = \min_{y \in L, j \leq i} |y - f_j|$;*
- *if $\min_{x \in J, j \leq i} |x - e_j| > 2^{k-i}$, then $\min_{y \in L, j \leq i} |y - f_j| > 2^{k-i}$.*

Proof of Claim 5.12. Since we know that $(M_s, J) \equiv_{k+2} (M_s, L)$, it suffices to show that for any x, y, $p \leq k$, and any $r \leq 2^p$, there is a formula of quantifier rank $p + 1$ that tests if $|x - y| = r$, and there is a formula of quantifier rank $p + 2$ that tests if the minimum distance from x to an element of the set (interpreted as J and L in the models) is exactly r. We prove the first statement; the second is an easy exercise for the reader.

We define $\alpha_0(x, y) \equiv (x = y)$; this tests if the distance is zero. To test if the distance is one, we see if x is the successor of y or y is the successor of x:

$$\alpha_1(x, y) \equiv \big(x < y \wedge \neg\exists z\ (x < z \wedge z < y)\big) \vee \big(y < x \wedge \neg\exists z\ (y < z \wedge z < x)\big).$$

Now, suppose for each $r \leq 2^p$, we have a formula $\alpha_r(x, y)$ in FO$[p+1]$ testing if the distance is r. We now show how to test distances up to 2^{p+1} using FO$[p+2]$ formulae. Suppose $2^p < r \leq 2^{p+1}$. The formula α_r is of the form $((x < y) \wedge \alpha'_r(x, y)) \vee ((y < x) \wedge \alpha''_r(x, y))$. We present $\alpha'_r(x, y)$ below. Let $r_1, r_2 \leq 2^p$ be such that $r_1 + r_2 = r$. Then

$$\alpha'_r(x, y) \equiv \exists z\ \Big((x < z) \wedge (z < y) \wedge \alpha_{r_1}(x, z) \wedge \alpha_{r_2}(z, y)\Big).$$

Clearly, this increases the quantifier rank by 1. This proves the claim.

Given $x \in S_r(a)$ and $y \in S_{r'}(a)$, define $\delta(x,y)$ as $r-r'$. Given $x_1, \ldots, x_m$ in $B_d(a)$, and $u \geq 0$, we define a structure $\mathfrak{S}_u[x_1, \ldots, x_m]$ as follows. Its universe is $\{x \mid -u \leq \delta(x, x_i) \leq u,\ i \leq m\}$. It inherits binary relations E and $\prec$ from $B_d(a)$. Note that the universe of $\mathfrak{S}_u[x_1, \ldots, x_m]$ is a union of spheres. Suppose these are spheres $S_{r_1}(a), \ldots, S_{r_w}(a)$, with $r_1 < \ldots < r_w$. Then the vocabulary of $\mathfrak{S}_u[x_1, \ldots, x_m]$ contains w unary predicates $U_1, \ldots, U_w$, interpreted as $S_{r_1}(a), \ldots, S_{r_w}(a)$.

Furthermore, $\mathfrak{S}_u^J[x_1, \ldots, x_m]$ and $\mathfrak{S}_u^L[x_1, \ldots, x_m]$ extend $\mathfrak{S}_u[x_1, \ldots, x_m]$ by means of an extra unary relation U interpreted as the union of spheres $S_{r_i}(a)$ with $r_i \in J$ ($r_i \in L$, respectively).

We shall be interested in the parameter u of the form $2^{k-i}, i \leq k$, and now define a relation $\mathfrak{S}_{2^{k-i}}^J[x_1, \ldots, x_m] \sim_{k-i} \mathfrak{S}_{2^{k-i}}^L[y_1, \ldots, y_m]$. The first condition is as follows:

$$\text{If the universe of } \mathfrak{S}_{2^{k-i}}^J[x_1, \ldots, x_m] \text{ is a union of } w \text{ spheres, } S_{r_1}(a) \cup \ldots \cup S_{r_w}(a) \text{, then the universe of } \mathfrak{S}_{2^{k-i}}^L[y_1, \ldots, y_m] \text{ is a union of } w \text{ spheres, } S_{r'_1}(a) \cup \ldots \cup S_{r'_w}(a) \text{, and } r_j \in J \text{ iff } r'_j \in L. \tag{5.6}$$

Define $\Delta_u(r_1, \ldots, r_w)$ as $\{j > 1 \mid r_{j+1} - r_j > u\}$. The second condition is:

$$\Delta_{2^{k-i}}(r_1, \ldots, r_w) = \Delta_{2^{k-i}}(r'_1, \ldots, r'_w). \tag{5.7}$$

For $1 \leq j < j' \leq w+1$, define the restriction $\mathfrak{S}_u^J[x_1, \ldots, x_m]_j^{j'}$ to include only the spheres from $S_{r_j}(a)$ up to $S_{r_{j'-1}}(a)$ (and likewise for $\mathfrak{S}_u^L[y_1, \ldots, y_m]_j^{j'}$). The next condition is:

$$\begin{array}{l}\text{For each consecutive } j, j' \in \{1, w+1\} - \Delta_{2^{k-i}}(r_1, \ldots, r_w),\\ \mathfrak{S}_{2^{k-i}}^J[x_1, \ldots, x_m]_j^{j'} \equiv_i \mathfrak{S}_{2^{k-i}}^L[y_1, \ldots, y_m]_j^{j'}.\end{array} \tag{5.8}$$

We now write $\mathfrak{S}_{2^{k-i}}^J[x_1, \ldots, x_m] \sim_{k-i} \mathfrak{S}_{2^{k-i}}^L[y_1, \ldots, y_m]$ if (5.6), (5.7), and (5.8) hold.

Our goal is to show that the duplicator can play in such a way that, after i moves,

$$\mathfrak{S}_{2^{k-i}}^J[p_0, p_1, \ldots, p_i] \sim_{k-i} \mathfrak{S}_{2^{k-i}}^L[q_0, q_1, \ldots, q_i], \tag{5.9}$$

where $p_0 = q_0 = a$.

The proof is by induction on i. The case of $i = 0$ (i.e., $\mathfrak{S}_{2^{k-i}}^J[p_0] \sim_k \mathfrak{S}_{2^{k-i}}^L[q_0]$) is immediate from the sparseness of J and L. We also set $e_0 = f_0 = 1$.

Now suppose (5.9) holds, and the spoiler plays $p_{i+1} \in N_{d,J}(a)$, such that $p_{i+1} \in S_r(a)$ (the case of the move $q_{i+1} \in N_{d,L}(a)$ is symmetric). The duplicator sets $e_{i+1} \in \{1, \ldots, d+1\}$ to be $r+1$, and finds the response f_{i+1} to e_{i+1}

in the game on (M_s, J) and (M_s, L), from position $((e_0, \ldots, e_i), (f_0, \ldots, f_i))$. Let $f_{i+1} = r' + 1$; then the response q_{i+1} will be found in $S_{r'}(a)$.

Assume that $\mathfrak{S}^J_{2^{k-i}}[p_0, p_1, \ldots, p_i]$ is the union of spheres $S_{r_1}(a) \cup \ldots \cup S_{r_w}(a)$, and $\mathfrak{S}^L_{2^{k-i}}[q_0, q_1, \ldots, q_i]$ is the union of spheres $S_{r'_1}(a) \cup \ldots \cup S_{r'_w}(a)$.

We distinguish two cases.

Case 1. In this case $|\, \delta(p_{i+1}, p_j) \,| > 2^{k-(i+1)}$ for all $j \leq w$ (i.e., $|\, e_{i+1} - e_j \,| > 2^{k-(i+1)}$). From Claim 5.12, we conclude $|\, \delta(q_{i+1}, q_j) \,| > 2^{k-(i+1)}$ for all j. Since e_{i+1} and f_{i+1} satisfy all the same unary predicates over (M_s, J) and (M_s, L), we see that there is an element q_{i+1} in $S_{r'}(a)$ such that $\mathfrak{S}_{2^k}[p_{i+1}] \equiv_{k+1} \mathfrak{S}_{2^k}[q_{i+1}]$ and hence

$$\mathfrak{S}_{2^{k-(i+1)}}[p_{i+1}] \equiv_{k-(i+1)} \mathfrak{S}_{2^{k-(i+1)}}[q_{i+1}].$$

Moreover, by Claim 5.12, $r \pm l \in J$ iff $r' \pm l \in L$, for every $l \leq 2^{k-(i+1)}$, and hence

$$\mathfrak{S}^J_{2^{k-(i+1)}}[p_{i+1}] \equiv_{k-(i+1)} \mathfrak{S}^L_{2^{k-(i+1)}}[q_{i+1}].$$

From here

$$\mathfrak{S}^J_{2^{k-(i+1)}}[p_0, p_1, \ldots, p_{i+1}] \sim_{k-(i+1)} \mathfrak{S}^L_{2^{k-(i+1)}}[q_0, q_1, \ldots, q_{i+1}]$$

follows easily. This implies (5.8), and (5.6), (5.7) follow from the construction. The final note to make about this case is that if $d - r \leq 2^{k-(i+1)}$, then q_{i+1} can be chosen to be equal to p_{i+1}, while preserving (5.9).

Case 2. In this case $|\, \delta(p_{i+1}, p_{j_0}) \,| \leq 2^{k-(i+1)}$ for some $j_0 \leq w$. Find two consecutive $j, j' \in \Delta_{2^{k-i}}(r_1, \ldots, r_w)$ such that p_{i+1} is in $\mathfrak{S}^J_{2^{k-i}}[p_0, \ldots, p_i]^{j'}_j$. From Claim 5.12, $|\, \delta(q_{i+1}, q_j) \,| \leq 2^{k-(i+1)}$. We then use (5.8) and find q_{i+1} in $S_{r'}(a)$ so that

$$\mathfrak{S}^J_{2^{k-i}}[p_0, \ldots, p_i, p_{i+1}]^{j'}_j \equiv_{k-(i+1)} \mathfrak{S}^L_{2^{k-i}}[q_0, \ldots, q_i, q_{i+1}]^{j'}_j . \tag{5.10}$$

Conditions (5.6) and (5.7) for $2^{k-(i+1)}$ now follow from Claim 5.12, and condition (5.8) then follows from (5.10), since for every sphere which is a part of one of the structures mentioned in (5.10), there is a unary predicate interpreted as that sphere.

Finally, if $d + 1 - e_{i+1} \leq 2^{k-(i+1)}$, then $d + 1 - e_{j_0} \leq 2^{k-i}$, and thus $p_{j_0} = q_{j_0}$ and the structures $\mathfrak{S}^J_{2^{k-i}}[p_0, \ldots, p_i]^{j'}_j$ and $\mathfrak{S}^L_{2^{k-i}}[q_0, \ldots, q_i]^{j'}_j$ are actually isomorphic. Hence, responding to p_{i+1} with $q_{i+1} = p_{i+1}$ will preserve the isomorphism of structures of the form $\mathfrak{S}^J_{2^{k-(i+1)}}[p_0, \ldots, p_i, p_{i+1}]^{l'}_l$ and $\mathfrak{S}^L_{2^{k-(i+1)}}[q_0, \ldots, q_i, q_{i+1}]^{l'}_l$ containing the sphere with $p_{i+1} = q_{i+1}$.

This finally shows that the duplicator plays in such a way that (5.9) is preserved. After k moves, the moves of the game $(\vec{p}, \vec{q})$ form a partial isomorphism. Indeed, if p_{i_1}, p_{i_2} are in different structures $\mathfrak{S}^J_1[\vec{p}]^{j'}_j$ and $\mathfrak{S}^J_1[\vec{p}]^{l'}_l$, then q_{i_1}, q_{i_2} are in different structures $\mathfrak{S}^L_1[\vec{q}]^{j'}_j$ and $\mathfrak{S}^L_1[\vec{q}]^{l'}_l$, and hence there is no

E-relation between them. Furthermore, since $e_{i_1} < e_{i_2}$ iff $f_{i_1} < f_{i_2}$, we see that $p_{i_1} \prec p_{i_2}$ iff $q_{i_1} \prec q_{i_2}$. If p_{i_1}, p_{i_2} are in the same structure $\mathfrak{S}_1^J[\vec{p}]_j^{j'}$, then q_{i_1}, q_{i_2} are in $\mathfrak{S}_1^L[\vec{q}]_j^{j'}$, and hence by (5.8), the E and $\prec$ relations between them are preserved. Finally, since $e_i \in J$ iff $f_i \in L$, we have $p_i \in U$ iff $q_i \in U$. This shows that $(\vec{p}, \vec{q})$ is a partial isomorphism between $N_{d,J}(a)$ and $N_{d,L}(a)$, and thus finishes the proof of Lemma 5.11 and Proposition 5.9. □

5.4 Bibliographic Notes

While the concept of invariant queries is extremely important in finite model theory, over arbitrary models it is not interesting, as Exercise 5.1 shows.

The separating example of Theorem 5.3 is due to Gurevich, although he never published it (it appeared as an exercise in [3]). Another separating example is given in Exercise 5.2.

Locality of invariant FO-definable queries is due to Grohe and Schwentick [113]. Their original proof is the subject of Exercises 5.8 and 5.9; the proof presented here is a slight simplification of that proof. It uses the concept of weak locality, introduced in Libkin and Wong [170].

Sources for exercises:

Exercise 5.1:	Ebbinghaus and Flum [60]
Exercise 5.2:	Otto [192]
Exercises 5.3 and 5.4:	Libkin and Wong [170]
Exercises 5.7–5.9:	Grohe and Schwentick [113]
Exercise 5.11:	Rossman [210]

5.5 Exercises

Exercise 5.1. Prove that over arbitrary structures, FO = (FO+ $<$)$_{\text{inv}}$.
Hint: use the interpolation theorem.

Exercise 5.2. The goal of this exercise it to give another separation example for FO $\subsetneq$ (FO+ $<$)$_{\text{inv}}$. We consider structures in the vocabulary $\sigma = (U_1, U_2, E, R, S)$ where U_1, U_2 are unary and E, R, S are binary. We consider a class $\mathcal{C}$ of structures $\mathfrak{A} \in \text{STRUCT}[\sigma]$ that satisfy the following conditions:

1. U_1 and U_2 partition the universe A.
2. $E \subseteq U_1 \times U_1$ and $S \subseteq U_2 \times U_2$.
3. The restriction of $\mathfrak{A}$ to $\langle U_2, S\rangle$ is a Boolean algebra (we refer to its set of atoms as X).
4. $|X| = |U_1| = 2m$; moreover, if $U_1 = \{u_1, \ldots, u_{2m}\}$ and $X = \{x_1, \ldots, x_{2m}\}$, then

$$R = \bigcup_{i=1}^{m} \{u_{2i-1}, u_{2i}\} \times \{x_{2i-1}, x_{2i}\}.$$

First, prove that the class $\mathcal{C}$ is FO-definable. Next, consider the following Boolean query Q on $\mathcal{C}$:

$$Q(\mathfrak{A}) = \mathit{true} \quad \text{iff} \quad \langle U_1, E \rangle \text{ is connected.}$$

Prove that $Q \in (\text{FO}+<)_{\text{inv}}$ on $\mathcal{C}$, but that Q is not FO-definable on $\mathcal{C}$.

Exercise 5.3. Give an example of a query that is weakly local, but is not Gaifman-local.

Exercise 5.4. Prove that weak locality implies the BNDP for binary queries. Does this implication hold for m-ary queries, where $m > 2$?

Exercise 5.5. Using Proposition 5.9, prove that acyclicity and k-colorability are not definable in FO+<.

Exercise 5.6. Prove Lemma 5.10.

Exercise 5.7. In the proof of weak locality of invariant queries presented in this chapter, we only dealt with nonoverlapping neighborhoods. To deal with the case of overlapping $B_d(a)$ and $B_d(b)$, prove the following.

Let $d' = 5d+1$, and let $a \approx^{\mathfrak{A}}_{d'} b$. Then there exists a set X containing $\{a, b\}$ and an automorphism g on $N^{\mathfrak{A}}_d(X)$ such that $g(a) = b$.

Exercise 5.8. Prove that every unary query in $(\text{FO}+<)_{\text{inv}}$ is Gaifman-local.

The main ingredients have already been presented in this chapter, but for the case of nonoverlapping neighborhoods. To deal with the case of overlapping neighborhoods $N_d(a)$ and $N_d(b)$, define d', g, and X as in Exercise 5.7.

Now note that each sphere $S_r(X)$ is a union of g-orbits; that is, sets of the form $\{g^i(v) \mid i \in \mathbb{Z}\}$. For each orbit O, we fix a node c_O and define a linear ordering $\leq_O$ on O by $c_O \leq_O g(c_O) \leq_O g^2(c_O) \leq_O \ldots$. Let $\leq_m$ be the image of $\leq_O$ under g^m.

The definition of $\leq_a$ and $\leq_b$ is almost the same as the definition we used in the proof of Proposition 5.9. We start with a fixed order on orbits that respects distance from X. It generates a preorder on $B_d(X)$, which we refine to two different orders in the following way. On $S_0(X)$, we let $\leq_a$ be $\leq_0$ and $\leq_b$ be $\leq_1 = g(\leq_0)$. Then, for suitably defined J and L (cf. the proof of Proposition 5.9), we do the following. Let $J = \{j_1, \ldots, j_m\}, j_1 < \ldots < j_m$. For all spheres $S_r(X), r < j_1$, the order on each orbit is $\leq_0$, but on $S_{j_1}(X)$ we use $\leq_1$ instead. We continue to use $\leq_1$ until $S_{j_2-1}(X)$, and on $S_{j_2}(X)$ we switch to $\leq_2$, and so on. For $\leq_b$, we do the same, except that we use the set L instead. We choose J and L so that $|J| = |L| + 1$, which means that on $S_d(X)$, both $\leq_a$ and $\leq_b$ coincide.

The goal of the exercise is then to turn this sketch (together with the proof of Proposition 5.9) into a proof of locality of unary queries in $(\text{FO}+<)_{\text{inv}}$.

Exercise 5.9. The goal of this exercise is to complete the proof of Theorem 5.8. Using Exercise 5.8, show that every m-ary query in $(\text{FO}+<)_{\text{inv}}$, for $m > 1$, is Gaifman-local.

Exercise 5.10. Calculate the locality rank of an order-invariant query produced in the proof of Theorem 5.8. You will probably have to use Exercise 3.10.

Exercise 5.11. We know that $\mathrm{FO} \subsetneq (\mathrm{FO}+<)_{\mathrm{inv}}$. What about $(\mathrm{FO}+\textsc{Succ})_{\mathrm{inv}}$? Clearly

$$\mathrm{FO} \subseteq (\mathrm{FO}+\textsc{Succ})_{\mathrm{inv}} \subseteq (\mathrm{FO}+<)_{\mathrm{inv}},$$

and at least one containment must be proper. Find the exact relationship between these three classes of queries.

Exercise 5.12.* Consider again the vocabulary $\sigma_{<,+}$ and a class $\mathcal{C}_{<,+}$ of $\sigma_{<,+}$-structures where $<$ is interpreted as a linear ordering, and $+$ as the addition corresponding to $<$. Prove that every query in $(\mathrm{FO}+\mathcal{C}_{<,+})_{\mathrm{inv}}$ is local.

6

Complexity of First-Order Logic

The goal of this chapter is to study the complexity of queries expressible in FO. We start with the general definition of different ways of measuring the complexity of a logic over finite structures: these are *data*, *expression*, and *combined* complexity. We then connect FO with Boolean circuits and establish some bounds on the data complexity. We also consider the issue of *uniformity* for a circuit model, and study it via logical definability. We then move to the combined complexity of FO, and show that it is much higher than the data complexity. Finally, we investigate an important subclass of FO queries – *conjunctive queries* – which play a central role in database theory.

6.1 Data, Expression, and Combined Complexity

Let us first consider the complexity of the *model-checking* problem: that is, given a sentence Φ in a logic $\mathcal{L}$ and a structure $\mathfrak{A}$, does $\mathfrak{A}$ satisfy Φ? There are two parameters of this question: the sentence Φ, and the structure $\mathfrak{A}$. Depending on which of them are considered parameters of the problem, and which are fixed, we get three different definitions of complexity for a logic.

Complexity theory defines its main concepts via acceptance of string languages by computational devices such as Turing machines. To talk about complexity of logics on finite structures, we need to encode finite structures and logical formulae as strings. For formulae, we shall assume some natural encoding: for example, $\text{enc}(\varphi)$, the encoding of a formula φ, could be its syntactic tree (represented as a string). For the notion of data complexity, defined below, the choice of a particular encoding of formulae does not matter.

There are several different ways to encode structures. The one we use here is the one most often used, but others are possible, and sometimes provide additional useful information about the running time of query-evaluation algorithms.

Suppose we have a structure $\mathfrak{A} \in \text{STRUCT}[\sigma]$. Let $A = \{a_1, \ldots, a_n\}$. For encoding a structure, we always assume an ordering on the universe. In some structures, the order relation is a part of the vocabulary; in others, it is not, and then we arbitrarily choose one. The order in this case will have no effect on the result of queries, but we need it to represent the encoding of a structure on the tape of a Turing machine, to be able to talk about computability and complexity of queries.

Thus, we choose an order on the universe, say, $a_1 < a_2 < \ldots < a_n$. Each k-ary relation $R^{\mathfrak{A}}$ will be encoded by an n^k-bit string $\mathrm{enc}(R^{\mathfrak{A}})$ as follows. Consider an enumeration of all k-tuples over A, in the lexicographic order (i.e., $(a_1, \ldots, a_1), (a_1, \ldots, a_1, a_2), \ldots, (a_n, \ldots, a_n, a_{n-1}), (a_n, \ldots, a_n)$). Let $\vec{a}_j$ be the jth tuple in this enumeration. Then the jth bit of $\mathrm{enc}(R^{\mathfrak{A}})$ is 1 if $\vec{a}_j \in R^{\mathfrak{A}}$, and 0 if $\vec{a}_j \notin R^{\mathfrak{A}}$. We shall assume without any loss of generality that σ contains only relation symbols, since a constant can be encoded as a unary relation containing one element.

If $\sigma = \{R_1, \ldots, R_p\}$, then the basic encoding of a structure is the concatenation of the encodings of relations: $\mathrm{enc}(R_1^{\mathfrak{A}}) \cdots \mathrm{enc}(R_p^{\mathfrak{A}})$. In some computational models (e.g., circuits), the length of the input is a parameter of the model and thus $|A|$ can easily be calculated from the basic encoding; in others (e.g., Turing machines), $|A|$ must be known by the device in order to use the encoding of a structure. For that purpose, we define an $\mathrm{enc}(\mathfrak{A})$ which is simply the concatenation of 0^n1 and all the $\mathrm{enc}(R_i^{\mathfrak{A}})$'s:

$$\mathrm{enc}(\mathfrak{A}) \;=\; 0^n 1 \cdot \mathrm{enc}(R_1^{\mathfrak{A}}) \cdots \mathrm{enc}(R_p^{\mathfrak{A}}). \tag{6.1}$$

The length of this string, denoted by $\|\mathfrak{A}\|$, is

$$\|\mathfrak{A}\| \;=\; (n+1) + \sum_{i=1}^{p} n^{\mathrm{arity}(R_i)} \tag{6.2}$$

Definition 6.1. *Let $\mathcal{K}$ be a complexity class, and $\mathcal{L}$ a logic. We say that*

- *the* data complexity *of $\mathcal{L}$ is $\mathcal{K}$ if for every sentence Φ of $\mathcal{L}$, the language*

$$\{\mathit{enc}(\mathfrak{A}) \mid \mathfrak{A} \models \Phi\}$$

  belongs to $\mathcal{K}$;

- *the* expression complexity *of $\mathcal{L}$ is $\mathcal{K}$ if for every finite structure $\mathfrak{A}$, the language*

$$\{\mathit{enc}(\Phi) \mid \mathfrak{A} \models \Phi\}$$

  belongs to $\mathcal{K}$; and

- *the* combined complexity *of $\mathcal{L}$ is $\mathcal{K}$ if the language*

$$\{(\mathit{enc}(\mathfrak{A}), \mathit{enc}(\Phi)) \mid \mathfrak{A} \models \Phi\}$$

  belongs to $\mathcal{K}$.

- *Furthermore, we say that the combined complexity of $\mathcal{L}$ is hard for $\mathcal{K}$ (or $\mathcal{K}$-hard) if the language $\{(enc(\mathfrak{A}), enc(\Phi)) \mid \mathfrak{A} \models \Phi\}$ is a $\mathcal{K}$-hard problem. The data complexity is $\mathcal{K}$-hard if for some Φ, $\{enc(\mathfrak{A}) \mid \mathfrak{A} \models \Phi\}$ is a hard problem for $\mathcal{K}$, and the expression complexity is $\mathcal{K}$-hard if for some $\mathfrak{A}$, $\{enc(\Phi) \mid \mathfrak{A} \models \Phi\}$ is $\mathcal{K}$-hard.*
- *A problem that is both in $\mathcal{K}$ and $\mathcal{K}$-hard is complete for $\mathcal{K}$, or $\mathcal{K}$-complete. Thus, we can talk about data/expression/combined complexity being $\mathcal{K}$-complete.*

Given our standard choice of encoding, we shall sometimes omit the notation $\text{enc}(\cdot)$, instead writing $\{\mathfrak{A} \mid \mathfrak{A} \models \Phi\} \in \mathcal{K}$, etc.

The notion of data complexity is most often used in the database context: the structure $\mathfrak{A}$ corresponds to a large relational database, and the much smaller sentence Φ is a query that has to be evaluated against $\mathfrak{A}$; hence Φ is ignored in this definition. The notions of expression and combined complexity are often used in verification and model-checking, where a complex specification needs to be evaluated on a description of a finite state machine; in this case the specification Φ may actually be more complex than the structure $\mathfrak{A}$. We shall also see that for most logics of interest, all the hardness results for the combined complexity will be shown on very simple structures, thereby giving us matching bounds for the combined and expression complexity. Thus, we shall concentrate on the *data* and *combined* complexity.

We defined the notion of complexity for sentences only. The notion of data complexity has a natural extension to formulae with free variables defining non-Boolean queries. Suppose an m-ary query Q is definable by a formula $\varphi(x_1, \ldots, x_m)$. Then the data complexity of Q is the complexity of the language $\{(\text{enc}(\mathfrak{A}), \text{enc}(\{\vec{a}\})) \mid \vec{a} \in Q(\mathfrak{A})\}$. This is the same as the data complexity of the sentence $(\exists! \vec{x}\ S(\vec{x})) \wedge (\forall \vec{x}\ (S(\vec{x}) \rightarrow \varphi(\vec{x})))$, where S is a new m-ary relation symbol not in σ (we assume that the logic $\mathcal{L}$ is closed under the Boolean connectives and first-order quantification). Recall that the quantifier $\exists! x$ means "there exists a unique x". Thus, as long as $\mathcal{L}$ has the right closure properties, we can only consider data complexity with respect to sentences.

6.2 Circuits and FO Queries

In this section we show how to code FO sentences over finite structures by Boolean circuits. This coding will give us bounds for both the data and combined complexity of FO.

Definition 6.2. *A* Boolean circuit *with n inputs $x_1, \ldots, x_n$ is a tuple*

$$\mathsf{C} = (V, E, \lambda, o),$$

where

1. (V, E) *is a directed acyclic graph with the set of nodes* V *(which we call* gates*) and the set of edges* E.
2. λ *is a function from* V *to* $\{x_1, \ldots, x_n\} \cup \{\vee, \wedge, \neg\}$ *such that:*
   - $\lambda(v) \in \{x_1, \ldots, x_n\}$ *implies that* v *has in-degree* 0;
   - $\lambda(v) = \neg$ *implies that* v *has in-degree* 1.
3. $o \in V$.

The in-degree of a node is called its fan-in. *The* size *of* C *is the number of nodes in* V*; the* depth *of* C *is the length of the longest path from a node of in-degree* 0 *to* o.

A circuit C computes a Boolean function with n inputs $x_1, \ldots, x_n$ as follows. Suppose we are given values of $x_1, \ldots, x_n$. Initially, we compute the values associated with each node of in-degree 0: for a node labeled x_i, it is the value of x_i; for a node labeled $\vee$ it is *false*; and for a node labeled $\wedge$ it is *true*. Next, we compute the value of each node by induction: if we have a node v with incoming edges from $v_1, \ldots, v_l$, and we know the values of $a_1, \ldots, a_l$ associated with $v_1, \ldots, v_l$, then the value a associated with v is:

- $a_1 \vee \ldots \vee a_l$ if $\lambda(v) = \vee$;
- $a_1 \wedge \ldots \wedge a_l$ if $\lambda(v) = \wedge$;
- $\neg a_1$ if $\lambda(v) = \neg$ (in this case we know that $l = 1$).

The output of the circuit is the value assigned to the node o. An example of a circuit computing the Boolean function $(x_1 \wedge \neg x_2 \wedge x_3) \vee \neg(x_3 \wedge \neg x_4)$ is shown in Fig. 6.1; the output node is depicted as a double circle.

Note that a circuit with no inputs is possible, and its in-degree zero gates are labeled $\vee$ or $\wedge$. Such a circuit always outputs a constant (i.e., *true* or *false*).

We next define families of circuits and languages in $\{0,1\}^*$ they accept.

Definition 6.3. *A family of circuits is a sequence* $\mathbf{C} = (\mathsf{C}_n)_{n \geq 0}$ *where each* C_n *is a circuit with* n *inputs. It accepts the language* $L(\mathbf{C}) \subseteq \{0,1\}^*$ *defined as follows. Let* s *be a string of length* n*. It can be viewed as a Boolean vector* $\vec{x}_s$ *such that the* i*th component of* $\vec{x}_s$ *is the* i*th symbol in* s*. Then* $s \in L(\mathbf{C})$ *iff* C_n *outputs* 1 *on* $\vec{x}_s$.

A family of circuits $\mathbf{C}$ *is said to be of polynomial size if there is a polynomial* $p : \mathbb{N} \to \mathbb{N}$ *such that the size of each* C_n *is at most* $p(n)$*. For a function* $f : \mathbb{N} \to \mathbb{N}$*, we say that* $\mathbf{C}$ *is of depth* $f(n)$ *if the depth of* C_n *is at most* $f(n)$*. We say that* $\mathbf{C}$ *is of constant depth if there is* $d > 0$ *such that for all* n*, the depth of* C_n *is at most* d.

The class of languages accepted by polynomial-size constant-depth families of circuits is called nonuniform AC^0.

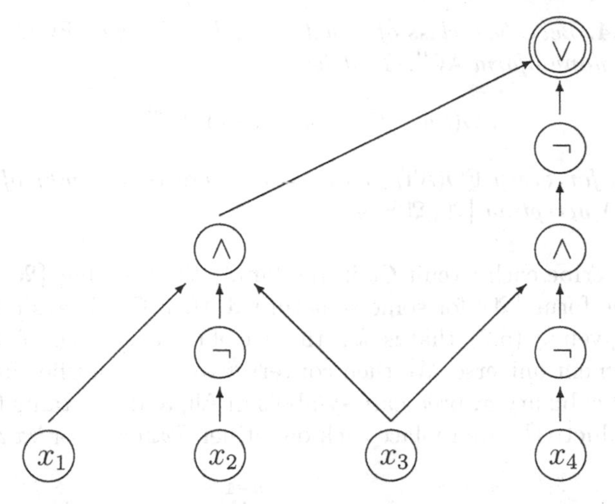

Fig. 6.1. Boolean circuit computing $(x_1 \wedge \neg x_2 \wedge x_3) \ \vee \ \neg(x_3 \wedge \neg x_4)$

For example, the language that consists of strings containing at least two ones is in nonuniform AC^0: each circuit C_n, $n > 1$, has $\wedge$-gates for every pair of inputs x_i and x_j, and then the outputs of those $\wedge$-gates form the input for one $\vee$-gate.

A class of structures $\mathcal{C} \subseteq \mathrm{STRUCT}[\sigma]$ is in nonuniform AC^0 if so is the language $\{\mathrm{enc}(\mathfrak{A}) \mid \mathfrak{A} \in \mathcal{C}\}$. An example of a class of structures that is not FO-definable, but belongs to nonuniform AC^0, is the class EVEN of structures of the empty vocabulary: that is, $\{\langle A, \emptyset \rangle \mid \mid A \mid \mod 2 = 0\}$. The coding of such a structure with $|A|= n$ is simply $0^n 1$; hence C_k always returns *true* for odd k (as it corresponds to structures of even cardinality), and *false* for even k.

Next, we extend FO as follows. Let $\mathcal{P}$ be a collection, finite or infinite, of numerical predicates; that is, subsets of $\mathbb{N}^k$. For example, they may include $<$, $+$ considered as a ternary predicate $\{(i, j, l) \mid i + j = l\}$, etc. For $\mathcal{P}$ including the linear order, we define $\mathrm{FO}(\mathcal{P})$ as an extension of FO with atomic formulae of the form $P(x_1, \ldots, x_k)$, for a k-ary $P \in \mathcal{P}$. The semantics is defined as follows. Suppose $\mathfrak{A}$ is a σ-structure, and its universe A is ordered by $<$ as $a_0 < \ldots < a_{n-1}$. Then $\mathfrak{A} \models P(a_{i_1}, \ldots, a_{i_k})$ iff the tuple of numbers $(i_1, \ldots, i_k)$ belongs to P.

For example, let $P_2 \subset \mathbb{N}$ consist of the even numbers. Then the query EVEN is expressed as an $\mathrm{FO}(\{<, P_2\})$ sentence as follows:

$$\forall x \ (\forall y \ (y \leq x) \ \rightarrow \ P_2(x)).$$

We are now interested in the class FO(All) where All stands for the family of *all* numerical predicates; that is, all subsets of $\mathbb{N}, \mathbb{N}^2, \mathbb{N}^3$, etc. We now show the connection between FO(All) and nonuniform AC^0.

Theorem 6.4. *Let C be a class of structures definable by an* FO(All) *sentence. Then C is in nonuniform* AC^0. *That is,*

$$\mathrm{FO}(\mathsf{All}) \subseteq \text{nonuniform } \mathrm{AC}^0.$$

Furthermore, for every FO(All) *sentence Φ, there is a family of circuits of depth $O(\|\Phi\|)$ accepting $\{\mathfrak{A} \mid \mathfrak{A} \models \Phi\}$.*

Proof. We describe each circuit C_k in the family $\mathbf{C}$ accepting $\{\mathfrak{A} \mid \mathfrak{A} \models \Phi\}$. If k is not of the form $\|\mathfrak{A}\|$ for some structure $\mathfrak{A}$, then C_k always returns *false*. Assume k is given by (6.2); that is, k is the size of the encodings of structures $\mathfrak{A}$ with an n-element universe. We then convert Φ into a quantifier-free sentence Φ' over the vocabulary σ, predicate symbols in All, and constants $0, \ldots, n-1$ as follows. Inductively, we replace each quantifier $\exists x \varphi(x, \vec{y})$ or $\forall x \varphi(x, \vec{y})$ with

$$\bigvee_{c=0}^{n-1} \varphi(c, \vec{y}) \quad \text{and} \quad \bigwedge_{c=0}^{n-1} \varphi(c, \vec{y}),$$

respectively. Notice that the number of connectives $\vee, \wedge, \neg, \bigvee, \bigwedge$ in Φ' is exactly the same as the number of connectives $\vee, \wedge, \neg$ and quantifiers $\exists, \forall$ in Φ.

We now build the circuit to evaluate Φ'. Note that Φ' is a Boolean combination (using connectives $\vee, \wedge, \neg, \bigvee, \bigwedge$) of formulae of the form $P(i_1, \ldots, i_k)$, where P is a numerical predicate, and $R(i_1, \ldots, i_m)$, where R is an m-ary symbol in σ. The former is replaced by its truth value (which is either a $\vee$ or a $\wedge$ gate with zero inputs), and the latter corresponds to one bit in $\mathrm{enc}(\mathfrak{A})$; that is, the input of the circuit. The depth of the resulting circuit is bounded by the number of connectives $\vee, \wedge, \neg, \bigvee, \bigwedge$ in Φ', and hence depends only on Φ, and not on k. The size of the circuit C_k is clearly polynomial in k, which completes the proof. □

Corollary 6.5. *The data complexity of* FO(All) *is nonuniform* AC^0.

We conclude this section with another bound on the complexity of FO queries. This time we determine the running time of such a query in terms of the sizes of encodings of a query and a structure.

Given an FO formula φ, its *width* is the maximum number of free variables in a subformula of φ.

Proposition 6.6. *Let Φ be an* FO *sentence in vocabulary σ, and let $\mathfrak{A} \in$* STRUCT$[\sigma]$. *If the width of Φ is k, then checking whether $\mathfrak{A} \models \Phi$ can be done in time*

$$O(\|\Phi\| \times \|\mathfrak{A}\|^k).$$

Proof. Assume, without loss of generality, that Φ uses $\wedge, \neg$, and $\exists$ but not $\vee$ and $\forall$. Let $\varphi_1, \ldots, \varphi_m$ enumerate all the subformulae of Φ; we know that they contain at most k free variables. We now inductively construct $\varphi_i(\mathfrak{A})$. If φ_i has k_i free variables, then $\varphi_i(\mathfrak{A}) \subseteq A^{k_i}$. It will be represented by a Boolean vector of length n^{k_i}, where $n = |A|$, in exactly the same way as we code relations in $\mathfrak{A}$.

If φ_i is an atomic formula $R(x_1, \ldots, x_{k_i})$, then $\varphi_i(\mathfrak{A})$ is simply the encoding of R in enc($\mathfrak{A}$). If φ_i is $\neg\varphi_j(\mathfrak{A})$, we simply flip all the bits in the representation of $\varphi_j(\mathfrak{A})$. If φ_i is $\varphi_j \wedge \varphi_l$, there are two cases. If the free variables of φ_j and φ_l are the same, then $\varphi_i(\mathfrak{A})$ is obtained as the bit-wise conjunction of $\varphi_j(\mathfrak{A})$ and $\varphi_l(\mathfrak{A})$. Otherwise, $\varphi_i(\vec{x}, \vec{y}, \vec{z}) = \varphi_j(\vec{x}, \vec{y}) \wedge \varphi_l(\vec{x}, \vec{z})$, and $\varphi_i(\mathfrak{A})$ is the join of $\varphi_j(\mathfrak{A})$ and $\varphi_l(\mathfrak{A})$, obtained by finding, for all tuples over $\vec{a} \in A^{|\vec{x}|}$, tuples $\vec{b} \in A^{|\vec{y}|}$ and $\vec{c} \in A^{|\vec{z}|}$ such that the bits corresponding to $(\vec{a}, \vec{b})$ in $\varphi_j(\mathfrak{A})$ and to $(\vec{a}, \vec{c})$ in $\varphi_l(\mathfrak{A})$ are set to 1, and then setting the bit corresponding to $(\vec{a}, \vec{b}, \vec{c})$ in $\varphi_i(\mathfrak{A})$ to 1. Finally, if $\varphi_i(\vec{x}) = \exists z \varphi_j(z, \vec{x})$, we simply go over $\varphi_j(\mathfrak{A})$, and if the bit corresponding to $(a, \vec{a})$ is set to 1, then we set the bit corresponding to $\vec{a}$ in $\varphi_i(\mathfrak{A})$ to 1.

The reader can easily check that the above algorithm can be implemented in time $O(\| \Phi \| \times \| \mathfrak{A} \|^k)$, since none of the formulae φ_i has more than k free variables. □

6.3 Expressive Power with Arbitrary Predicates

In the previous section, we introduced a powerful extension of FO – the logic FO(All). Since this logic can use arbitrary predicates on the natural numbers, it can express noncomputable queries: for example, we can test if the size of the universe of $\mathfrak{A}$ is a number n which codes a pair (k, m) such that the kth Turing machine halts on the mth input (assuming some standard enumeration of Turing machines and their inputs). Nevertheless, we can prove some strong bounds on the expressiveness of FO(All): although we saw that EVEN is FO(All)-expressible, it turns out that the closely related query, PARITY, is not.

Recall that PARITY_U is a query on structures whose vocabulary σ contains one unary relation symbol U. Then

$$\text{PARITY}_U(\mathfrak{A}) \quad \Leftrightarrow \quad |U^{\mathfrak{A}}| \bmod 2 = 0.$$

We shall omit the subscript U if it is understood from the context.

To show that PARITY is not FO(All)-expressible, we consider the Boolean function *parity* with n arguments (for each n) defined as follows:

$$parity(x_1, \ldots, x_n) \quad = \quad \begin{cases} 1 & \text{if} \quad |\{i \mid x_i = 1\}| \bmod 2 = 0, \\ 0 & \text{otherwise.} \end{cases}$$

We shall need the following deep result in circuit complexity.

Theorem 6.7 (Furst-Saxe-Sipser, Ajtai). *There is no constant-depth polynomial-size family of circuits that computes parity.* □

Corollary 6.8. PARITY *is not expressible in* FO(All).

Proof. Assume, to the contrary, that PARITY is expressible. By Theorem 6.4, there is a polynomial-size constant-depth circuit family **C** that computes PARITY on encodings of structures. Such an encoding of a structure $\mathfrak{A}$ with $|A| = n$ is $0^n 1 \cdot s$, where s is the string of length n whose ith element is 1 iff the ith element of A is in $U^{\mathfrak{A}}$.

We now use **C** to construct a new family of circuits defining *parity*. The circuit with n inputs $x_1, \ldots, x_n$ works as follows. For each x_i, it adds an in-degree 0 gate g_i labeled $\vee$, and for x_n it also adds an in-degree 0 gate g'_n labeled $\wedge$. Then it puts C_{2n+1}, the circuit with $2n+1$ inputs from **C** on the outputs of $g_1, \ldots, g_n, g'_n$ followed by $x_1, \ldots, x_n$, as shown below:

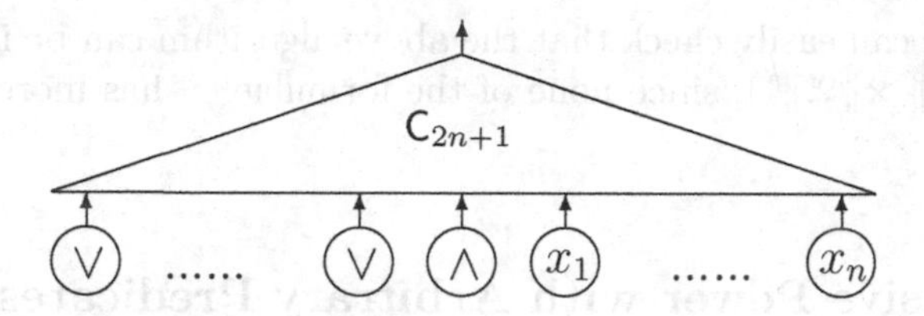

Clearly this circuit computes *parity*$(x_1, \ldots, x_n)$, and by Theorem 6.4 the resulting family of circuits is of polynomial size and bounded depth. This contradicts Theorem 6.7. □

As another example of inexpressibility in FO(All), we show the following.

Corollary 6.9. *Graph connectivity is not expressible in* FO(All).

Proof. We shall follow the idea of the proof of Corollary 3.19; however, that proof used inexpressibility of the query EVEN, which of course is definable in FO(All). We modify the proof to make use of Corollary 6.8 instead.

First, we show that for a graph $G = (V, E)$, where E is a successor relation on a set $U \subseteq V$ of nodes, FO(All) cannot test if the cardinality of U is even. Indeed, suppose to the contrary that it can; then this can be done in nonuniform AC^0, by a family of circuits **C**. We now show how to use **C** to test PARITY. Suppose an encoding $0^n 1 \cdot s$ of a unary relation U is given, where $U = \{i_1, \ldots, i_k\} \subseteq \{1, \ldots, n\}$. We transform U into a successor relation $S_U = \{(i_1, i_2), \ldots, (i_{k-1}, i_k)\}$. We leave it to the reader to show how to use bounded-depth circuits to transform $0^n 1 \cdot s$ into $0^n 1 \cdot s'$ where s' of length n^2 codes S_U. Then using the circuit C_{n^2+n+1} from **C** on $0^n 1 \cdot s'$ we can test if U is even.

Finally, using inexpressibility of parity of a successor relation, we show inexpressibility of connectivity in FO(All) using the same proof as in Corollary 3.19. □

6.4 Uniformity and AC^0

We have noticed that nonuniform AC^0 is not truly a complexity class: in fact, the function that computes the circuit C_n from n need not even be recursive. It is customary to impose some *uniformity* conditions that postulate how C_n is obtained. While it is possible to formulate these conditions purely in terms of circuits, we prefer to follow the logic connection, and instead put restrictions on the choice of available predicates in FO(All).

We now associate a finite n-element universe of a structure with the set $\{0, \ldots, n-1\}$, and consider an extension of FO over σ-structures by adding two ternary predicates, $+$ and $\times$, which are graphs of addition and multiplication. That is,

$$+ = \{(i, j, k) \mid i + j = k\} \quad \text{and} \quad \times = \{(i, j, k) \mid i \cdot j = k\}.$$

Note that we have to use $+$ and $\times$ as ternary relations rather than binary functions, to ensure that the result of addition or multiplication is always in the universe of the structure. The resulting logic is denoted by $\mathrm{FO}(+, \times)$.

Definition 6.10. *The class of structures definable in* $\mathrm{FO}(+, \times)$ *is called* uniform AC^0.

We shall normally omit the word *uniform*; hence, by referring to just AC^0, we mean uniform AC^0. Note that many examples of AC^0 queries seen so far only use the standard arithmetic on the natural numbers; for example, EVEN is in AC^0.

It turns out that AC^0 is quite powerful and can define several interesting numerical relations on the domain $\{0, \ldots, n-1\}$. One of them, which we shall see quite often, is the *bit* relation:

$$\mathrm{BIT}(x, y) \text{ is true} \quad \Leftrightarrow \quad \text{the } y\text{th bit of the binary expansion of } x \text{ is } 1.$$

For example, the binary expansion of $x = 18$ is 10010, and hence $\mathrm{BIT}(x, y)$ is true if y is 1 or 4, and $\mathrm{BIT}(x, y)$ is false if y is 0, 2, or 3.

We now start building the family of functions definable in $\mathrm{FO}(+, \times)$. Whenever we say that a k-ary function is definable, we actually mean that the graph of this function, a $k + 1$-ary relation, is definable. However, to make formulae more readable, we often use functions instead of their graphs. First, we note that the linear order is definable by $x \le y \Leftrightarrow \exists z\, +(x, z, y)$ (i.e., $\exists z\ (x + z = y)$), and thus the minimum element 0, and the maximum element, denoted by $\underline{\max}$, are definable.

Lemma 6.11. *The integer division* $\lfloor x/y \rfloor$ *and* $(x \bmod y)$ *are definable in* $\mathrm{FO}(+, \times)$.

Proof. If $y \neq 0$, then

$$u = \lfloor x/y \rfloor \;\Leftrightarrow\; ((u \cdot y) \leq x \wedge (\exists v < y\ (x = u \cdot y + v))).$$

Furthermore,

$$u = (x \bmod y) \;\Leftrightarrow\; \exists v\ ((v = \lfloor x/y \rfloor) \wedge (u + y \cdot v = x)). \qquad \square$$

In particular, we can express divisibility $x \mid y$ as $(x \bmod y) = 0$.

Now our goal is to show the following.

Theorem 6.12. BIT *is expressible in* FO$(+,\times)$.

Proof. We shall prove this in several stages. First, note that the following tests if x is a power of 2:

$$pow_2(x) \;\equiv\; \forall u, v\ \big((x = u \cdot v) \wedge (v \neq 1)\big) \rightarrow \big(\exists z\ (v = z + z)\big).$$

This is because $pow_2(x)$ asserts that 2 is the only prime factor of x. Next, we define the predicate

$$\mathrm{BIT}'(x, y) \;\equiv\; (\lfloor x/y \rfloor \bmod 2) \;=\; 1.$$

Note that if $y = 2^z$, then $\mathrm{BIT}'(x, y)$ is true iff the zth bit of x is 1. Assume that we can define the predicate $y = 2^z$. Then

$$\mathrm{BIT}(x, y) \;\equiv\; \exists u\ \big(u = 2^y \wedge \mathrm{BIT}'(x, u)\big).$$

Thus, it remains to show how to express the binary predicate $x = 2^y$. We do so by coding an iterative computation of 2^y. The codes of such computations will be numbers, and as we shall see, those numbers can be as large as x^4. Since we only quantify over $\{0, \ldots, n-1\}$, where n is the size of the finite structure, we show below how to express the predicate

$$P_2(x, y) \;\equiv\; x = 2^y \;\wedge\; x^4 \leq n - 1.$$

With P_2, we can define $x = 2^y$ as follows:

$$\exists u \exists v \begin{pmatrix} (y = 4v \quad\;\;\; \wedge P_2(u, v) \wedge x = u^4) \\ \vee\ (y = 4v + 1 \wedge P_2(u, v) \wedge x = 2 \cdot u^4) \\ \vee\ (y = 4v + 2 \wedge P_2(u, v) \wedge x = 4 \cdot u^4) \\ \vee\ (y = 4v + 3 \wedge P_2(u, v) \wedge x = 8 \cdot u^4) \end{pmatrix}.$$

We now show how to express $P_2(x, y)$. Let $y = \sum_{i=0}^{k-1} y_i \cdot 2^i$, so that y is $y_{k-1} y_{k-2} \cdots y_1 y_0$ in binary (we assume that the most significant bit y_{k-1} is 1). Then $2^y = \prod_{i=0}^{k-1} 2^{y_i \cdot 2^i}$. We now define the following recurrences for $i < k$:

$$\begin{array}{lll} p_0 = 1 & a_0 = 0 & b_0 = 1 \\ p_{i+1} = 2p_i & a_{i+1} = a_i + y_i \cdot 2^i & b_{i+1} = b_i \cdot 2^{y_i \cdot 2^i} \end{array}$$

Thus, $p_i = 2^i$, a_i is the number whose binary representation is $y_{i-1} \cdots y_0$, and $b_i = 2^{a_i}$. We define sequences $\vec{p} = (p_0, \ldots, p_k), \vec{a} = (a_0, \ldots, a_k), \vec{b} = (b_0, \ldots, b_k)$.

Next, we explain how to code these sequences. Notice that in all three of them, the ith element needs at most 2^i bits to be represented in binary. Suppose we have an arbitrary sequence $\vec{c} = (c_0, \ldots, c_k)$, where each c_i has at most 2^i bits in binary. Such a sequence will be coded by a number $\boldsymbol{c}$ such that its 2^i bits from 2^i to $2^{i+1} - 1$ form the binary representation of c_i. These codes, when applied to $\vec{p}, \vec{a}$, and $\vec{b}$, result in numbers $\boldsymbol{p}$, $\boldsymbol{a}$, and $\boldsymbol{b}$, respectively. These numbers turn out to be relatively small. Since the length of the binary representation of y is k, we know that $y \geq 2^{k-1}$. If $x = 2^y$, then $x \geq 2^{2^{k-1}}$ and $x^4 \geq 2^{2^{k+1}}$. The binary representation of $\vec{p}, \vec{a}$, and $\vec{b}$ has at most $2^{k+1} - 1$ bits, and hence the maximum value of those codes is $2^{2^{k+1}-1} - 1$, which is bounded above by x^4. Hence, for defining P_2, codes of all the sequences will be bounded by the size of the universe.

How can one extract numbers c_i from the code $\boldsymbol{c}$ of $\vec{c}$? Notice that

$$\lfloor x/2^{2^i} \rfloor \bmod 2^{2^i}$$

is c_i. In general, we define $\text{extract}(x, u) \equiv \lfloor x/u \rfloor \bmod u$, and thus $c_i = \text{extract}(\boldsymbol{c}, 2^{2^i})$. Notice that since $(2^{2^i})^2 = 2^{2^{i+1}}$, for $u = 2^{2^i}$ we have $c_i = \text{extract}(\boldsymbol{c}, u)$ and $c_{i+1} = \text{extract}(\boldsymbol{c}, u^2)$.

Assume now that we have an extra predicate $\boldsymbol{ppow}_2(u)$ which holds iff u is of the form 2^{2^i}. With this, we express $P_2(x, y)$ by stating the existence of $\boldsymbol{a}, \boldsymbol{b}, \boldsymbol{p}$ (coding $\vec{a}, \vec{b}, \vec{p}$) such that:

- $\text{extract}(\boldsymbol{p}, 2) = 1$, $\text{extract}(\boldsymbol{a}, 2) = 0$, $\text{extract}(\boldsymbol{b}, 2) = 1$ (the initial conditions of the recurrences hold).
- If $u < x$ and $\boldsymbol{ppow}_2(u)$, then $\text{extract}(\boldsymbol{p}, u^2) = 2 \cdot \text{extract}(\boldsymbol{p}, u)$ (the recurrence for $\vec{p}$ is correct).
- If $u < x$ and $\boldsymbol{ppow}_2(u)$, then either
  1. $\text{extract}(\boldsymbol{a}, u^2) = \text{extract}(\boldsymbol{a}, u)$ and $\text{extract}(\boldsymbol{b}, u^2) = \text{extract}(\boldsymbol{b}, u)$, or
  2. $\text{extract}(\boldsymbol{a}, u^2) = \text{extract}(\boldsymbol{a}, u) + \text{extract}(\boldsymbol{p}, u)$ and $\text{extract}(\boldsymbol{b}, u^2) = u \cdot \text{extract}(\boldsymbol{b}, u)$.

  That is, the recurrences for $\vec{a}$ and $\vec{b}$ are coded correctly: the first case corresponds to $y_i = 0$, and hence $a_{i+1} = a_i$ and $b_{i+1} = b_i$; the second case corresponds to $y_i = 1$, and hence $a_{i+1} = a_i + p_i$ and $b_{i+1} = b_i \cdot 2^{2^i} = b_i \cdot u$.
- There is u such that $\boldsymbol{ppow}_2(u)$ holds, $\text{extract}(\boldsymbol{a}, u) = y$, and $\text{extract}(\boldsymbol{b}, u) = x$. That is, the sequences show that $2^y = x$.

Clearly, the above can be expressed as an FO formula.

All that remains is to show how to express the predicate $\boldsymbol{ppow}_2(u)$. This in turn is done in two steps. First, we define a predicate $P_1(v)$ that holds iff v is of the form $\sum_{i=1}^{s} 2^{2^i}$ (i.e., in its binary representation, ones appear only in positions corresponding to powers of 2). With this predicate, we define

$$\boldsymbol{ppow}_2(u) \equiv pow_2(u) \wedge \exists w\ P_1(w) \wedge \mathrm{BIT}'(w,u).$$

Note that if $\boldsymbol{ppow}_2(u)$ holds, one can find w with $\mathrm{BIT}'(w,u)$ such that $w \le u^2$; given that all numbers for which $\boldsymbol{ppow}_2(\cdot)$ is checked are below $\sqrt[4]{n-1}$, the $\exists w$ is guaranteed to range over the finite universe.

To express P_1, we need an auxiliary formula $pow_4(u) = pow_2(u) \wedge (u \bmod 3 = 1)$ testing if u is a power of 4. Now $P_1(u)$ is the conjunction of $\neg\mathrm{BIT}'(u,1) \wedge \mathrm{BIT}'(u,2)$ and the following formula:

$$\forall v \Big(2 < v \le u \rightarrow (\mathrm{BIT}'(u,v) \leftrightarrow (pow_4(v) \wedge \exists w\ [(w \cdot w = v) \wedge \mathrm{BIT}'(u,w)]))\Big).$$

This formula states that 1-bits in the binary representation of u are 2 and others given by the sequence $e_1 = 2, e_2 = 4, \ldots, e_{i+1} = e_i^2$; that is, bits in positions of the form 2^{2^i}. This defines P_1, and thus completes the proof of the theorem. □

The BIT predicate turns out to be quite powerful. First note the following.

Lemma 6.13. *Addition is definable in* FO($<$, BIT).

Proof. We use the standard carry-lookahead algorithm. Given x, y, and u, we define carry(x,y,u) to be true if, while adding x, y given as binary numbers, the carry bit with number u is 1:

$$\exists v \begin{pmatrix} (v < u \wedge \mathrm{BIT}(x,v) \wedge \mathrm{BIT}(y,v)) \\ \wedge \forall w\ ((w < u \wedge w > v) \rightarrow (\mathrm{BIT}(x,w) \vee \mathrm{BIT}(y,w))) \end{pmatrix}.$$

Then $x + y = z$ iff

$$\forall u \Big(\mathrm{BIT}(z,u) \leftrightarrow ((\mathrm{BIT}(x,u) \oplus \mathrm{BIT}(y,u)) \oplus \mathrm{carry}(x,y,u))\Big),$$

where $\varphi \oplus \psi$ is an abbreviation for $\varphi \leftrightarrow \neg\psi$. □

A more complicated result (see Exercise 6.5) states the following.

Lemma 6.14. *Multiplication is definable in* FO($<$, BIT).

We thus obtain:

Corollary 6.15. FO($<$, BIT) = FO($+, \times$).

Hence, uniform AC^0 can be characterized as the class of structures definable in FO($<$, BIT).

6.5 Combined Complexity of FO

We have seen that the data complexity of FO(All) is nonuniform AC^0, and the data complexity of FO is AC^0. What about the combined and expression complexity of FO? It turns out that they belong to a much larger class than AC^0.

Theorem 6.16. *The combined complexity of* FO *is* PSPACE*-complete.*

Proof. The membership in PSPACE follows immediately from the evaluation method used in the proof of Proposition 6.6. To show hardness, recall the problem QBF, satisfiability of quantified Boolean formulae.

Problem: QBF
Input: A formula $\Phi = Q_1 x_1 \ldots Q_n x_n \; \alpha(x_1, \ldots, x_n)$, where:
each Q_i is either $\exists$ or $\forall$, and
α is a propositional formula in $x_1, \ldots, x_n$.
Question: If all x_i's range over $\{\mathit{true}, \mathit{false}\}$, is Φ true?

It is known that QBF is PSPACE-hard (see the bibliographic notes at the end of the chapter). We now prove PSPACE-hardness of FO by reduction from QBF.

Given a formula $\Phi = Q_1 x_1 \ldots Q_n x_n \; \alpha(x_1, \ldots, x_n)$, construct a structure $\mathfrak{A}$ whose vocabulary includes one unary relation U as follows: $A = \{0, 1\}$, and $U^{\mathfrak{A}} = \{1\}$. Then modify α by changing each occurrence of x_i to $U(x_i)$, and each occurrence of $\neg x_i$ to $\neg U(x_i)$. Let α^U be the resulting formula. For example, if $\alpha(x_1, x_2, x_3) = (x_1 \wedge x_2) \vee (\neg x_1 \wedge x_3)$, then α^U is $(U(x_1) \wedge U(x_2)) \vee (\neg U(x_1) \wedge U(x_3))$. Then

$$\Phi \text{ is true} \quad \Leftrightarrow \quad \mathfrak{A} \models Q_1 x_1 \ldots Q_n x_n \; \alpha^U(x_1, \ldots, x_n),$$

which proves PSPACE-hardness. □

Since the structure $\mathfrak{A}$ constructed in the proof of Theorem 6.16 is fixed, we obtain:

Corollary 6.17. *The expression complexity of* FO *is* PSPACE*-complete.*

For most of the logics we study, the expression and combined complexity coincide; however, this need not be the case in general.

6.6 Parametric Complexity and Locality

Proposition 6.6 says that checking whether $\mathfrak{A} \models \Phi$ can be done in time $O(\|\Phi\| \cdot \|\mathfrak{A}\|^k)$, where k is the width of Φ: the maximum number of free variables of a subformula of Φ. In particular, this gives a polynomial time algorithm for evaluating FO queries on finite structures, for a fixed sentence Φ. Although polynomial time is good, in many cases it is not sufficient: for example, in the database context where $\| \mathfrak{A} \|$ is very large, even for small k the running time $O(\| \mathfrak{A} \|^k)$ may be prohibitively expensive (in fact, the goal of most join algorithms in database systems is to reduce the running time from the impractical $O(n^2)$ to $O(n \log n)$ – at least if the result of the join is not too large – and running time of the order n^{10} is completely out of the question).

The question is, then, whether sometimes (or always) one can find better algorithms for evaluating FO queries on finite structures. In particular, it would be ideal if one could always guarantee time linear in $\| \mathfrak{A} \|$. Since the combined complexity of FO queries is PSPACE-complete, something must be exponential, so in that case we would expect the complexity to be $O\big(g(\| \Phi \|) \cdot \| \mathfrak{A} \|\big)$, where $g : \mathbb{N} \to \mathbb{N}$ is some function.

This is the setting of *parameterized complexity*, where the standard input of a problem is split into the input part and the parameter part, and one looks for *fixed parameter tractable* problems that admit algorithms with running time $O(g(\pi) \cdot n^p)$ for a fixed p; here π is the size of the parameter, and n is the size of the input. It is known that even some NP-hard problems become fixed parameter tractable if the parameters are chosen correctly. For example, SET COVER is the problem whose input is a set V, a family $\mathcal{F}$ of its subsets, and a number k, and the output is "yes" if there is a subset of V of size at most k that intersects every member of $\mathcal{F}$. This problem is NP-complete, but if we choose $\pi = k + \max_{F \in \mathcal{F}} |F|$ to be the parameter, it becomes solvable in time $O(\pi^{\pi+1} \cdot |\mathcal{F}|)$, thus becoming linear in what is likely the largest part of the input.

We now formalize the concept of fixed-parameter tractability.

Definition 6.18. *Let $\mathcal{L}$ be a logic, and $\mathcal{C}$ a class of structures. The* model-checking problem for $\mathcal{L}$ on $\mathcal{C}$ *is the problem to check, for a given structure $\mathfrak{A} \in \mathcal{C}$ and an $\mathcal{L}$-sentence Φ, whether $\mathfrak{A} \models \Phi$.*

We say that the model-checking problem for $\mathcal{L}$ on $\mathcal{C}$ is fixed-parameter tractable, *if there is a constant p and a function $g : \mathbb{N} \to \mathbb{N}$ such that for every $\mathfrak{A} \in \mathcal{C}$ and every $\mathcal{L}$-sentence Φ, checking whether $\mathfrak{A} \models \Phi$ can be done in time*

$$g(\| \Phi \|) \cdot \| \mathfrak{A} \|^p.$$

We say that the model-checking problem for $\mathcal{L}$ on $\mathcal{C}$ is fixed-parameter linear, *if $p = 1$; that is, if there is a function $g : \mathbb{N} \to \mathbb{N}$ such that for every $\mathfrak{A} \in \mathcal{C}$ and every $\mathcal{L}$-sentence Φ, checking whether $\mathfrak{A} \models \Phi$ can be done in time*

$$g(\| \Phi \|) \cdot \| \mathfrak{A} \|.$$

We now prove that on structures of bounded degree, model-checking for FO is fixed-parameter linear. The proof is based on Hanf-locality of FO.

Theorem 6.19. *Fix $l > 0$. Then the model-checking problem for* FO *on* $\text{STRUCT}_l[\sigma]$ *is fixed-parameter linear.*

Proof. We use threshold equivalence and Theorem 4.24. Given l and Φ, we can find numbers d and m such that for every $\mathfrak{A}, \mathfrak{B} \in \text{STRUCT}_l[\sigma]$, it is the case that $\mathfrak{A} \leftrightarrows^{thr}_{d,m} \mathfrak{B}$ implies that $\mathfrak{A}$ and $\mathfrak{B}$ agree on Φ.

We know that for structures of fixed degree l, the upper bound on the number of isomorphism types of radius d neighborhoods of a point is determined by d, l, and σ. We assume that $\tau_1, \ldots, \tau_M$ enumerate isomorphism types of all the structures of the form $N_d^{\mathfrak{A}}(a)$ for $\mathfrak{A} \in \text{STRUCT}_l[\sigma]$.

Let $n_i(\mathfrak{A}) = \mid \{a \mid N_d^{\mathfrak{A}}(a) \text{ of type } \tau_i\} \mid$. With each structure $\mathfrak{A}$, we now associate an M-tuple $\vec{t}(\mathfrak{A}) = (t_1, \ldots, t_M)$ such that

$$t_i = \begin{cases} n_i(\mathfrak{A}), & \text{if } n_i(\mathfrak{A}) \leq m, \\ * & \text{otherwise.} \end{cases}$$

Let T be the set of all M-tuples whose elements come from $\{1, \ldots, m\} \cup \{*\}$. Note that the number of such tuples is $(m+1)^M$, which depends only on l and Φ, and that each $\vec{t}(\mathfrak{A})$ is a member of T.

From Theorem 4.24, $\vec{t}(\mathfrak{A}) = \vec{t}(\mathfrak{B})$ implies that $\mathfrak{A}$ and $\mathfrak{B}$ agree on Φ. Let T_0 be the set of $\vec{t} \in T$ such that for some structure $\mathfrak{A} \in \text{STRUCT}_l[\sigma]$, we have $\mathfrak{A} \models \Phi$ and $\vec{t}(\mathfrak{A}) = \vec{t}$. We leave it as an exercise for the reader (see Exercise 6.7) to show that T_0 is computable.

The idea of the algorithm then is to compute, for a given structure $\mathfrak{A}$, the tuple $\vec{t}(\mathfrak{A})$ in linear time. Once this is done, we check if $\vec{t} \in T_0$. The computation of T_0 depends entirely on Φ and l, but not on $\mathfrak{A}$; hence the resulting algorithm has linear running time.

For simplicity, we present the algorithm for computing $\vec{t}(\mathfrak{A})$ for the case when $\mathfrak{A}$ is an undirected graph; extension to the case of arbitrary $\mathfrak{A}$ is straightforward. We compute, for each node i (assuming that nodes are numbered $0, \ldots, n-1$), $\tau(i)$, the isomorphism type of its d-neighborhood. For this, we first do a pass over the code of $\mathfrak{A}$, and construct an array that, for each node i, has the list of all nodes j such that there is an edge (i, j). Note that the size of any such list is at most l. Next, we construct the radius d neighborhood of each node by looking up its neighbors, then the neighbors of its neighbors, etc., in the array constructed in the first step. After d iterations, we have radius d neighborhood, whose size is bounded by a number that depends on the Φ and l but not on $\mathfrak{A}$. Now for each i, we find $j \leq M$ such that $\tau(i) = \tau_j$; since the enumeration $\tau_1, \ldots, \tau_M$ does not depend on $\mathfrak{A}$, each such step takes constant time. Finally, we do one extra pass over $(\tau(i))_i$ and compute $\vec{t}(\mathfrak{A})$. Hence, $\vec{t}(\mathfrak{A})$ is computed in linear time. As we already explained, to check if $\mathfrak{A} \models \Phi$, we check if $\vec{t} \in T_0$, which takes constant time. Hence, the entire algorithm has linear running time. ☐

Can one prove a similar result for FO queries on arbitrary structures? The answer is most likely *no*, assuming some separation results in complexity theory (see Exercise 6.9). In fact, these results show that even fixed-parameter tractability is very unlikely for arbitrary structures.

Nevertheless, fixed-parameter tractability can be shown for some interesting classes of structures.

Recall that a graph H is a *minor* of a graph G if H can be obtained from a subgraph of G by contracting edges. A class $\mathcal{C}$ of graphs is called *minor-closed* if for any $G \in \mathcal{C}$ and H a minor of G, we have $H \in \mathcal{C}$.

Theorem 6.20. *If $\mathcal{C}$ is a minor-closed class of graphs which does not include all the graphs, then model-checking for* FO *on $\mathcal{C}$ is fixed-parameter tractable.*

The proof of this (hard) theorem is not given here (see Exercise 6.10).

Corollary 6.21. *Model-checking for* FO *on the class of planar graphs is fixed-parameter tractable.* □

6.7 Conjunctive Queries

In this section we introduce a subclass of FO queries that plays a central role in database theory. This is the class of *conjunctive* queries. These are the queries most commonly asked in relational databases; in fact any SQL `SELECT-FROM-WHERE` query that only uses conjunction of attribute equalities in the `WHERE` clause is such. Logically this class has a simple characterization.

Definition 6.22. *A first-order formula $\varphi(\vec{x})$ over a relational vocabulary σ is called a* conjunctive query *if it is built from atomic formulae using only conjunction $\wedge$ and existential quantification $\exists$.*

By renaming variables and pushing existential quantifiers outside, we can see that every conjunctive query can be expressed as

$$\varphi(\vec{x}) \quad = \quad \exists \vec{y} \bigwedge_{i=1}^{k} \alpha_i(\vec{x}, \vec{y}), \tag{6.3}$$

where each α_i is either of the form $R(\vec{u})$, where $R \in \sigma$ and $\vec{u}$ is a tuple of variables from $\vec{x}, \vec{y}$, or $u = v$, where u, v are variables from $\vec{x}, \vec{y}$ or constant symbols.

We have seen an example of a conjunctive query in Chap. 1: to test if there is a path of length $k+1$ between x and x' in a graph E, one can write

$$\exists y_1, \ldots, y_k \; R(x, y_1) \wedge R(y_1, y_2) \wedge \ldots \wedge R(y_{k-1}, y_k) \wedge R(y_k, x').$$

To see how conjunctive queries can be evaluated, we introduce the concept of a *join* of two relations. Suppose we have a formula $\varphi(x_1, \ldots, x_m)$ over vocabulary σ. For each $\mathfrak{A} \in \text{STRUCT}[\sigma]$, this formula defines an m-ary relation $\varphi(\mathfrak{A}) = \{\vec{a} \mid \mathfrak{A} \models \varphi(\vec{a})\}$. We can view $\varphi(\mathfrak{A})$ as an m-ary relation with attributes $x_1, \ldots, x_m$: that is, a set of finite mappings $\{x_1, \ldots, x_m\} \to A$. Viewing $\varphi(\mathfrak{A})$ as a relation with columns and rows lets us name individual columns.

Suppose now we have two relations over A: an m-ary relation S and an l-ary relation R, such that R is viewed as a set of mappings $t : X \to A$ and S is viewed as a set of mappings $t : Y \to A$. Then the join of R and S is defined as

$$R \bowtie S \quad = \quad \{t : X \cup Y \to A \mid t|_X \in R,\ t|_Y \in S\}. \tag{6.4}$$

Suppose that R is $\varphi(\mathfrak{A})$ where φ has free variables $(\vec{x}, \vec{z})$, and S is $\psi(\mathfrak{A})$ where ψ has free variables $(\vec{y}, \vec{z})$. How can one construct $R \bowtie S$? According to (6.4), it consists of tuples $(\vec{a}, \vec{b}, \vec{c})$ such that $\varphi(\vec{a}, \vec{c})$ and $\psi(\vec{b}, \vec{c})$ hold. Thus, $R \bowtie S = [\varphi \wedge \psi](\mathfrak{A})$.

As another operation corresponding to conjunctive queries, consider again a relation R viewed as a set of finite mappings $t : X \to A$, and let $Y \subseteq X$. Then the *projection* of R on Y is defined as

$$\pi_Y(R) \quad = \quad \{t : Y \to A \mid \exists t' \in R :\ t'|_Y = t\}. \tag{6.5}$$

Again, if R is $\varphi(\mathfrak{A})$, where φ has free variables $(\vec{x}, \vec{y})$, then $\pi_{\vec{y}}(R)$ is simply $[\exists \vec{x}\ \varphi(\vec{x}, \vec{y})](R)$.

Now suppose we have a conjunctive query

$$\varphi(\vec{y}) \quad \equiv \quad \exists \vec{x}\ (\alpha_1(\vec{u}_1) \wedge \ldots \wedge \alpha_n(\vec{u}_n)), \tag{6.6}$$

where each $\alpha_i(\vec{u}_i)$ is an atomic formula $S(\vec{u}_i)$ for some $S \in \sigma$, and $\vec{u}_i$ is a list of variables among $\vec{x}, \vec{y}$. Then for any structure $\mathfrak{A}$,

$$\varphi(\mathfrak{A}) \quad = \quad \pi_{\vec{y}}(\alpha_1(\mathfrak{A}) \bowtie \ldots \bowtie \alpha_n(\mathfrak{A})). \tag{6.7}$$

A slight extension of the correspondence between conjunctive queries and the join and projection operations involves queries of the form

$$\varphi(\vec{y}) \quad \equiv \quad \exists \vec{x}\ (\alpha_1(\vec{u}_1) \wedge \ldots \wedge \alpha_n(\vec{u}_n) \wedge \beta(\vec{x}, \vec{u})), \tag{6.8}$$

where β is a conjunction of formulae $u_1 = u_2$, where u_1 and u_2 are variables occurring among $\vec{u}_1, \ldots, \vec{u}_n$.

Suppose we have a relation R, again viewed as a set of finite mappings $t : X \to A$, and a set C of conditions $x_i = x_j$, for $x_i, x_j \in X$. Then the *selection* operation, $\sigma_C(R)$, is defined as

$$\{t : X \to A \mid t \in R,\ t(x_i) = t(x_j) \text{ for all } x_i = x_j \in C\}.$$

If R is $\varphi(\mathfrak{A})$, then $\sigma_C(R)$ is simply $[\varphi \wedge \beta](R)$, where β is the conjunction of all the conditions $x_i = x_j$ that occur in C.

For β being as in (6.8), let C_β be the list of all equalities listed in β. Then, using the selection operation, the most general form of a conjunctive query above can be translated into

$$\pi_{\vec{y}}\Big(\sigma_{C_\beta}(\alpha_1(\mathfrak{A}) \bowtie \ldots \bowtie \alpha_n(\mathfrak{A}))\Big)\,. \tag{6.9}$$

Many common database queries are of the form (6.9): they compute the join of two or more relations, select some tuples from them, and output only certain elements of those tuples. These can be expressed as conjunctive queries.

The data complexity of conjunctive queries is the same as for general FO queries: uniform AC^0. For the combined and expression complexity, we can lower the PSPACE bound of Theorem 6.16.

Theorem 6.23. *The combined and expression complexity of conjunctive queries are* NP*-complete (even for Boolean conjunctive queries).*

Proof. It is easy to see that the combined complexity is NP: for the query given by (6.3) and a tuple $\vec{a}$, to check if $\varphi(\vec{a})$ holds, one has to guess a tuple $\vec{b}$ and then check in polynomial time if $\bigwedge_i \alpha_i(\vec{a}, \vec{b})$ holds.

For completeness, we use reduction from 3-colorability, defined in Chap. 1 (and known to be NP-complete). Define a structure $\mathfrak{A} = \langle\{0,1,2\}, N\rangle$, where N is the binary inequality relation: $N = \{(0,1),(0,2),(1,0),(1,2),(2,0),(2,1)\}$. Suppose we are given a graph with the set of nodes $U = \{a_1, \ldots, a_n\}$, and a set of edges $E \subseteq U \times U$. We then define the following Boolean conjunctive query:

$$\exists x_1 \ldots \exists x_n \bigwedge_{(a_i,a_j)\in E} N(x_i, x_j). \tag{6.10}$$

Note that for a given graph $\langle U, E\rangle$, the query Φ can be constructed in deterministic logarithmic time.

For the query Φ given by (6.10), $\mathfrak{A} \models \Phi$ iff there is an assignment of variables $x_i, 1 \leq i \leq n$, to $\{0,1,2\}$ such that for every edge (a_i, a_j), the corresponding values x_i and x_j are different. That is, $\mathfrak{A} \models \Phi$ iff $\langle U, E\rangle$ is 3-colorable, which provides the desired reduction, and thus proves NP-completeness for the combined (and expression, since $\mathfrak{A}$ is fixed) complexity of conjunctive queries. ☐

As for the data complexity of conjunctive queries, so far we have seen no results that would distinguish it from the data complexity of FO. We shall now see one result that lowers the complexity of conjunctive query evaluation rather significantly, under certain assumptions on the structure of queries. Unlike Theorem 6.19, this result will apply to arbitrary structures.

Recall that in general, an FO sentence Φ can be evaluated on a structure $\mathfrak{A}$ in time $O(\|\Phi\| \cdot \|\mathfrak{A}\|^k)$, where k is the width of Φ. We shall now lower this to $O(\|\Phi\| \cdot \|\mathfrak{A}\|)$ for the class of *acyclic* conjunctive queries. That is, for a certain class of queries, we shall prove that they are fixed-parameter linear on the class of all finite structures. To define this class of queries, we need a few preliminary definitions.

Let $\mathcal{H}$ be a *hypergraph*: that is, a set U and a set E of *hyper-edges*, or subsets of U. A *tree decomposition* of $\mathcal{H}$ is a tree $\mathcal{T}$ together with a set $B_t \subseteq U$ for each node t of $\mathcal{T}$ such that the following two conditions hold:

1. For every $a \in U$, the set $\{t \mid a \in B_t\}$ is a subtree of $\mathcal{T}$.
2. Every hyper-edge of $\mathcal{H}$ is contained in one of the B_t's.

A hypergraph $\mathcal{H}$ is *acyclic* if there exists a tree decomposition of $\mathcal{H}$ such that each B_t, $t \in \mathcal{T}$, is a hyper-edge of $\mathcal{H}$.

Definition 6.24. *Given a conjunctive query* $\varphi(\vec{y}) \equiv \exists \vec{x}\ \big(\alpha_1(\vec{u}_1) \wedge \ldots \wedge \alpha_n(\vec{u}_n)\big)$, *its hypergraph* $\mathcal{H}(\varphi)$ *is defined as follows. Its set of nodes is the set of all variables used in* φ, *and its hyper-edges are precisely* $\vec{u}_1, \ldots, \vec{u}_n$.

We say that φ *is* acyclic *if the hypergraph* $\mathcal{H}(\varphi)$ *is acyclic.*

For example, let $\Phi \equiv \exists x \exists y \exists z\ R(x,y) \wedge R(y,z)$. Then $\mathcal{H}(\Phi)$ is a hypergraph on $\{x,y,z\}$ with edges $\{(x,y),(y,z)\}$. A tree decomposition of $\mathcal{H}(\Phi)$ would have two nodes, say t_1 and t_2, with an edge from t_1 to t_2, and $B_{t_1} = \{x,y\}, B_{t_2} = \{y,z\}$. Hence, Φ is acyclic.

As a different example, let $\Phi' \equiv \exists x \exists y \exists z\ R(x,y) \wedge R(y,z) \wedge R(z,x)$. Then $\mathcal{H}(\Phi')$ is a hypergraph on $\{x,y,z\}$ with edges $\{(x,y),(y,z),(z,x)\}$. Assume it is acyclic. Then there is some tree decomposition of $\mathcal{H}(\Phi')$ in which the sets B_t include $\{x,y\},\{y,z\},\{x,z\}$. By a straightforward inspection, there is no way to assign these sets to nodes of a tree so that condition 1 of the definition of tree decomposition would hold. Hence, Φ' is not acyclic.

In general, for binary relations, hypergraph and graph acyclicity coincide. To give an example involving hyper-edges, consider a query

$$\Psi \equiv \exists x \exists y \exists z \exists u \exists v\ \big(R(x,y,z) \wedge R(z,u,v) \wedge S(u,z) \wedge S(x,y) \wedge S(v,w)\big).$$

Its hypergraph has hyper-edges $\{x,y,z\},\{z,u,v\},\{u,z\},\{x,y\},\{v,w\}$. The maximal edges of this hypergraph are shown in Fig. 6.2 (a). This hypergraph is acyclic. Indeed, consider a tree with three nodes, t_1,t_2,t_3, and edges (t_1,t_2) and (t_1,t_3). Define B_{t_1} as $\{z,u,v\}$, B_{t_2} as $\{x,y,z\}$, and B_{t_3} as $\{v,w\}$ (see Fig. 6.2 (b)). This defines an acyclic tree decomposition of $\mathcal{H}(\Psi)$.

If, on the other hand, we consider a query

$$\Psi' \equiv \exists x \exists y \exists z \exists u \exists v\ \big(R(x,y,z) \wedge R(z,u,v) \wedge R(x,v,w)\big)$$

then one can easily check that $\mathcal{H}(\Phi')$ (shown in Fig. 6.2 (c)) is not acyclic.

We now show that acyclic conjunctive queries are fixed-parameter tractable (in fact, fixed-parameter linear) over arbitrary structures. The result below is given for Boolean conjunctive queries; for extension to queries with free variables, see Exercise 6.13.

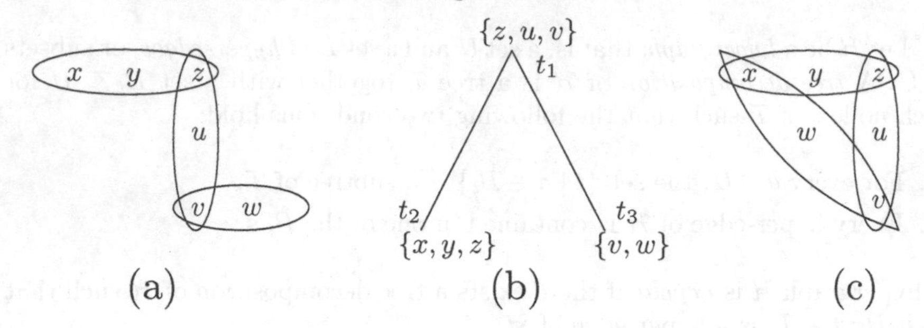

Fig. 6.2. Cyclic and acyclic hypergraphs

Theorem 6.25. *Let Φ be a Boolean acyclic conjunctive query over σ-structures, and let $\mathfrak{A} \in \text{STRUCT}[\sigma]$. Then checking whether $\mathfrak{A} \models \Phi$ can be done in time $O(\| \Phi \| \cdot \| \mathfrak{A} \|)$.*

Proof. Let Φ be

$$\Phi \equiv \exists x_1 \ldots x_m \bigwedge_{i=1}^{n} \alpha_i(\vec{u}_i),$$

where each $\alpha_i(\vec{u}_i)$ is of the form $S(\vec{u}_i)$ for $S \in \sigma$, and $\vec{u}_i$ contains some variables from $\vec{x}$. The case when some of the α_i's are variable equalities can be shown by essentially the same argument, by adding one selection over the join of all $\alpha_i(\mathfrak{A})$'s.

We use a known result that if $\mathcal{H}$ is acyclic, then its tree decomposition satisfying the condition that each B_t is a hyper-edge of $\mathcal{H}$ can be computed in linear time. Furthermore, one can construct this decomposition so that $B_{t_1} \not\subseteq B_{t_2}$ for any $t_1 \neq t_2$. Hence, we assume that we have such a decomposition $(\mathcal{T}, (B_t)_{t \in \mathcal{T}})$ for $\mathcal{H}(\Phi)$, computed in time $O(\| \Phi \|)$. Let $\prec$ denote the partial order of $\mathcal{T}$, with the root being the smallest node.

From the acyclicity of $\mathcal{H}$, it follows that there is a bijection between maximal, with respect to $\subseteq$, sets $\vec{u}_i$, and nodes t of $\mathcal{T}$. For each i, let ν_i be the node t such that $\vec{u}_i$ is contained in B_t. This node is unique: we look for the maximal $\vec{u}_j$ that contains $\vec{u}_i$, and find the unique node t such that $B_t = \vec{u}_j$.

We now define

$$R_t = \bowtie_{\substack{i \in [1,n] \\ \nu_i = t}} \alpha_i(\mathfrak{A}). \tag{6.11}$$

Our goal is now to compute the join of all R_t's, since (6.7) implies that

$$\mathfrak{A} \models \Phi \Leftrightarrow \bowtie_{t \in \mathcal{T}} R_t \neq \emptyset. \tag{6.12}$$

To show that (6.11) and (6.12) yield a linear time algorithm, we need two complexity bounds on computing projections and joins: $\pi_X(R)$ can be computed in time $O(\| R \|)$, and $R \bowtie S$ can be computed in time $O(\| R \| + \| S \| + \| R \bowtie S \|)$ (see Exercise 6.12).

To see that each R_t can be computed in linear time, let i_t be such that $\vec{u}_{i_t} = B_t$ (it exists since the query is acyclic). Then

$$R_t = \alpha_{i_t}(\vec{u}_{i_t}) \bowtie \alpha_{i_1}(\vec{u}_{i_1}) \bowtie \ldots \bowtie \alpha_{i_k}(\vec{u}_{i_k}),$$

where all $\vec{u}_{i_j} \subseteq \vec{u}_{i_t}$, $j \leq k$. Hence $R_t \subseteq \alpha_{i_t}(\mathfrak{A})$. Using the above bounds for computing joins and projections, we conclude that the entire family $R_t, t \in \mathcal{T}$, can be computed in time $O(\| \Phi \| \cdot \| \mathfrak{A} \|)$.

We define

$$P_t = \bowtie_{v \succeq t} R_v,$$

where $\succeq$ is the partial order of $\mathcal{T}$, with the root r being the smallest element. If t is a leaf of $\mathcal{T}$, then $P_t = R_t$. Otherwise, let t be a node with children $t_1, \ldots, t_l$. Then

$$P_t = R_t \bowtie \Big(\bowtie_{1 \leq i \leq l} \big(\bowtie_{v \succeq t_i} R_v \big) \Big) = R_t \bowtie \Big(\bowtie_{1 \leq i \leq l} P_{t_i} \Big). \tag{6.13}$$

Using (6.13) inductively, we compute $P_r = \bowtie_t R_t$ in time $O(\| \mathcal{T} \| \cdot \max_t \| R_t \|)$. We saw that $\| R_t \| \leq \| \mathfrak{A} \|$ for each t, and, furthermore, $\mathcal{T}$ can be computed from Φ in linear time. Hence, P_r can be found in time $O(\| \Phi \| \cdot \| \mathfrak{A} \|)$, which together with (6.12) implies that $\mathfrak{A} \models \Phi$ can be tested with the same bounds. This completes the proof. □

There is another interesting way to connect tree decompositions with tractability of conjunctive queries. Suppose we have a conjunctive query $\varphi(\vec{x})$ given by (6.3). We define its graph $\mathcal{G}(\varphi)$, whose set of vertices is the set of variables used in φ, with an edge between two variables u and v if there is an atom α_i such that both u and v are its free variables. For example, if $\varphi(x, y) \equiv \exists z \exists v \; R(x, y, z) \wedge S(z, v)$, then $\mathcal{G}(\varphi)$ has undirected edges $(x, y), (x, z), (y, z)$, and (z, v).

A *tree decomposition* of $\mathcal{G}(\varphi)$ is a tree decomposition, as defined earlier, when we view $\mathcal{G}(\varphi)$ as a hypergraph. In other words, it consists of a tree $\mathcal{T}$, and a set B_t of nodes of $\mathcal{G}(\varphi)$ for each $t \in \mathcal{T}$, such that

1. $\{t \mid v \in B_t\}$ forms a subtree of $\mathcal{T}$ for each v, and
2. for every edge (u, v), both u and v are in one of the B_t's.

The *width* of a tree decomposition is $\max_t |B_t| - 1$. The *treewidth* of $\mathcal{G}(\varphi)$ is the minimum width of a tree decomposition of $\mathcal{G}(\varphi)$. It is easy to see that the treewidth of a tree is 1.

For $k > 0$, let CQ_k be the class of conjunctive queries φ such that the treewidth of $\mathcal{G}(\varphi)$ is at most k. Then the following can be shown.

Theorem 6.26. *Let $k > 0$ be fixed, and let φ be a query from CQ_k. Then, for every structure $\mathfrak{A}$, one can compute $\varphi(\mathfrak{A})$ in polynomial time in $\| \Phi \| + \| \mathfrak{A} \| + \| \varphi(\mathfrak{A}) \|$. In particular, Boolean queries from CQ_k can be evaluated in polynomial time in $\| \Phi \| + \| \mathfrak{A} \|$.* □

In other words, conjunctive-query evaluation becomes tractable for queries whose graphs have bounded treewidth. Exercise 6.15 shows that the converse holds, under certain complexity-theoretic assumptions.

6.8 Bibliographic Notes

The notions of data, expression, and combined complexity are due to Vardi [244], see also [3].

Representation of first-order formulae by Boolean circuits is fairly standard, see, e.g., books [133] and [247]. Proposition 6.6 was explicitly shown by Vardi [245].

Theorem 6.7 is perhaps the deepest result in circuit complexity. It was proved by Furst, Saxe, and Sipser [86] (see also Ajtai [10] and Denenberg, Gurevich, and Shelah [55]).

The notion of uniformity and its connection with logical descriptions of complexity classes was studied by Barrington, Immerman, and Straubing [16]. Proofs of FO(<, BIT) = FO(+, ×) are given in [133] and – partially – in [247]. The proof of expressibility of BIT (Theorem 6.12) follows closely the presentation in Buss [29] and Cook [40].

PSPACE-completeness of FO (expression complexity) and of QBF is due to Stockmeyer [222].

The idea of using parameterized complexity as a refinement of the notions of the data and expression complexity was proposed by Yannakakis [250], and developed by Papadimitriou and Yannakakis [196]. Parameterized complexity is treated in a book by Downey and Fellows [58]; see also surveys by Grohe [109, 111]. Theorem 6.19 is from Seese [219], Theorem 6.20 is from Flum and Grohe [81].

The notion of conjunctive queries is a fundamental one in database theory, see [3]. NP-completeness of conjunctive queries (combined complexity) is due to Chandra and Merlin [34]. Fixed-parameter linearity of acyclic conjunctive queries is due to Yannakakis [249]; the presentation here follows closely Flum, Frick, and Grohe [80]. A linear time algorithm for producing tree decompositions of hypergraphs, used in Theorem 6.25, is due to Tarjan and Yannakakis [228]. Flum, Frick, and Grohe [80] show how to extend the notion of acyclicity to FO formulae. Theorem 6.26 and Exercise 6.15 are from Grohe, Schwentick, and Segoufin [114]. See also Gottlob, Leone, and Scarcello [96] for additional results on the complexity of acyclic conjunctive queries.

Sources for exercises:

Exercise 6.4:	Dawar et al. [50]
Exercise 6.5:	Immerman [133], Vollmer [247]
Exercise 6.9:	Papadimitriou and Yannakakis [196]
Exercise 6.12:	Flum, Frick, and Grohe [80]
Exercises 6.13 and 6.14:	Flum, Frick, and Grohe [80]
	Yannakakis [249]
Exercise 6.15:	Grohe, Schwentick, and Segoufin [114]
Exercise 6.16:	Flum and Grohe [81]
Exercise 6.18:	Gottlob, Leone, and Scarcello [96]
Exercise 6.19:	Chandra and Merlin [34]

6.9 Exercises

Exercise 6.1. Show that none of the following is expressible in FO(All): transitive closure of a graph, testing for planarity, acyclicity, 3-colorability.

Exercise 6.2. Prove that $\lfloor\sqrt{x}\rfloor$ is expressible in FO$(+,\times)$.

Exercise 6.3. Consider two countable undirected graphs. For the first one, the universe is $\mathbb{N}$, and we have an edge between i and j iff BIT(i,j) or BIT(j,i) is true. In the other graph, the universe is $\mathbb{N}_+ = \{n \in \mathbb{N} \mid n > 0\}$ and there is an edge between n and m, for $n > m$, iff n is divisible by p_m, the mth prime. Prove that these graphs are isomorphic.

Hint: if you find it hard to do all the calculations required for the proof, you may want to wait until Chap. 12, which introduces some powerful logical tools that let you prove results of this kind without using any number theory at all (see Exercise 12.9, part a).

Exercise 6.4. Show that the standard linear order is expressible in FO(BIT). Conclude that FO$(+,\times)$ = FO(BIT).

Exercise 6.5. Prove Lemma 6.14.

You may find it useful to show that the following predicate is expressible in FO$(+,\times)$: BitSum(x,y) iff the number of ones in the binary representation of x is y.

Exercise 6.6. Prove that QBF is PSPACE-complete.

Exercise 6.7. We stated in the proof of Theorem 6.19 that the set of tuples $\vec{t} \in T$ for which there exists a structure $\mathfrak{A}$ with $\vec{t}(\mathfrak{A}) = \vec{t}$ and $\mathfrak{A} \models \Phi$ is computable. Prove this statement, using the assumption that $\mathfrak{A}$ is of bounded degree. Derive bounds on the constant in the $O(\|\mathfrak{A}\|)$ running time.

Exercise 6.8. Give an example of a two-element structure over which the expression complexity of conjunctive queries is NP-hard. Recall that in the proof of Theorem 6.23, we used a structure whose universe had three elements.

Exercise 6.9. In this exercise, we refer to parameterized complexity class $W[1]$ whose definition can be found in [58, 81]. This class is believed to contain problems which are not fixed-parameter tractable.

Prove that checking $\mathfrak{A} \models \Phi$, with Φ being the parameter, is $W[1]$-hard, even if Φ is a conjunctive query. Thus, it is unlikely that FO (or even conjunctive queries) are fixed-parameter tractable.

Exercise 6.10. Derive Theorem 6.20 from the following facts. H is an excluded minor of a class of graphs $\mathcal{C}$ if no $G \in \mathcal{C}$ has H as a minor. If such an H exists, then $\mathcal{C}$ is called a class of graphs with an excluded minor.

- If $\mathcal{C}$ is a minor-closed class of graphs, membership in $\mathcal{C}$ can be verified in PTIME (see Robertson and Seymour [205]).
- If $\mathcal{C}$ is a PTIME-decidable class of graphs with an excluded minor, then checking Boolean FO queries on $\mathcal{C}$ is fixed-parameter tractable (see Flum and Grohe [81]).

Exercise 6.11. Prove that an order-invariant conjunctive query is FO-definable without the order relation. That is, $(\text{CQ}+<)_{\text{inv}} \subseteq \text{FO}$.

Exercise 6.12. Prove that $R \bowtie S$ can be evaluated in $O(\|R\| + \|S\| + \|R \bowtie S\|)$.

Exercise 6.13. Extend the proof of Theorem 6.25 to deal conjunctive queries with free variables, by showing that $\varphi(\mathfrak{A})$, for an acyclic φ, can be computed in time $O(\|\varphi\| \cdot \|\mathfrak{A}\| \cdot \|\varphi(\mathfrak{A})\|)$. Also show that if the set of free variables of φ is contained in one of the B_t's, for a tree decomposition of $\mathcal{H}(\varphi)$, then the evaluation can be done in time $O(\|\varphi\| \cdot \|\mathfrak{A}\|)$.

Exercise 6.14. Extend Theorem 6.25 and Exercise 6.13 to conjunctive queries with negation; that is, conjunctive queries in which some atoms are of the form $x \neq y$, where x and y are variables.

Exercise 6.15. Under the complexity-theoretic assumption that $W[1]$ contains problems which are not fixed-parameter tractable (see Exercise 6.9), the converse to Theorem 6.26 holds: if for a class of graphs $\mathcal{C}$, it is the case that every conjunctive query φ with $\mathcal{G}(\varphi) \in \mathcal{C}$ can be evaluated in time polynomial in $\|\Phi\| + \|\mathfrak{A}\| + \|\varphi(\mathfrak{A})\|$, then $\mathcal{C}$ has bounded treewidth (i.e., there is a constant $k > 0$ such that every graph in $\mathcal{C}$ has treewidth at most k).

Exercise 6.16. We say that a class of structures $\mathcal{C} \subseteq \text{STRUCT}[\sigma]$ has bounded treewidth if there is $k > 0$ such that for every $\mathfrak{A} \in \mathcal{C}$, the treewidth of its Gaifman graph is at most k. Prove that FO is fixed-parameter tractable on classes of structures of bounded treewidth.

Exercise 6.17. Give an example of a conjunctive query which is of treewidth 2 but not acyclic. Also, give an example of a family of acyclic conjunctive queries that has queries of arbitrarily large treewidth.

Exercise 6.18. Given a hypergraph $\mathcal{H}$, its *hypertree decomposition* is a triple $(T, (B_t)_{t \in T}, (C_t)_{t \in T})$ such that $(T, (B_t)_{t \in T})$ is a tree decomposition of $\mathcal{H}$, and each C_t is a set of hyper-edges. It is required to satisfy the following two properties for every $t \in T$:

1. $B_t \subseteq \bigcup C_t$;
2. $\bigcup C_t \cap \bigcup_{v \succeq t} B_v \subseteq B_t$.

The *hypertree width* of $\mathcal{H}$ is defined as the minimum value of $\max_{t \in T} \mid C_t \mid$, taken over all hypertree decompositions of $\mathcal{H}$.

Prove the following:

(a) A hypergraph is acyclic iff its hypertree width is 1.
(b) For each fixed k, conjunctive queries whose hypergraphs have hypertree width at most k can be evaluated in polynomial time.

Note that this does not contradict the result of Exercise 6.15 which refers to graph-based (as opposed to hypergraph-based) classes of conjunctive queries.

Exercise 6.19. Suppose $\varphi_1(\vec{x})$ and $\varphi_2(\vec{x})$ are two conjunctive queries. We write $\varphi_1 \subseteq \varphi_2$, if $\varphi_1(\mathfrak{A}) \subseteq \varphi_2(\mathfrak{A})$ for all $\mathfrak{A}$ (in other words, $\forall \vec{x}\ \varphi_1(\vec{x}) \rightarrow \varphi_2(\vec{x})$ is valid in all finite structures). We write $\varphi_1 = \varphi_2$ if both $\varphi_1 \subseteq \varphi_2$ and $\varphi_2 \subseteq \varphi_1$ hold.

Prove that testing both $\varphi_1 \subseteq \varphi_2$ and $\varphi_1 = \varphi_2$ is NP-complete.

Exercise 6.20.* Use Ehrenfeucht-Fraïssé games to prove that PARITY is not expressible in FO$(+, \times)$.

7

Monadic Second-Order Logic and Automata

We now move to extensions of first-order logic. In this chapter we introduce second-order logic, and consider its often used fragment, *monadic second-order logic*, or MSO, in which one can quantify over subsets of the universe. We study the expressive power of this logic over graphs, proving that its existential fragment expresses some NP-complete problems, but at the same time cannot express graph connectivity. Then we restrict our attention to strings and trees, and show that, over them, MSO captures regular string and tree languages. We explore the connection with automata to prove further definability and complexity results.

7.1 Second-Order Logic and Its Fragments

We have seen a few examples of second-order formulae in Chap. 1. The idea is that in addition to quantification over the elements of the universe, we can also quantify over subsets of the universe, as well as binary, ternary, etc., relations on it. For example, to express the query EVEN, we can say that there are two disjoint subsets U_1 and U_2 of the universe A such that $A = U_1 \cup U_2$ and there is a one-to-one mapping $F : U_1 \to U_2$. This is expressed by a formula

$$\exists U_1 \, \exists U_2 \, \exists F \; \varphi,$$

where φ is an FO formula in the vocabulary (U_1, U_2, F) stating that U_1 and U_2 form a partition of the universe ($\forall x \; (U_1(x) \leftrightarrow \neg U_2(x))$), and that $F \subseteq U_1 \times U_2$ is functional, onto, and one-to-one.

Note that the formula φ in this example has three *second-order free variables* U_1, U_2, and F. We now formally define second-order logic.

Definition 7.1 (Second-order logic). *The definition of* second-order logic, SO, *extends the definition of* FO *with second-order variables, ranging over subsets and relations on the universe, and quantification over such variables. We* *assume that for every $k > 0$, there are infinitely many variables $X_1^k, X_2^k, \ldots$, ranging over k-ary relations. A formula of* SO *can have both first-order and second-order free variables; we write $\varphi(\vec{x}, \vec{X})$ to indicate that $\vec{x}$ are free first-order variables, and $\vec{X}$ are free second-order variables.*

Given a vocabulary σ that consists of relation and constant symbols, we define SO *terms and formulae, and their free variables, as follows:*

- *Every first-order variable x, and every constant symbol c, are first-order terms. The only free variable of a term x is the variable x, and c has no free variables.*
- *There are three kinds of atomic formulae:*
  - FO *atomic formulae; that is, formulae of the form*
    - $t = t'$, *where t, t' are terms, and*
    - $R(\vec{t})$, *where $\vec{t}$ is a tuple of terms, and $R \in \sigma$, and*
  - $X(t_1, \ldots, t_k)$, *where $t_1, \ldots, t_k$ are terms, and X is a second-order variable of arity k. The free first-order variables of this formula are free first-order variables of $t_1, \ldots, t_k$; the free second-order variable is X.*
- *The formulae of* SO *are closed under the Boolean connectives $\vee, \wedge, \neg$, and first-order quantification, with the usual rules for free variables.*
- *If $\varphi(\vec{x}, Y, \vec{X})$ is a formula, then $\exists Y\ \varphi(\vec{x}, Y, \vec{X})$ and $\forall Y\ \varphi(\vec{x}, Y, \vec{X})$ are formulae, whose free variables are $\vec{x}$ and $\vec{X}$.*

The semantics is defined as follows. Suppose $\mathfrak{A} \in$ STRUCT$[\sigma]$. For each formula $\varphi(\vec{x}, \vec{X})$, we define the notion $\mathfrak{A} \models \varphi(\vec{b}, \vec{B})$, where $\vec{b}$ is a tuple of elements of A of the same length as $\vec{x}$, and for $\vec{X} = (X_1, \ldots, X_l)$, with each X_i being of arity n_i, $\vec{B} = (B_1, \ldots, B_l)$, where each B_i is a subset of A^{n_i}.

We give the semantics only for constructors that are different from those for FO*:*

- *If $\varphi(\vec{x}, X)$ is $X(t_1, \ldots, t_k)$, where X is k-ary and $t_1, \ldots, t_k$ are terms, with free variables among $\vec{x}$, then $\mathfrak{A} \models \varphi(\vec{b}, B)$ iff the tuple $(t_1^{\mathfrak{A}}(\vec{b}), \ldots, t_k^{\mathfrak{A}}(\vec{b}))$ is in B.*
- *If $\varphi(\vec{x}, \vec{X})$ is $\exists Y\ \psi(\vec{x}, Y, \vec{X})$, where Y is k-ary, then $\mathfrak{A} \models \varphi(\vec{b}, \vec{B})$ if for some $C \subseteq A^k$, it is the case that $\mathfrak{A} \models \psi(\vec{b}, C, \vec{B})$.*
- *If $\varphi(\vec{x}, \vec{X})$ is $\forall Y\ \psi(\vec{x}, Y, \vec{X})$, and Y is k-ary, then $\mathfrak{A} \models \varphi(\vec{b}, \vec{B})$ if for all $C \subseteq A^k$, we have $\mathfrak{A} \models \psi(\vec{b}, C, \vec{B})$.*

We know that every FO formula can be written in the prenex normal form $Q_1 x_1 \ldots Q_n x_n\ \psi$, where Q_i's are $\exists$ or $\forall$, and ψ is quantifier-free. Likewise, every SO formula can be written as a sequence of first- and second-order quantifiers, followed by a quantifier-free formula. Furthermore, note the following equivalences:

$$\exists x\ \mathbf{Q}\ \varphi(x,\cdot) \ \leftrightarrow\ \exists X\ \mathbf{Q}\ \big(\exists x\ (X(x) \land \varphi(x,\cdot))\big) \tag{7.1}$$

$$\forall x\ \mathbf{Q}\ \varphi(x,\cdot) \ \leftrightarrow\ \forall X\ \mathbf{Q}\ \big(\exists! x\ X(x) \to \forall x\ (X(x) \to \varphi(x,\cdot))\big), \tag{7.2}$$

where $\mathbf{Q}$ stands for an arbitrary sequence of first- and second-order quantifiers. Using those inductively, we can see that every SO formula is equivalent to a formula in the form

$$\mathbf{Q}_1 X_1 \ldots \mathbf{Q}_n X_n Q_1 x_1 \ldots Q_l x_l\ \psi, \tag{7.3}$$

where $\mathbf{Q}_i X_i$ are second-order quantifiers, $Q_j x_j$ are first-order quantifiers, and ψ is quantifier-free.

We now define some restrictions of the full SO logic of interest to us. The first one is the central notion studied in this chapter.

Definition 7.2. *Monadic* SO *logic, or* MSO, *is defined as the restriction of* SO *where all second-order variables have arity* 1.

In other words, in MSO, second-order variables range over subsets of the universe.

Rules (7.1) and (7.2) do not take us out of MSO, and hence every MSO formula is equivalent to one in the normal form (7.3), where the second-order quantifiers precede the first-order quantifiers.

Definition 7.3. *Existential* SO *logic, or* $\exists$SO, *is defined as the restriction of* SO *that consists of the formulae of the form*

$$\exists X_1 \ldots \exists X_n\ \varphi,$$

where φ does not have any second-order quantification. If, furthermore, all X_i's have arity 1, *the resulting restriction is called* existential monadic SO, *or* $\exists$MSO.

If the second-order quantifier prefix consists only of universal quantifiers, we speak of the universal SO *logic, or* $\forall$SO, *and its further restriction to monadic quantifiers is referred to as* $\forall$MSO.

In other words, an $\exists$SO formula starts with a second-order existential prefix $\exists X_1 \ldots \exists X_n$, and what follows is an FO formula φ (in the original vocabulary expanded with $X_1, \ldots, X_n$).

Formula (1.2) from Chap. 1 stating the 3-colorability of a graph is an example of an $\exists$MSO formula, while (1.3) stating the existence of a clique of a given size is an example of an $\exists$SO formula.

Definition 7.4. *The* quantifier rank *of an* SO *formula is defined as the maximum depth of quantifier-nesting, including both first-order and second-order quantifiers. That is, the rules for the quantifier rank for* FO *are augmented with*

- $\mathsf{qr}(\exists X\ \varphi) = \mathsf{qr}(\forall X\ \varphi) = \mathsf{qr}(\varphi) + 1$.

7.2 MSO Games and Types

MSO can be characterized by a type of Ehrenfeucht-Fraïssé game, which is fairly close to the game we have used for FO. As in the case of FO, the game is also closely connected with the notion of *type*.

Let MSO$[k]$ consist of all MSO formulae of quantifier-rank at most k. An MSO *rank-k m,l-type* is a consistent set S of MSO$[k]$ formulae with m free first-order variables and l free second-order variables such that for every $\varphi(x_1,\ldots,x_m,X_1,\ldots,X_l)$ from MSO$[k]$, either $\varphi \in S$ or $\neg\varphi \in S$.

Given a structure $\mathfrak{A}$, an m-tuple $\vec{a} \in A$, and an l-tuple V of subsets of A, the MSO *rank-k type* of $(\vec{a},\vec{U})$ in $\mathfrak{A}$ is the set

$$\text{mso-tp}_k(\mathfrak{A},\vec{a},\vec{V}) \;=\; \{\varphi(\vec{x},\vec{X}) \in \text{MSO}[k] \;\mid\; \mathfrak{A} \models \varphi(\vec{a},\vec{V})\}.$$

Clearly, $\text{mso-tp}_k(\mathfrak{A},\vec{a},\vec{V})$ is an MSO rank-k type.

When both $\vec{a}$ and $\vec{V}$ are empty, $\text{mso-tp}_k(\mathfrak{A})$ is the set of all MSO$[k]$ sentences that are true in $\mathfrak{A}$.

Just as for FO, a simple inductive argument shows that for each m and l, up to logical equivalence, there are only finitely many different formulae $\varphi(x_1,\ldots,x_m,X_1,\ldots,X_l)$ in MSO$[k]$. Hence, MSO rank-k m,l types (where m and l stand for the number of free first-order and second-order variables, respectively) are essentially finite objects. In fact, just as for FO, one can show the following result for MSO.

Proposition 7.5. *Fix k,l,m.*

- *There exist only finitely many* MSO *rank-k m,l types.*
- *Let $T_1,\ldots,T_s$ enumerate all the* MSO *rank-k m,l types. There exist* MSO$[k]$ *formulae $\alpha_i(\vec{x},\vec{X})$, $i=1,\ldots,s$, such that for every structure $\mathfrak{A}$, every m-tuple $\vec{a}$ of elements of A, and every l-tuple $\vec{U}$ of subsets of A, it is the case that $\mathfrak{A} \models \alpha_i(\vec{a},\vec{U})$ iff* $\text{mso-tp}_k(\mathfrak{A},\vec{a},\vec{U}) = T_i$.

  Furthermore, each MSO$[k]$ *formula with m free first-order variables and l free second-order variables is equivalent to a disjunction of some of the α_i's.*

Hence, just as in the case of FO, we shall associate rank-k types with their defining formulae, which are also of quantifier rank k.

We now present the modification of Ehrenfeucht-Fraïssé games for MSO.

Definition 7.6. *An* MSO *game is played by two players, the spoiler and the duplicator, on two structures $\mathfrak{A}$ and $\mathfrak{B}$ of the same vocabulary σ. The game has two different kinds of moves:*

Point move *This is the same move as in the Ehrenfeucht-Fraïssé game for* FO*: the spoiler chooses a structure,* $\mathfrak{A}$ *or* $\mathfrak{B}$*, and an element of that structure; the duplicator responds with an element in the other structure.*

Set move *The spoiler chooses a structure,* $\mathfrak{A}$ *or* $\mathfrak{B}$*, and a subset of that structure. The duplicator responds with a subset of the other structure.*

Let $a_1, \ldots, a_p \in A$ *and* $b_1, \ldots, b_p \in B$ *be the point moves played in the* k*-round game, with* $V_1, \ldots, V_s \subseteq A$ *and* $U_1, \ldots, U_s \subseteq B$ *being the set moves (i.e.,* $p+s=k$*, and the moves of the same round have the same index). Then the duplicator wins the game if* $(\vec{a}, \vec{b})$ *is a partial isomorphism of* $(\mathfrak{A}, \vec{V})$ *and* $(\mathfrak{B}, \vec{U})$*. If the duplicator has a winning strategy in the* k*-round* MSO *game on* $\mathfrak{A}$ *and* $\mathfrak{B}$*, we write* $\mathfrak{A} \equiv_k^{\mathrm{MSO}} \mathfrak{B}$*.*

Furthermore, we write $(\mathfrak{A}, \vec{a}_0, \vec{V}_0) \equiv_k^{\mathrm{MSO}} (\mathfrak{B}, \vec{b}_0, \vec{U}_0)$ *if the duplicator has a winning strategy in the* k*-round* MSO *game on* $\mathfrak{A}$ *and* $\mathfrak{B}$ *starting with position* $((\vec{a}_0, \vec{V}_0), (\vec{b}_0, \vec{U}_0))$*. That is, when* k *rounds of the game* $\vec{a}, \vec{b}, \vec{V}, \vec{U}$ *are played,* $(\vec{a}_0\vec{a}, \vec{b}_0\vec{b})$ *is a partial isomorphism between* $(\mathfrak{A}, \vec{V}_0, \vec{V})$ *and* $(\mathfrak{B}, \vec{U}_0, \vec{U})$*.*

This game captures the expressibility in MSO[k].

Theorem 7.7. *Given two structures* $\mathfrak{A}$ *and* $\mathfrak{B}$*, two* m*-tuples* $\vec{a}_0$*,* $\vec{b}_0$ *of elements of* A *and* B*, and two* l*-tuples* $\vec{V}_0, \vec{U}_0$ *of subsets of* A *and* B*, we have*

$$\text{mso-tp}_k(\mathfrak{A}, \vec{a}_0, \vec{V}_0) = \text{mso-tp}_k(\mathfrak{B}, \vec{b}_0, \vec{U}_0) \quad \Leftrightarrow \quad (\mathfrak{A}, \vec{a}_0, \vec{V}_0) \equiv_k^{\mathrm{MSO}} (\mathfrak{B}, \vec{b}_0, \vec{U}_0).$$

That is, $(\mathfrak{A}, \vec{a}_0, \vec{V}_0) \equiv_k^{\mathrm{MSO}} (\mathfrak{B}, \vec{b}_0, \vec{U}_0)$ *iff for every* MSO[k] *formula* $\varphi(\vec{x}, \vec{X})$*,*

$$\mathfrak{A} \models \varphi(\vec{a}_0, \vec{V}_0) \quad \Leftrightarrow \quad \mathfrak{B} \models \varphi(\vec{b}_0, \vec{U}_0).$$

The proof is essentially the same as the proof of Theorem 3.9, and is left to the reader as an exercise (see Exercise 7.1).

In the case of sentences, Theorem 7.7 gives us the following.

Corollary 7.8. *If* $\mathfrak{A}$ *and* $\mathfrak{B}$ *are two structures of the same vocabulary, then* $\mathfrak{A} \equiv_k^{\mathrm{MSO}} \mathfrak{B}$ *iff* $\mathfrak{A}$ *and* $\mathfrak{B}$ *agree on all the sentences of* MSO[k].

As for FO, the method of games is complete for expressibility in MSO.

Proposition 7.9. *A property* $\mathcal{P}$ *of* σ*-structures is expressible in* MSO *iff there is a number* k *such that for every two* σ*-structures* $\mathfrak{A}, \mathfrak{B}$*, if* $\mathfrak{A}$ *has the property* $\mathcal{P}$ *and* $\mathfrak{B}$ *does not, then the spoiler wins the* k*-round* MSO *game on* $\mathfrak{A}$ *and* $\mathfrak{B}$*.*

Proof. Assume $\mathcal{P}$ is expressible by a sentence Φ of quantifier rank k. Let $\alpha_1, \ldots, \alpha_s$ enumerate all the MSO rank k types (without free variables). Then $\mathcal{P}$ is equivalent to a disjunction of some of the α_i's. Hence, if $\mathfrak{A}$ has $\mathcal{P}$ and $\mathfrak{B}$ does not, there is some i such that $\mathfrak{A} \models \alpha_i$ and $\mathfrak{B} \models \neg\alpha_i$, and thus $\mathfrak{A} \not\equiv_k^{\mathrm{MSO}} \mathfrak{B}$.

Conversely, suppose that we can find $k \geq 0$ such that for every $\mathfrak{A}$ having $\mathcal{P}$ and $\mathfrak{B}$ not having $\mathcal{P}$, we have $\mathfrak{A} \not\equiv_k^{\text{MSO}} \mathfrak{B}$. Now take any two structures $\mathfrak{A}_1$ and $\mathfrak{A}_2$ such that $\mathfrak{A}_1 \equiv_k^{\text{MSO}} \mathfrak{A}_2$. Suppose $\mathfrak{A}_1$ has $\mathcal{P}$. If $\mathfrak{A}_2$ does not have $\mathcal{P}$, we would conclude $\mathfrak{A}_1 \not\equiv_k^{\text{MSO}} \mathfrak{A}_2$, which contradicts the assumption; hence $\mathfrak{A}_2$ has $\mathcal{P}$ as well. Thus, $\mathcal{P}$ is a union of rank-k MSO types. Since there are finitely many of them, and each is definable by a rank-k MSO sentence, we conclude that $\mathcal{P}$ is MSO[k]-definable. □

Most commonly, we use the contrapositive of this proposition, which tells us when some property is *not* expressible in MSO.

Corollary 7.10. *A property $\mathcal{P}$ of σ-structures is not expressible in* MSO *iff for every $k \geq 0$, one can find $\mathfrak{A}_k, \mathfrak{B}_k \in \text{STRUCT}[\sigma]$ such that:*

- *$\mathfrak{A}_k$ has the property $\mathcal{P}$,*
- *$\mathfrak{B}_k$ does not have the property $\mathcal{P}$, and*
- *$\mathfrak{A}_k \equiv_k^{\text{MSO}} \mathfrak{B}_k$.*

Our next goal it to use games to study expressibility in MSO. A useful technique is the *composition* of MSO games, which allows us to construct more complex games from simpler ones. Similarly to Exercise 3.15, we can show the following.

Lemma 7.11. *Let $\mathfrak{A}_1, \mathfrak{A}_2, \mathfrak{B}_1, \mathfrak{B}_2$ be σ-structures, and let $\mathfrak{A}$ be the disjoint union of $\mathfrak{A}_1$ and $\mathfrak{A}_2$, and $\mathfrak{B}$ the disjoint union of $\mathfrak{B}_1$ and $\mathfrak{B}_2$. Assume $\mathfrak{A}_1 \equiv_k^{\text{MSO}} \mathfrak{B}_1$ and $\mathfrak{A}_2 \equiv_k^{\text{MSO}} \mathfrak{B}_2$. Then $\mathfrak{A} \equiv_k^{\text{MSO}} \mathfrak{B}$.*

Proof sketch. Assume the spoiler makes a point move, say a in $\mathfrak{A}$. Then a is in $\mathfrak{A}_1$ or $\mathfrak{A}_2$. Suppose a is in $\mathfrak{A}_1$; then the duplicator selects a response b in $\mathfrak{B}_1$ according to his winning strategy on $\mathfrak{A}_1$ and $\mathfrak{B}_1$.

Assume the spoiler makes a set move, say $U \subseteq A$. The universe A is the disjoint union of A_1 and A_2, the universes of $\mathfrak{A}_1$ and $\mathfrak{A}_2$. Let $U_i = U \cap A_i, i = 1, 2$. Let V_i be the response of the duplicator to U_i in $\mathfrak{B}_i$, $i = 1, 2$, according to the winning strategy. Then the response to U is $V = V_1 \cup V_2$. It is routine to verify that, using this strategy, the duplicator wins in k rounds. □

As an application of the composition argument, we prove the following.

Proposition 7.12. *Let $\sigma = \emptyset$. Then* EVEN *is not* MSO*-expressible.*

Proof. We claim that for every $\mathfrak{A}$ and $\mathfrak{B}$ with $|A|, |B| \geq 2^k$, it is the case that $\mathfrak{A} \equiv_k^{\text{MSO}} \mathfrak{B}$. Clearly this implies that EVEN is not MSO-definable. Since $\sigma = \emptyset$, we shall write $U \equiv_k^{\text{MSO}} V$ instead of the more formal $(U, \emptyset) \equiv_k^{\text{MSO}} (V, \emptyset)$.

We prove the statement by induction on k. The cases of $k = 0$ and $k = 1$ are easy, so we show how to go from k to $k + 1$.

Suppose $\mathfrak{A}$ and $\mathfrak{B}$ with $|A|, |B| \geq 2^{k+1}$ are given. We only consider a set move by the spoiler, since any point move a can be identified with the set move $\{a\}$. Assume that in the first move, the spoiler plays $U \subseteq A$. We distinguish the following cases:

1. $|U| \leq 2^k$. Then pick an arbitrary set $V \subseteq B$ such that $|V|=|U|$. We have $U \cong V$ (and thus $U \equiv_k^{\text{MSO}} V$), and $A - U \equiv_k^{\text{MSO}} B - V$ – the latter is by the induction hypothesis, since $|A - U|, |B - V| \geq 2^k$. Combining the two games, we see that from the position (U, V) on $\mathfrak{A}$ and $\mathfrak{B}$, the duplicator can continue the game for k rounds, and hence $\mathfrak{A} \equiv_{k+1}^{\text{MSO}} \mathfrak{B}$.
2. $|A - U| \leq 2^k$. This case is treated in exactly the same way as the previous one.
3. $|U| > 2^k$ and $|A - U| > 2^k$. Since $|B| \geq 2^{k+1}$, we can find a subset $V \subseteq B$ such that both $|V|$ and $|B - V|$ are at least 2^k. By the induction hypothesis, we know that $U \equiv_k^{\text{MSO}} V$ and $A - U \equiv_k^{\text{MSO}} B - V$, and hence from (U, V), the duplicator can play for k more rounds, thus proving $\mathfrak{A} \equiv_{k+1}^{\text{MSO}} \mathfrak{B}$. □

Suppose now that the vocabulary is expanded by one binary symbol $<$ interpreted as a linear ordering; that is, we deal with finite linear orders. Then EVEN *is* expressible in MSO. To see this, we let our MSO sentence guess the set that consists of alternating elements $a_1, a_3, \ldots, a_{2n+1}, \cdots$ in the ordering $a_1 < a_2 < a_3 < \ldots$, such that the first element is in this set, and the last element is not:

$$\exists X \begin{pmatrix} \forall x \; (\text{first}(x) \rightarrow X(x)) \\ \wedge \; \forall x \; (\text{last}(x) \rightarrow \neg X(x)) \\ \wedge \; \forall x \forall y \; \text{succ}_<(x, y) \rightarrow (X(x) \leftrightarrow \neg X(y)) \end{pmatrix},$$

where $\text{first}(x)$ stands for $\forall y \; (y \geq x)$, $\text{last}(x)$ stands for $\forall y \; (y \leq x)$, and $\text{succ}_<(x, y)$ stands for $(x < y) \wedge \neg\exists z \; (x < z \wedge z < y)$.

Thus, as for FO, we have a separation between the ordered and unordered case. Noticing that EVEN is an order-invariant query, we obtain the following.

Corollary 7.13. $\text{MSO} \subsetneq (\text{MSO+} <)_{\text{inv}}$. □

Note the close connection between Corollary 7.13 and Theorem 5.3: the latter showed that $\text{FO} \subsetneq (\text{FO+} <)_{\text{inv}}$, and the separating example was the parity of the number of atoms of a Boolean algebra. We used the Boolean algebra to simulate monadic second-order quantification; in MSO it comes for free, and hence EVEN worked as a separating query.

7.3 Existential and Universal MSO on Graphs

In this section we study two restrictions of MSO: existential MSO, or ∃MSO, and universal MSO, or ∀MSO, whose formulae are respectively of the form

$$\exists X_1 \ldots \exists X_n \; \varphi$$

and

$$\forall X_1 \ldots \forall X_n \, \varphi,$$

where φ is first-order.

These also are commonly found in the literature under the names *monadic* Σ_1^1 for ∃MSO and *monadic* Π_1^1 for ∀MSO, where monadic, of course, refers to second-order quantification over sets. In general, Σ_k^1 consists of formulae whose prefix of second-order quantifiers consists of k blocks, with the first block being existential. For example, a formula $\exists X_1 \exists X_2 \forall Y_1 \exists Z_1 \psi$ is a Σ_3^1-formula. The class Π_k^1 is defined likewise, except that the first block of quantifiers is universal.

Another name for ∃MSO is monadic NP, and ∀MSO is referred to as monadic coNP. The reason for these names will become clear in Chap. 9, when we prove Fagin's theorem.

We now give an example of a familiar property that separates monadic Π_1^1 from monadic Σ_1^1 (i.e., ∀MSO from ∃MSO).

Proposition 7.14. *Graph connectivity is expressible in* ∀MSO, *but is not expressible in* ∃MSO.

Proof. A graph is not connected if its nodes can be partitioned into two nonempty sets with no edges between them:

$$\exists X \begin{pmatrix} \exists x \; X(x) \;\; \wedge \;\; \exists x \; \neg X(x) \\ \wedge \; (\forall x \forall y \; (X(x) \wedge \neg X(y) \rightarrow \neg E(x,y))) \end{pmatrix} \tag{7.4}$$

Since (7.4) is an ∃MSO sentence, its negation, expressing graph connectivity, is a universal MSO sentence.

For the converse, we use Hanf-locality. Suppose that connectivity is definable by an ∃MSO sentence $\Phi \equiv \exists X_1 \ldots \exists X_m \varphi$. Assume without loss of generality that $m > 0$. Since φ is a first-order sentence (over structures of vocabulary σ extended with $X_1, \ldots, X_n$), it is Hanf-local. Let $d = \mathsf{hlr}(\varphi)$, the Hanf-locality rank of φ. That is, if $(G, U_1, \ldots, U_m) \leftrightarrows_d (G', V_1, \ldots, V_m)$, where G, G' are graphs and the U_i's and the V_i's interpret X_i's over them, then $(G, U_1, \ldots, U_m)$ and $(G', V_1, \ldots, V_m)$ agree on φ.

We now set $K = 2^{m(2d+1)}$ and $r = (4d+4)K$. We claim the following: if G is an m-colored graph (i.e., a graph on which m unary predicates are defined), which is a cycle of length at least r, then there exist two nodes a and b such that the distance between them is at least $2d+2$, and their d-neighborhoods are isomorphic.

Indeed, for a long enough cycle, the d-neighborhood of each node a is a chain of length $2d+1$ with a being the middle node. Each node on the chain can belong to some of the U_i's, and there are 2^m possibilities for choosing a subset of indexes $1, \ldots, m$ of U_i's such that $a \in U_i$. Hence, there are at most K different isomorphism types of d-neighborhoods. If the length of the cycle is at least $(4d+4)K$, then there is one type of d-neighborhoods which

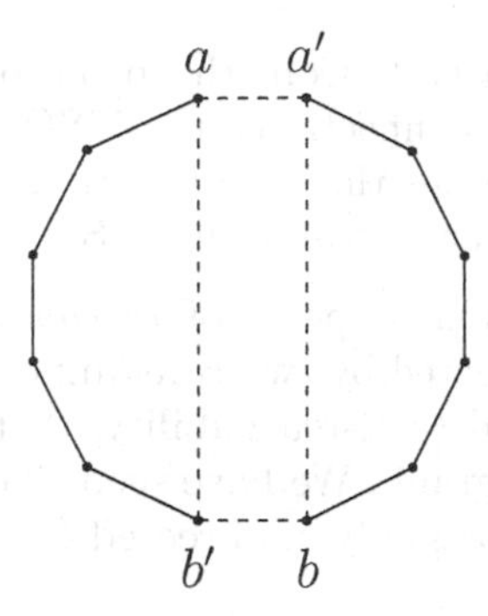

Fig. 7.1. Illustration for the proof of Proposition 7.14

is realized by at least $4d+4$ elements, and hence two of those elements will be at distance at least $2d+2$ from each other.

Now let G be a cycle of length at least r. Since G is a connected graph, we have $G \models \Phi$. Let $U_1, \ldots, U_m$ witness it; that is, $(G, U_1, \ldots, U_m) \models \varphi$. Choose a, b such that $a \approx_d^{(G,U_1,\ldots,U_m)} b$ and $d(a,b) > 2d+1$, and let a', b' be their successors (in an arbitrarily chosen orientation of G; the one shown in Fig. 7.1 is the clockwise orientation).

We now construct a new graph G' by removing edges (a, a') and (b, b') from G, and adding edges (a, b') and (b, a'). We claim that for every node c,

$$N_d^{(G,U_1,\ldots,U_m)}(c) \cong N_d^{(G',U_1,\ldots,U_m)}(c). \tag{7.5}$$

First, since a and b are at the distance at least $2d+2$, the d-neighborhood of any point in G or G' is a chain of length $2d+1$. If c is at the distance d or greater from a and b, its d-neighborhood is the same in $(G, U_1, \ldots, U_m)$ and $(G', U_1, \ldots, U_m)$, which means that (7.5) holds.

Suppose now that the distance between c and a is $d_0 < d$, and assume that c precedes a in the clockwise orientation of G. Then the d predecessors of c are the same in both structures. Furthermore, since $a \approx_d^{(G,U_1,\ldots,U_m)} b$, in both structures the $d - d_0$ successors of a agree on all the U_i's. Hence, (7.5) holds for c. The remaining cases (again, viewing G in the clockwise order) are those of c preceding b, or following a or a' and being at the distance less than d from them. In all of those cases the same argument as above proves (7.5).

We have thus established a bijection f between the universes of $(G, U_1, \ldots, U_m)$ and $(G', U_1, \ldots, U_m)$ (which is in fact the identity) that witnesses

$$(G, U_1, \ldots, U_m) \leftrightarrows_d (G', U_1, \ldots, U_m).$$

Since $d = \mathsf{hlr}(\varphi)$, we conclude that $(G', U_1, \ldots, U_m) \models \varphi$, and hence $G' \models \exists X_1 \ldots \exists X_m\, \varphi$; that is, $G' \models \Phi$. But G' is not a connected graph, which contradicts our assumption that Φ is an $\exists$MSO sentence defining graph connectivity. $\square$

Notice that the formula (7.4) from the proof of Proposition 7.14 shows that the negation of graph connectivity is ∃MSO-expressible, which means that ∃MSO can express queries that are not Hanf-local. One can also show that other forms of locality are violated in ∃MSO (see Exercise 7.6).

We now consider a related property of *reachability*. We assume that the language of graphs is augmented by two constants, s and t, and we are interested in the property, called (s,t)-reachability, that asks whether there is a path from s to t in a given graph. We have seen that undirected connectivity is not ∃MSO-definable; surprisingly, undirected (s,t)-reachability is!

Proposition 7.15. *For undirected graphs without loops, (s,t)-reachability is expressible in* ∃MSO.

Proof. Consider the sentence φ in the language of graphs expanded with one unary relation X that says the following:

1. both s and t are in X,
2. both s and t have an edge to exactly one member of X, and
3. every member of X except s and t has edges to precisely two members of X.

Let Φ be $\exists X\ \varphi$. We claim that $G \models \Phi$ iff there is a path from s to t in G. Indeed, if there is a path from s to t, we can take X to be the shortest path from s to t. Conversely, if $(G, X) \models \varphi$, then X is a path that starts in s; since the graph G is finite, X must contain the last node on the path, which could be only t. □

The approach of Proposition 7.15 does not work for directed graphs, because of back edges. Consider, for example, a directed graph which consists of a chain $\{(s,a_1),(a_1,a_2),(a_2,a_3),(a_3,t)\}$ together with the edge (a_3,a_1). The only path between s and t consists of edges s, a_1, a_2, a_3, t; however, if we let $X = \{s, a_1, a_2, a_3, t\}$, the sentence φ from the proof of Proposition 7.15 is false, since a_3 has one incoming edge, and two outgoing edges. It seems that the approach of Proposition 7.15 could be generalized if there is a bound on degrees in the input graph, and this is indeed the case (Exercise 7.7). However, in general, one can show a negative result.

Theorem 7.16. *Reachability for directed graphs is not expressible in* ∃MSO.

We conclude this section by showing that there are games that characterize expressibility in ∃MSO, much in the same way as Ehrenfeucht-Fraïssé games and MSO games characterize expressibility in FO and MSO.

Definition 7.17. *The* l,k-Fagin game *on two structures* $\mathfrak{A}, \mathfrak{B} \in \mathrm{STRUCT}[\sigma]$ *is played as follows. The spoiler selects l subsets $U_1, \ldots, U_l$ of A. Then the duplicator selects l subsets $V_1, \ldots, V_l$ of B. After that, the spoiler and the* *duplicator play k rounds of the Ehrenfeucht-Fraïssé game on* $(\mathfrak{A}, U_1, \ldots, U_l)$ *and* $(\mathfrak{B}, V_1, \ldots, V_l)$.

The winning condition for the duplicator is that after k rounds of the Ehrenfeucht-Fraïssé game, the elements played on $(\mathfrak{A}, U_1, \ldots, U_l)$ *and* $(\mathfrak{B}, V_1, \ldots, V_l)$ *form a partial isomorphism between these two structures.*

A fairly simple generalization of the previous game proofs shows the following.

Proposition 7.18. *A property* $\mathcal{P}$ *of* σ*-structures is* ∃MSO*-definable iff there exist* l *and* k *such that for every* $\mathfrak{A} \in \mathrm{STRUCT}[\sigma]$ *having* $\mathcal{P}$*, and for every* $\mathfrak{B} \in \mathrm{STRUCT}[\sigma]$ *not having* $\mathcal{P}$*, the spoiler wins the* l, k*-Fagin game on* $\mathfrak{A}$ *and* $\mathfrak{B}$. □

This game, however, is often rather inconvenient for the duplicator to play (after all, we use games to show that a certain property is inexpressible in a logic, so we need the win for the duplicator). A somewhat surprising result (see Exercise 7.9) shows that a different game that is easier for the duplicator to win, also characterizes the expressiveness of ∃MSO.

Definition 7.19. *Let* $\mathcal{P}$ *be a property of* σ*-structures (that is, a class of* σ*-structures closed under isomorphism). The* $\mathcal{P}, l, k$-Ajtai-Fagin game *is played as follows:*

1. *The duplicator selects a structure* $\mathfrak{A} \in \mathcal{P}$.
2. *The spoiler selects* l *subsets* $U_1, \ldots, U_l$ *of* A.
3. *The duplicator selects a structure* $\mathfrak{B} \notin \mathcal{P}$*, and* l *subsets* $V_1, \ldots, V_l$ *of* B.
4. *The spoiler and the duplicator play k rounds of the Ehrenfeucht-Fraïssé game on* $(\mathfrak{A}, U_1, \ldots, U_l)$ *and* $(\mathfrak{B}, V_1, \ldots, V_l)$.

The winning condition for the duplicator is that after k rounds of the Ehrenfeucht-Fraïssé game, the elements played on $(\mathfrak{A}, U_1, \ldots, U_l)$ *and* $(\mathfrak{B}, V_1, \ldots, V_l)$ *form a partial isomorphism between these two structures.*

Intuitively, this game is easier for the duplicator to win, because he selects the second structure $\mathfrak{B}$ and the coloring of it only after he has seen how the spoiler chose to color the first structure $\mathfrak{A}$.

Proposition 7.20. *A property* $\mathcal{P}$ *of* σ*-structures is* ∃MSO*-definable iff there exist* l *and* k *such that the spoiler has a winning strategy in the* $\mathcal{P}, l, k$*-Ajtai-Fagin game.* □

Hence, to show that a certain property $\mathcal{P}$ is not expressible in ∃MSO, it suffices to construct, for every l and k, a winning strategy for the *duplicator* in the $\mathcal{P}, l, k$-Ajtai-Fagin game. This is easier than a winning strategy in the l, k-Fagin game, since the duplicator sees the sets U_i's before choosing the second structure $\mathfrak{B}$ for the game. An example is given in Exercise 7.10.

7.4 MSO on Strings and Regular Languages

We now study MSO on strings. Recall that a string over a finite alphabet can be represented as a first-order structure. For example, the string $s = abaab$ is represented as $\langle\{1,2,3,4,5\},<,P_a,P_b\rangle$, where $<$ is the usual ordering, and P_a and P_b contain positions in s where a (or b, respectively) occurs: that is, $P_a = \{1,3,4\}$ and $P_b = \{2,5\}$.

In general, for a finite alphabet Σ, we define the vocabulary σ_Σ that contains a binary symbol $<$ and unary symbols P_a for each $a \in \Sigma$. A string $s \in \Sigma^*$ of length n is then represented as a structure $\mathfrak{M}_s \in \text{STRUCT}[\sigma_\Sigma]$ whose universe is $\{1,\ldots,n\}$, with $<$ interpreted as the order on the natural numbers, and P_a being the set of positions where the letter a occurs, for each a in Σ.

Suppose we have a sentence Φ of some logic $\mathcal{L}$, in the vocabulary σ_Σ. Such a sentence defines a *language*, that is, a subset of Σ^*, given by

$$L(\Phi) \quad = \quad \{s \in \Sigma^* \mid \mathfrak{M}_s \models \Phi\}. \tag{7.6}$$

We say that a language L is definable in a logic $\mathcal{L}$ if there exists an $\mathcal{L}$-sentence Φ such that $L = L(\Phi)$.

The following is a fundamental result that connects MSO-definability and regular languages.

Theorem 7.21 (Büchi). *A language is definable in* MSO *iff it is regular.*

Proof. We start by showing how to define every regular language L in MSO. If L is regular, then its strings are accepted by a deterministic finite automaton $\mathcal{A} = (Q, q_0, F, \delta)$, where $Q = \{q_0, \ldots, q_{m-1}\}$ is the set of states, $q_0 \in Q$ is the initial state, $F \subseteq Q$ is the set of final states, and $\delta : Q \times \Sigma \to Q$ is the transition function. We take Φ to be the MSO sentence

$$\exists X_0 \ldots \exists X_{m-1}\ \varphi_{\text{part}} \ \wedge\ \varphi_{\text{start}} \ \wedge\ \varphi_{\text{trans}} \wedge\ \varphi_{\text{accept}}. \tag{7.7}$$

In this sentence, we are guessing m sets $X_0, \ldots, X_{m-1}$ that correspond to elements of the universe of $\mathfrak{M}_s$ (i.e., positions of s) where the automaton $\mathcal{A}$ is in the state $q_0, q_1, \ldots, q_{m-1}$, respectively, and the remaining three first-order formulae ensure that the behavior of $\mathcal{A}$ is simulated correctly. That is:

- φ_{part} asserts that $X_0, \ldots, X_{m-1}$ partition the universe of $\mathfrak{M}_s$. This is easy to express in FO:

$$\forall x \bigvee_{i=0}^{m-1} \Big(X_i(x) \wedge \bigwedge_{j \neq i} \neg X_j(x) \Big).$$

- φ_{start} asserts that the automaton starts in state q_0:

$$\forall x \bigwedge_{a\in\Sigma} \Big((P_a(x) \wedge \forall y\ (y \geq x)) \ \rightarrow\ X_{\delta(q_0,a)}(x)\Big) .$$

  Note some abuse of notation: $\delta(q_0, a) = q_i$ for some i, but we write $X_{\delta(q_0,a)}$ instead of X_i.
- φ_{trans} asserts that transitions are simulated correctly:

$$\forall x \forall y \bigwedge_{i=0}^{m-1} \bigwedge_{a\in\Sigma} \Big(((x \prec y) \wedge X_i(x) \wedge P_a(y)) \ \rightarrow\ X_{\delta(q_i,a)}(y)\Big) ,$$

  where $x \prec y$ means that y is the successor of x.
- φ_{accepts} asserts that at the end of the string, $\mathcal{A}$ enters an accepting state:

$$\forall x \ \Big((\forall y\ (y \leq x)) \ \rightarrow\ \bigvee_{q_i\in F} X_i(x)\Big) .$$

Hence, (7.7) captures the behavior of $\mathcal{A}$, and thus $L(\Phi) = L$.

For the converse, let Φ be an MSO sentence in the vocabulary σ_Σ, and let $k = \mathsf{qr}(\Phi)$. Let $\tau_0, \ldots, \tau_m$ enumerate all the rank-k MSO types of σ_Σ structures (more precisely, rank-k $0, 0$ types, with zero free first- and second-order variables, or, in other words, sentences).

Let Ψ_i be an MSO sentence of quantifier rank k defining the type τ_i. That is,

$$M_s \models \Psi_i \ \Leftrightarrow\ \text{mso-tp}_k(M_s) = \tau_i .$$

Since $\mathsf{qr}(\Phi) = k$, the sentence Φ is a disjunction of some of the Ψ_i's. We define $F \subseteq \{\tau_0, \ldots, \tau_m\}$ to be the set of types consistent with Φ. Then Φ is equivalent to $\bigvee_{\tau_i\in F} \Psi_i$.

We further assume that τ_0 is the type of M_ϵ, where ϵ denotes the empty string. That is, this is the only type among the τ_i's that is consistent with $\neg\exists x\ (x = x)$.

We now define the automaton

$$\mathcal{A}_\Phi \ =\ (\{\tau_0, \ldots, \tau_m\}, \tau_0, F, \delta_\Phi), \tag{7.8}$$

with the set of states $S = \{\tau_0, \ldots, \tau_m\}$, the initial state τ_0, the set of final states F, and the transition function $\delta_\Phi : S \times \Sigma \rightarrow 2^S$ defined as follows:

$$\tau_j \in \delta_F(\tau_i, a) \ \Leftrightarrow\ \exists s \in \Sigma^* \begin{pmatrix} \text{mso-tp}_k(M_s) = \tau_i \\ \text{and}\ \ \text{mso-tp}_k(M_{s\cdot a}) = \tau_j \end{pmatrix} . \tag{7.9}$$

We now claim that the automaton $\mathcal{A}_\Phi$ is deterministic (i.e., for every τ_i and $a \in \Sigma$ there is exactly one τ_j satisfying (7.9)). For that, notice that by a composition argument similar to that of Lemma 7.11, if $s_1, s_2, t_1, t_2 \in \Sigma^*$ are such that $M_{s_1} \equiv_k^{\text{MSO}} M_{t_1}$ and $M_{s_2} \equiv_k^{\text{MSO}} M_{t_2}$, then $M_{s_1 \cdot s_2} \equiv_k^{\text{MSO}} M_{t_1 \cdot t_2}$.

Now suppose that $\text{mso-tp}_k(M_{s_1}) = \text{mso-tp}_k(M_{s_2}) = \tau_i$. In particular, $M_{s_1} \equiv_k^{\text{MSO}} M_{s_2}$. Then $M_{s_1 \cdot a} \equiv_k^{\text{MSO}} M_{s_2 \cdot a}$. Suppose also that we have $j_1 \neq j_2$ such that $\text{mso-tp}_k(M_{s_1 \cdot a}) = \tau_{j_1}$ and $\text{mso-tp}_k(M_{s_2 \cdot a}) = \tau_{j_2}$. Then $M_{s_1 \cdot a} \models \Psi_{j_1}$, but since $M_{s_2 \cdot a} \models \Psi_{j_2}$ and $\mathsf{qr}(\Psi_{j_2}) = k$, we obtain $M_{s_1 \cdot a} \models \Psi_{j_2}$, which implies $\text{mso-tp}_k(M_{s_1 \cdot a}) = \tau_{j_2} \neq \tau_{j_1}$. This contradiction proves that the automaton (7.8) is deterministic.

Now by a simple induction on the length of the string we prove that for any string s, after reading s the automaton $\mathcal{A}_\Phi$ ends in the state τ_i such that $\text{mso-tp}_k(M_s) = \tau_i$. For the empty string, this is our choice of τ_0. Suppose now that $\text{mso-tp}_k(M_s) = \tau_i$ and $\mathcal{A}_\Phi$ is in state τ_i after reading s. By the definition of the transition function δ_Φ and the fact that it is deterministic, if $\mathcal{A}_\Phi$ reads a, it moves to the state τ_j such that $\text{mso-tp}_k(M_{s \cdot a}) = \tau_j$, which proves the statement.

Therefore, $\mathcal{A}_\Phi$ accepts a string s iff $\text{mso-tp}_k(M_s)$ is in F, that is, is consistent with Φ. The latter happens iff $M_s \models \Phi$, which proves that the language accepted by $\mathcal{A}_\Phi$ is $L(\Phi)$. This completes the proof. □

We have seen that over graphs, there are universal MSO-sentences which are not expressible in ∃MSO. In contrast, over strings every MSO sentence can be represented by an automaton, and (7.7) shows that the behavior of every automaton can be captured by an ∃MSO sentence. Hence, we obtain the following.

Corollary 7.22. *Over strings,* MSO = ∃MSO. □

As an application of Theorem 7.21, we prove a few bounds on the expressive power of MSO. We have seen before that MSO over the empty vocabulary cannot express EVEN. What about the power of MSO on linear orderings? Recall that L_n denotes a linear ordering on n elements. From Theorem 7.21, we immediately derive the following.

Corollary 7.23. *Let $X \subseteq \mathbb{N}$. Then the set $\{L_n \mid n \in X\}$ is* MSO*-definable iff the language $\{a^n \mid n \in X\}$ is regular.* □

Thus, MSO can test, for example, if the size of a linear ordering is even, or – more generally – a multiple of k for any fixed k. On the other hand, one cannot test in MSO if the cardinality of a linear ordering is a square, or the kth power, for any $k > 1$; nor is it possible to test if such a cardinality is a power of $k > 1$.

As a more interesting application, we show the following.

Corollary 7.24. *It is impossible to test in* MSO *if a graph is Hamiltonian.*

Proof. Let $K_{n,m}$ denote the complete bipartite graph on sets of cardinalities n and m; that is, an undirected graph G whose nodes can be partitioned into two sets X, Y such that $|X| = n$, $|Y| = m$, and the set of edges is $\{(x,y),(y,x) \mid x \in X, y \in Y\}$. Notice that $K_{n,m}$ is Hamiltonian iff $n = m$.

Assume that Hamiltonicity is definable in MSO. Let $\Sigma = \{a, b\}$. Given a string s, we define, in FO, the following graph over the universe of M_s:

$$\varphi(x,y) \quad \equiv \quad (P_a(x) \wedge P_b(y)) \ \vee \ (P_b(x) \wedge P_a(y)).$$

That is, $\varphi(M_s)$ is $K_{n,m}$, where n is the number of a's in s, and m is the number of b's. Thus, if Hamiltonicity were definable in MSO, the language $\{s \in \Sigma^* \mid \text{the number of } a\text{'s in } s \text{ equals the number of } b\text{'s}\}$ would have been a regular language, but it is well known that it is not (by a pumping lemma argument). □

7.5 FO on Strings and Star-Free Languages

Since MSO on strings captures regular languages, what can be said about the class of languages captured by FO? It turns out that FO corresponds to a well-known class of languages, which we define below.

Definition 7.25. *A* star-free *regular expression over Σ is an expression built from the symbols $\emptyset$ and a, for each a in Σ, using the operations of union $(+)$, complement $(^-)$, and concatenation $(\cdot)$. Such a regular expression e denotes a language $L(e)$ over Σ as follows:*

- $L(\emptyset) = \emptyset$; $L(a) = \{a\}$ *for* $a \in \Sigma$.
- $L(e_1 + e_2) = L(e_1) \cup L(e_2)$.
- $L(\bar{e}) = \Sigma^* - L(e)$.
- $L(e_1 \cdot e_2) = \{s_1 \cdot s_2 \mid s_1 \in L(e_1), s_2 \in L(e_2)\}$.

A language denoted by a star-free expression is called a star-free language.

Note that some of the regular expressions that use the Kleene star $*$ are actually star-free, because in the definition of star-free expressions one can use the operation of complementation. For example, suppose $\Sigma = \{a, b\}$. Then $(a + b)^*$ defines a star-free language, denoted by the star-free expression $\bar{\emptyset}$. Likewise, $e = a^*b^*$ also denotes a star-free language, since it can be characterized as a language in which there is no b preceding an a. A language with a b preceding an a can be defined as $(a + b)^* \cdot ba \cdot (a + b)^*$, and hence $L(e)$ is defined by the star-free expression

$$\overline{\bar{\emptyset} \cdot b \cdot a \cdot \bar{\emptyset}}.$$

Theorem 7.26. *A language is definable in* FO *iff it is star-free.*

Proof. We show that every star-free language is definable in FO by induction on the star-free expression. The empty language is definable by *false*, the language $\{a\}$ is definable by $\exists! x\ (x = x) \wedge \forall x\ P_a(x)$. If $e = \bar{e}_1$ and $L(e_1)$ is definable by Φ, then $\neg\Phi$ defines $L(e)$. If $e = e_1 + e_2$, with $L(e_1)$ and $L(e_2)$ definable by Φ_1 and Φ_2 respectively, then $\Phi_1 \vee \Phi_2$ defines $L(e)$.

Now assume that $e = e_1 \cdot e_2$, and again $L(e_1)$ and $L(e_2)$ are definable by Φ_1 and Φ_2. Let x be a variable that does not occur in Φ_1 and Φ_2, and let $\varphi_i(x)$, $i = 1, 2$, be the formula obtained from Φ_1 by relativizing each quantifier to the set of positions $\{y \mid y \leq x\}$ for φ_1, and to $\{y \mid y > x\}$ for φ_2. More precisely, we inductively replace each subformula $\exists y \psi$ of Φ_1 by $\exists y\ (y \leq x) \wedge \psi$, and each such subformula of Φ_2 by $\exists y\ (y > x) \wedge \psi$. Then, for a string s and a position p, we have $M_s \models \varphi_1(p)$ iff $M_s^{\leq p} \models \Phi_1$, where $M_s^{\leq p}$ is the substructure of M_s with the domain $\{1, \ldots, p\}$. Furthermore, $M_s \models \varphi_2(p)$ iff $M_s^{>p} \models \Phi_2$, where $M_s^{>p}$ is the substructure of M_s whose universe is the complement of $\{1, \ldots, p\}$. Hence, $s \in L(e)$ iff $M_s \models \exists x\ \varphi_1(x) \wedge \varphi_2(x)$, which proves that every star-free language is FO-definable.

We now prove the other direction: every FO-definable language is star-free. For technical reasons (to get the induction off the ground), we expand σ_Σ with a constant $\underline{\max}$, to be interpreted as the largest element of the universe. Since $\underline{\max}$ is FO-definable, this does not affect the set of FO-definable languages.

The proof is now by induction on the quantifier rank k of a sentence Φ. Note that since star-free languages are closed under the Boolean operations, an arbitrary Boolean combination of sentences defining star-free languages also defines a star-free language.

For $k = 0$, we have Boolean combinations of the sentences of the form $P_a(\underline{\max})$, as well as *true* and *false*. The sentence $P_a(\underline{\max})$ defines the language denoted by $\bar{\emptyset} \cdot a$, *true* defines $L(\bar{\emptyset})$, and *false* defines $L(\emptyset)$.

Given the closure under Boolean combinations, for the inductive step it suffices to consider sentences $\Phi = \exists x \varphi(x)$, where $\mathsf{qr}(\varphi) = k$.

Let $\tau_0, \ldots, \tau_m$ enumerate all the rank-k FO-types (again, with respect to sentences: we do not have free variables). We define

$$S_\Phi = \left\{ (\tau_i, \tau_j) \;\middle|\; \begin{array}{l} \text{for some } s \text{ and a position } p,\ M_s \models \varphi(p), \\ \mathrm{tp}_k(M_s^{\leq p}) = \tau_i \ \text{ and } \ \mathrm{tp}_k(M_s^{>p}) = \tau_j \end{array} \right\}.$$

Our goal is now to show the following: for every string u, $M_u \models \Phi$ iff there exists a position p in u such that for some (τ_i, τ_j) in S_Φ, we have

$$\mathrm{tp}_k(M_u^{\leq p}) = \tau_i \ \text{ and } \ \mathrm{tp}_k(M_u^{>p}) = \tau_j. \tag{7.10}$$

First, we notice that this claim implies that the language $L(\Phi)$ is star-free. Indeed, each of τ_i is definable by an FO sentence Ψ_i of quantifier rank k, and hence by the induction hypothesis, each language $L(\Psi_i)$ is star-free. Thus,

$$L(\Phi) \;=\; \bigcup_{(\tau_i, \tau_j) \in S_\Phi} L(\Psi_i) \cdot L(\Psi_j).$$

That is, $L(\Phi)$ is a union of concatenations of star-free languages, and hence it is star-free.

If $M_u \models \Phi$, then the existence of p and a pair (τ_i, τ_j) follows from the definition of S_Φ. Conversely, suppose we have a string u and a position p such that (7.10) holds. Since $(\tau_i, \tau_j) \in S_\Phi$, we can find a string s with a position p' in it such that $M_s \models \varphi(p')$, $\text{tp}_k(M_s^{\leq p'}) = \tau_i$, and $\text{tp}_k(M_s^{>p}) = \tau_j$. Hence,

$$M_u^{\leq p} \equiv_k M_s^{\leq p'}, \quad M_u^{>p} \equiv_k M_s^{>p'},$$

and thus $(M_u, p) \equiv_k (M_s, p')$. Since $\mathsf{qr}(\varphi) = k$, it follows that $M_u \models \varphi(p)$, and hence $M_u \models \Phi$, as claimed. This completes the proof. □

Corollary 7.27. *There exist regular languages which are not star-free.*

Proof. The language denoted by $(aa)^*$ is regular, but clearly not star-free, since EVEN is not FO-definable over linear orders. □

7.6 Tree Automata

We now move from strings to trees. Our goal is to define trees as first-order structures, and study MSO over them. We shall connect MSO with the notion of tree automata. Tree automata play an important role in many applications, including rewriting systems, automated theorem proving, verification, and recently database query languages, especially in the XML context.

We consider two kinds of trees in this section. *Ranked* trees have the property that every node which is not a leaf has the same number of children (in fact we shall fix this number to be 2, but all the results can be generalized to any fixed $k > 1$). On the other hand, in *unranked* trees different nodes can have a different number of children. We shall start with ranked (binary) trees.

Definition 7.28. *A* tree domain *is a subset D of $\{1, 2\}^*$ that is prefix-closed; that is, if $s \in D$ and s' is a prefix of D, then $s' \in D$. Furthermore, if $s \in D$, then either both $s \cdot 1$ and $s \cdot 2$ are in D, or none of them is in D.*

A Σ-tree *T is a pair (D, f) where D is a tree domain and f is a function from D to Σ (the labeling function).*

We refer to the elements of D as the nodes of T. Every nonempty tree domain has the node ϵ, which is called the root. A node s such that $s \cdot 1, s \cdot 2 \notin D$ is called a leaf.

The first tree in Fig. 7.2 is a binary tree. We show both the nodes and the labeling in that picture. The nodes $111, 112, 12, 21, 22$ are the leaves.

We represent a tree $T = (D, f)$ as a first-order structure

$$M_T = \langle D, \prec, (P_a)_{a \in \Sigma}, \text{succ}_1, \text{succ}_2 \rangle$$

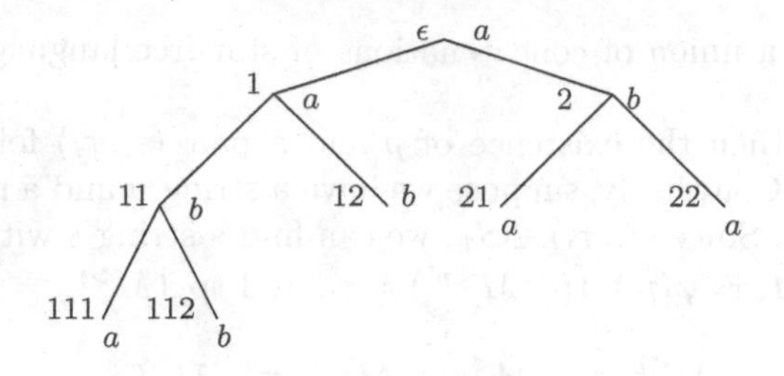

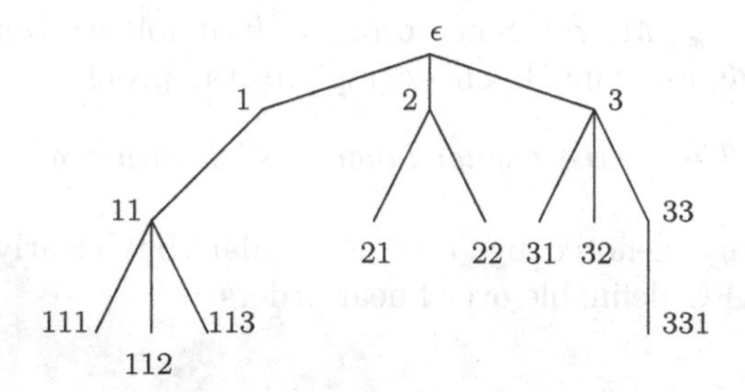

Fig. 7.2. Examples of a ranked and an unranked tree

of vocabulary σ_Σ expanded with two binary relations succ_1 and succ_2. Here $\prec$ is interpreted as the prefix relation on D (in particular, it is a partial order, rather than a linear order, as was the case with strings), P_a is interpreted as $\{s \in D \mid f(s) = a\}$, and succ_i is $\{(s, s \cdot i) \mid s, s \cdot i \in D\}$, for $i = 1, 2$.

We let $\mathrm{Trees}(\Sigma)$ be the set of all Σ-trees. If we have a sentence Φ of some logic, it defines the set of trees (also called a tree language)

$$L_T(\Phi) \quad = \quad \{T \in \mathrm{Trees}(\Sigma) \mid M_T \models \Phi\}.$$

Thus, we shall be talking about tree languages definable in various logics.

Definition 7.29 (Tree automata and regular tree languages). *A (nondeterministic)* tree automaton *is a tuple $\mathcal{A} = (Q, q_0, \delta, F)$, where Q is a finite set of states, $q_0 \in Q$, $F \subseteq Q$ is the set of final (accepting) states, and $\delta : Q \times Q \times \Sigma \to 2^Q$ is the transition function.*

Given a tree $T = (D, f)$, a run *of $\mathcal{A}$ on T is a function $r : D \to Q$ such that*

- *if s is a leaf labeled a, then $r(s) \in \delta(q_0, q_0, a)$;*
- *if $r(s \cdot 1) = q$, $r(s \cdot 2) = q'$ and $f(s) = a$, then $r(s) \in \delta(q, q', a)$.*

A run is called successful *if $r(\epsilon) \in F$ (the root is in the accepting state). The set of trees accepted by $\mathcal{A}$ is the set of all trees T for which there exists a successful run.*

A tree language is called regular *if it is accepted by a tree automaton.*

In a *deterministic* tree automaton, the transition function is $\delta : Q \times Q \times \Sigma \to Q$, and the definition of a run is modified as follows:

- if s is a leaf labeled a, then $r(s) = \delta(q_0, q_0, a)$;
- if $r(s \cdot 1) = q$, $r(s \cdot 2) = q'$ and $f(t) = a$, then $r(s) = \delta(q, q', a)$.

For example, consider a deterministic tree automaton $\mathcal{A}$ whose set of states is $\{q_0, q_a, q_b, q, q'\}$, with $F = \{q'\}$, and the transition function has the following:

$$\begin{aligned}
\delta(q_0, q_0, a) &= q_a \\
\delta(q_0, q_0, b) &= q_b \\
\delta(q_a, q_b, b) &= q \\
\delta(q_a, q_a, b) &= q' \\
\delta(q, q_b, a) &= q \\
\delta(q, q', a) &= q' .
\end{aligned}$$

Then this automaton accepts the ranked tree shown in Fig. 7.2: following the definition of the transition function, we define the run r such that:

- for the leaves, $r(111) = r(21) = r(22) = q_a$ and $r(112) = r(12) = q_b$;
- $r(11) = \delta(q_a, q_b, b) = q$;
- $r(1) = \delta(q, q_b, a) = q$;
- $r(2) = \delta(q_a, q_a, b) = q'$; and finally,
- $r(\epsilon) = \delta(q, q', a) = q'$, and since $q' \in F$, the automaton accepts.

We now establish the analog of Theorem 7.21 for trees, by showing that regular tree languages are precisely those definable in MSO.

Theorem 7.30. *A set of trees is definable in* MSO *iff it is regular.*

Proof. The proof is similar to that of Theorem 7.21. To find an MSO definition of the tree language accepted by an automaton $\mathcal{A}$, we guess, for each state q, the set X_q of nodes where the run of $\mathcal{A}$ is in state q, and then check, in FO, that each leaf labeled a is in X_q for some $q \in \delta(q_0, q_0, a)$, that transitions are modeled properly, and that the root is in one of the accepting states. The sentence looks very similar to (7.7), and is in fact an $\exists$MSO sentence.

The proof of the converse, i.e., that MSO only defines regular languages, again follows the proof in the string case. Suppose an MSO sentence Φ of quantifier rank k is given. We let $\tau_0, \ldots, \tau_m$ enumerate all the rank-k MSO types, with τ_0 being the type of the empty tree, and take $\{\tau_0, \ldots, \tau_m\}$ as the set of states of an automaton $\mathcal{A}_\Phi$. Since Φ is equivalent to a disjunction of types, we let $F = \{\tau_i \mid \tau_i \text{ is consistent with } \Phi\}$. Finally,

$$\tau_l \in \delta(\tau_i, \tau_j, a)$$

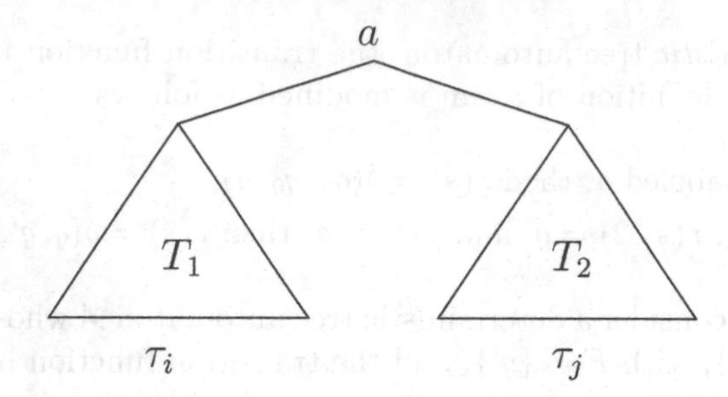

Fig. 7.3. Illustration for the proof of Theorem 7.30

if there are trees T_1 and T_2 whose rank-k MSO types are τ_i and τ_j, respectively, such that the rank-k MSO type of the tree obtained by hanging T_1 and T_2 as children of a root node labeled a is τ_l (see Fig. 7.3).

Again, similarly to the proof of Theorem 7.21, one can show that $\mathcal{A}_\Phi$ is a deterministic tree automaton accepting the tree language $\{T \mid T \models \Phi\}$. ☐

Corollary 7.31. *Every tree automaton is equivalent to a deterministic tree automaton, and every* MSO *sentence over trees is equivalent to an* ∃MSO *sentence.* ☐

The connection between FO-definability and star-free languages does not, however, extend to trees. There are several interesting logics between FO and MSO, and some of them will be introduced in exercises.

We next show how to extend these results to *unranked trees.*

Definition 7.32 (Unranked trees). *An* unranked tree domain *is a subset D of $\{1, 2, \ldots\}^*$ (finite words over positive integers) that is prefix-closed, and such that for $s \cdot i \in D$ and $j < i$, the string $s \cdot j$ is in D as well. An unranked tree is a pair (D, f), where D is an unranked tree domain, and f is the labeling function $f : D \to \Sigma$.*

Thus, a node in an unranked tree can have arbitrarily many children. An example is shown in Fig. 7.2 (the second tree). Some nodes – the root, nodes 11 and 3 – have three children; some have two (node 2), some have one (nodes 1 and 33).

The transition function for an automaton working on binary trees was of the form $\delta : Q \times Q \times \Sigma \to Q$, based on the fact that each nonleaf node has exactly two children. In an unranked tree, the number of children could be arbitrary. The idea of extending the notion of tree automata to the unranked case is then as follows: we have additional string automata that run on the children of each node, and the acceptance conditions of those automata determine the state of the parent node. This is formalized in the definition below.

Definition 7.33 (Unranked tree automata). *An* unranked tree automaton *is a triple* $\mathcal{A} = (Q, q_0, \delta)$*, where as before* Q *is the set of states,* q_0 *is an element of* Q*, and* δ *is the transition function* $\delta : Q \times \Sigma \to 2^{Q^*}$ *such that* $\delta(q, a)$ *is a regular language over* Q *for every* $q \in Q$ *and* $a \in \Sigma$*.*

Given an unranked tree $T = (D, f)$*, a* run *of* $\mathcal{A}$ *on* T *is defined as a function* $r : D \to Q$ *such that the following holds:*

- *if* s *is a node labeled* a*, with children* $s \cdot 1, \ldots, s \cdot n$*, then the string* $r(s \cdot 1)r(s \cdot 2) \ldots r(s \cdot n)$ *is in* $\delta(r(s), a)$*.*

In particular, if s *is a leaf, then* $r(s) = q$ *implies that the empty string belongs to* $\delta(q, a)$*.*

A run is successful *if* $r(\epsilon) = q_0$*, and* T *is accepted by* $\mathcal{A}$ *if there exists an accepting run. An unranked tree language* L *is called* regular *if it is accepted by an unranked tree automaton.*

To connect regular languages with MSO-definability, we have to represent unranked trees as structures. It is no longer sufficient to model just two successor relations, since a node can have arbitrarily many successors. Instead, we introduce an ordering on successor relations. That is, an unranked tree $T = (D, f)$ is represented as a structure

$$\langle D, \prec, (P_a)_{a \in \Sigma}, <_{\text{sibl}} \rangle, \tag{7.11}$$

where $\prec$, as before, is the prefix relation, P_a is interpreted as $\{s \in D \mid f(s) = a\}$, and $s' <_{\text{sibl}} s''$ iff there is a node s and $i, j \in \mathbb{N}, i < j$, such that $s' = s \cdot i$, $s'' = s \cdot j$. In other words, s' and s'' are siblings, and s' precedes s''.

Thus, when we talk about FO-definability, or MSO-definability over unranked trees, we mean definability over structures of the form (7.11).

Finally, the connection between automata and MSO-definability extends to unranked trees.

Theorem 7.34. *An unranked tree language is regular iff it is* MSO*-definable.*

The proof of this theorem is similar in spirit to the proofs of Theorems 7.21 and 7.30, and is left as an exercise for the reader.

7.7 Complexity of MSO

In this section we study complexity of MSO. We have seen that MSO, and even ∃MSO, are significantly more expressive than FO: ∃MSO can express NP-complete problems (3-colorability, for example), and by using negation, we can express CONP-complete problems in MSO.

This suggests a close connection between MSO and the polynomial hierarchy, PH, for which NP and coNP are the two lowest levels above polynomial time. Recall that the levels of the polynomial hierarchy are defined as $\Sigma_0^p = \Pi_0^p = \text{PTIME}$, $\Sigma_i^p = \text{NP}^{\Sigma_{i-1}^p}$, and Π_i^p is the set of problems whose complement is in Σ_i^p (see Sect. 2.3).

We next show that the data complexity of MSO is well approximated by the polynomial hierarchy (although MSO does not capture PH: for example, Hamiltonicity is not MSO-expressible).

Proposition 7.35. *For each level Σ_i^p or Π_i^p of the polynomial hierarchy, there exists a problem complete for that level which is expressible in* MSO.

Proof. We show how to express a variant of QBF (quantified Boolean formulae), which we used in the proof of PSPACE-completeness of the combined complexity of FO. We define the problem Σ_i-SAT as follows. Its input is a formula of the form

$$(\exists \ldots \exists)(\forall \ldots \forall)(\exists \ldots \exists) \ldots \varphi, \tag{7.12}$$

where φ is a propositional Boolean formula in conjunctive normal form, such that each conjunct contains three propositional variables. The quantifier prefix starts with a block of existential quantifiers, followed by a block of universal quantifiers, followed by a block of existential quantifiers, and so on – such that there are i blocks of quantifiers. The output is "yes" if the formula (7.12) is true.

The problem Π_i-SAT is defined similarly, except that in (7.12), the first block of quantifiers is universal. We use the known fact that Σ_i-SAT is complete for Σ_i^p, and Π_i-SAT is complete for Π_i^p.

We now show how to encode an instance Φ (7.12) of Σ_i-SAT as a structure $\mathfrak{A}_\Phi$. Its universe is the set of variables used in (7.12). It has four binary relations R_0, R_1, R_2, R_3, and $i+1$ unary relations $E_1, U_2, E_3, \ldots$. Each relation E_k or U_k is interpreted as the set of variables quantified by the kth block of quantifiers. Relations R_0, R_1, R_2, R_3 encode the formula φ: relation R_i corresponds to all the conjuncts of φ in which exactly i variables appear positively. That is, R_0 has all the triples (x, y, z) such that $(\neg x \vee \neg y \vee \neg z)$ is a conjunct of φ, R_1 has all the triples (x, y, z) such that $(x \vee \neg y \vee \neg z)$ is a conjunct of φ, and so on.

Next, we find an MSO sentence Ψ such that $\mathfrak{A}_\Phi \models \Psi$ iff Φ is true. This sentence is of the form

$$\exists X_1 \subseteq E_1 \forall X_2 \subseteq U_2 \exists X_3 \subseteq E_3 \ldots \quad \varphi',$$

where each X_i corresponds to the set of variables set to *true* in the ith quantifier block. The formula φ' says that the variable assignment of the quantifier prefix of Ψ makes φ true. For example, for each triple (x, y, z) in R_1, it would state that either y or z belongs to some of the X_i's, or x belongs to neither of them, and similarly for R_0, R_2, and R_3. We leave the details to the reader. The proof for Π_i-SAT is almost identical: the sentence Φ must start with a universal MSO quantifier. □

We shall return to complexity of SO in Chap. 9. For the combined complexity of MSO, see Exercise 7.21.

Even though the complexity of MSO is quite high, for many interesting structures, in particular, strings and trees, the connection with automata provides nice bounds in terms of parameterized complexity.

Corollary 7.36. *Over strings and trees (ranked and unranked), evaluating* MSO *sentences is fixed-parameter linear. In particular, over strings and trees, the data complexity of* MSO *is linear.*

Proof. Suppose we have a sentence Φ and a structure $\mathfrak{A}$ (string or tree). We convert Φ into a deterministic automaton, by Theorems 7.21, 7.30, and 7.34, and run that automaton over $\mathfrak{A}$, which takes linear time. □

Can Corollary 7.36 be extended to a larger class of structures? The answer to this is positive, and it uses the concept of bounded treewidth we first encountered in Section 6.7. Recall that a class $\mathcal{C}$ of σ-structures is said to be of bounded treewidth if there is a number k such that for every $\mathfrak{A} \in \mathcal{C}$, the treewidth of its Gaifman graph is at most k. (See Sect. 6.7 for the definition of treewidth.)

Theorem 7.37 (Courcelle). *Let $\mathcal{C}$ be a class of structures of bounded treewidth. Then evaluating* MSO *sentences over $\mathcal{C}$ is fixed-parameter linear. In particular, the data complexity of* MSO *over $\mathcal{C}$ is linear.*

Proof sketch. We outline the idea of the proof. For simplicity, assume that our input structures are graphs. Given a graph $\mathcal{G}$, compute, in linear time, its tree decomposition, consisting of a tree T and a set B_t for each node t of T. Since the treewidth is fixed, say k, each B_t is of size at most $k+1$, and thus all the graphs generated by B_t's can be explicitly enumerated. This allows us to express MSO quantification over the original graph $\mathcal{G}$ in terms of MSO quantification over T. Thus, we are now in the setting where MSO sentences have to be evaluated over trees, and this problem is fixed-parameter linear, which can be shown by converting MSO sentences into tree automata, as in Corollary 7.36. □

Fixed-parameter linearity implies that the complexity of the model-checking is of the form $f(\|\Phi\|)\cdot\|\mathfrak{A}\|$. What can be said about the function f? Even over strings, to achieve fixed-parameter linearity, one has to convert Φ to an automaton. We have seen this conversion in the proof of Theorem 7.21, and it was based on computing all rank-k MSO-types. One can also convert MSO sentences into automata directly, with existential quantifiers corresponding to nondeterministic guesses. For such a conversion, the main problem is negation, since complementing nondeterministic automata is not easy: one has to make them deterministic first, and then reverse the roles of accepting and rejecting states. Going from a nondeterministic automaton to a deterministic one entails an exponential blow-up.

When we try to apply this reasoning to an MSO sentence of the form

$$(\exists \ldots \exists)(\forall \ldots \forall)(\exists \ldots \exists) \ldots \varphi,$$

we see that at each quantifier alternation, one needs to make the automaton deterministic. Hence, the size of the resulting automaton will be bounded by (roughly)

$$\left.2^{2^{\cdot^{\cdot^{\cdot^{n}}}}}\right\}k \text{ times},$$

where n is the size of the automaton corresponding to φ, and k is the number of alternations of quantifiers. That is, the size of the automaton is actually *nonelementary* in terms of $\| \Phi \|$. We recall that a function $f : \mathbb{N} \to \mathbb{N}$ is elementary if for some fixed l,

$$f(n) \;<\; \left.2^{2^{\cdot^{\cdot^{\cdot^{n}}}}}\right\}l \text{ times} \quad \text{for all } n.$$

In fact, it is known that converting MSO formulae into automata is inherently nonelementary. Thus, even though over some classes of structures MSO is fixed-parameter linear, the function of the parameter (that depends on the MSO sentence) is extremely large. Exercise 7.22 shows that the complexity cannot be lowered unless NP collapses to PTIME.

7.8 Bibliographic Notes

Second-order logic is described in most logic textbooks. Monadic second order logic and its games can be found in Ebbinghaus and Flum [60].

Proposition 7.14 is from Fagin [71], the proof is from Fagin, Stockmeyer, and Vardi [76]. Expressibility of undirected reachability in ∃MSO is due to Kanellakis; the proof was published in [11]. Inexpressibility of directed reachability in ∃MSO is due to Ajtai and Fagin [11].

The Fagin game is from Fagin [71], and the Ajtai-Fagin game is from [11]. For additional results on ∃MSO and its relatives, see a survey by Fagin [75] and Ajtai, Fagin, and Stockmeyer [12], Janin and Marcinkowski [137], and Schwentick [216].

Theorem 7.21 is due to Büchi [27], and the proof presented here follows Ladner [160], see also Neven and Schwentick [187]. Corollary 7.24 is due to Turán [236] and de Rougemont [56]; the proof here follows Makowsky [176].

Theorem 7.26 was proved by McNaughton and Papert [182]; the proof based on games follows Thomas [233].

For connections between automata theory, logical definability, and circuit complexity, see Straubing [225].

In the proofs of Proposition 7.12 and Theorems 7.21 and 7.26 we used the composition method already encountered in Chap. 3. The composition techniques used here are a special case of the Feferman-Vaught Theorem [79]. For more on the composition method, see a recent survey by Makowsky [177], as well as Exercises 7.25 and 7.26.

Tree automata are treated in several books and surveys [38, 90]. Theorem 7.30 is due to Thatcher and Wright [230]. The corresponding result for unranked trees seems to be a part of folklore, and can be found in several papers dealing with querying XML, e.g., Neven [186].

Proposition 7.35 also appears to be folklore. Completeness of Σ_i^p-SAT and Π_i^p-SAT is due to Stockmeyer [223] (it is also known that the quantifier-free formula can always be taken to be 3-CNF [59]). Theorem 7.37 was proved by Courcelle [44]. Linearity of finding a tree decomposition of small treewidth is from Bodlaender [24]. The nonelementary complexity of the translation from MSO to automata is due to Stockmeyer and Meyer [224].

Sources for exercises:

Exercises 7.2–7.5:	Courcelle [45, 46]
Exercise 7.11:	Schwentick [215]
Exercise 7.12:	Cosmadakis [42]
Exercise 7.13:	Otto [190]
Exercise 7.14:	Matz, Schweikardt, and Thomas [180]
Exercise 7.16:	Thomas [231]
Exercises 7.18 and 7.19:	Thomas [232, 233]
Exercise 7.20:	Blumensath and Grädel [23]
	Bruyère et al. [26], Benedikt et al. [21]
	Benedikt and Libkin [20]
Exercise 7.22:	Frick and Grohe [85]
Exercise 7.23:	Grandjean and Olive [105]
	Schwentick [216], Lynch [174]
Exercise 7.25:	Makowsky [177]
Exercise 7.26:	Courcelle and Makowsky [47]
Exercise 7.27:	Seese [218]

7.9 Exercises

Exercise 7.1. Prove Theorem 7.7.

Exercise 7.2. Prove that the following properties of an undirected graph G are expressible in MSO:

- G is planar;
- G is a tree.

Exercise 7.3. Prove that the following properties of an undirected graph G are expressible in $\exists$MSO:

- G is not planar;
- G is not a tree;
- G is not chordal (recall that a chord of a cycle C is an edge (a, b) such that a, b are nodes of C, but the edge (a, b) is not in C; a graph without loops is chordal if it has no cycle of length at least 4 without a chord).

Exercise 7.4. Consider a different representation of graphs as first-order structures. Given a graph G, we create a structure $\mathfrak{A}_G = \langle A_G, P_G \rangle$ whose universe is the disjoint union of nodes and edges of G, and P_G is a ternary relation that consists of pairs (a, e, b), where e is the edge (a, b) in G.

Prove that over such a representation of graphs, Hamiltonicity is MSO-definable.

Exercise 7.5. Corollary 7.24 and Exercise 7.4 show that the expressive power of MSO is different over two representation of graphs: one with the universe consisting of nodes, and the other one with the universe consisting of both nodes and edges.

Prove that if we restrict the class of graphs to be one of the following:

- graphs of bounded degree, or
- planar graphs, or
- graphs of treewidth at most k, for a fixed k,

then the expressive power of MSO over the two different representations of graphs is the same.

Exercise 7.6. Prove that $\exists$MSO can express queries that are not Gaifman-local and violate the BNDP.

Exercise 7.7. Prove that for each fixed k, directed reachability is expressible in $\exists$MSO over graphs whose in-degrees and out-degrees do not exceed k.

Exercise 7.8. Prove Theorem 7.16.

Conclude that undirected reachability is in $\forall\text{MSO} \cap \exists\text{MSO}$, while directed reachability is in $\forall\text{MSO} - \exists\text{MSO}$.

Exercise 7.9. Prove Proposition 7.20.

Exercise 7.10. Use Ajtai-Fagin games to prove that there is no $\exists$MSO sentence Φ such that, if a graph G is a disjoint union of two cycles, then $G \models \Phi$ iff the cycles are of the same size.

Exercise 7.11. Prove that graph connectivity is not definable in $\exists\text{MSO}+<$.

Exercise 7.12. Prove that non-3-colorability of graphs cannot be expressed in $\exists$MSO.

Exercise 7.13. Prove that the number of second-order quantifiers in $\exists$MSO gives rise to a strict hierarchy.

Exercise 7.14. Prove that the alternation depth of second-order quantifiers in MSO gives rise to a strict hierarchy.

Exercise 7.15. Prove the composition result used in the proof of Theorem 7.21. That is, if $s_1, s_2, t_1, t_2 \in \Sigma^*$ are such that $M_{s_1} \equiv_k^{\text{MSO}} M_{t_1}$ and $M_{s_2} \equiv_k^{\text{MSO}} M_{t_2}$, then $M_{s_1 \cdot s_2} \equiv_k^{\text{MSO}} M_{t_1 \cdot t_2}$.

Exercise 7.16. Prove that over strings, every MSO sentence is equivalent to an ∃MSO sentence with a single second-order quantifier.

Exercise 7.17. Complete the proof of Theorem 7.30, and prove Theorem 7.34.

Exercise 7.18. Consider a restriction of MSO on binary trees, in which we only allow second-order quantifications over antichains: sets of nodes X such that for $s, s' \in X$, $s \neq s'$, neither $s \prec s'$ nor $s' \prec s$ holds. Such a logic is called the *antichain logic*.

Prove that every regular tree language is definable in the antichain logic.

Exercise 7.19. Next, consider a restriction of MSO on binary trees, in which we only allow second-order quantifications over chains: sets of nodes X such that for $s, s' \in X$, $s \neq s'$, either $s \prec s'$ or $s' \prec s$ holds.

Prove that there are regular tree languages that are not definable in this restriction of MSO.

Exercise 7.20. Let $s_1, \ldots, s_n \in \Sigma^*$. We construct a string $[\vec{s}]$ over the alphabet $(\Sigma \cup \{\#\})^n$, whose length is the maximum of the lengths of s_j's, and whose ith symbol is a tuple $(c_1, \ldots, c_n)$, where each c_k is the ith symbol of s_k, if the length of s_k is at least i, or $\#$ otherwise. We say that a set $S \subseteq (\Sigma^*)^n$ is regular if the set $\{[\vec{s}] \mid \vec{s} \in S\} \subseteq (\Sigma \cup \{\#\})^n$ is regular.

Consider the *infinite* first-order structure $\mathfrak{M}$ whose universe is Σ^*, and the predicates include $\prec$ (the prefix relation), a unary predicate L_a for each a in Σ, such that $L_a(x)$ holds iff the last symbol of x is a, and a binary predicate el such that $\text{el}(s, s')$ holds iff the length of s equals the length of s'.

We call a subset S of $(\Sigma^*)^n$ *definable* in $\mathfrak{M}$ if there is an FO formula $\varphi(x_1, \ldots, x_n)$ in the vocabulary of $\mathfrak{M}$ such that $S = \{\vec{s} \mid \mathfrak{M} \models \varphi(\vec{s})\}$.

Prove the following:

(a) A subset of $(\Sigma^*)^n$ is definable in $\mathfrak{M}$ iff it is regular.
(b) A subset of Σ^* is definable in $\mathfrak{M}$ by a formula that does not mention the equal length predicate iff it is star-free.
(c) Generalize (a) to binary trees.

Exercise 7.21. Prove that the combined complexity of MSO is PSPACE-complete.

Exercise 7.22. Prove that if the model-checking problem for MSO on strings can be solved in time $f(\|\Phi\|) \cdot p(|s|)$, for a polynomial p and an elementary function f, then PTIME = NP.

Exercise 7.23. Define complexity class NLIN as the class of problems accepted by nondeterministic RAMs in linear time. Consider a different encoding of strings as finite structures. A string $s = s_1 \ldots s_n \in \{0,1\}^*$ is encoded as follows. Partition s into m pieces of length $\frac{1}{2} \cdot \log n$, where $m = \lceil \frac{2n}{\log n} \rceil$. Let $g_s(i)$ be the number encoded by the ith piece of the partitioned string. We define a structure $\mathcal{M}_s$ whose universe is $\{1, \ldots, m\}$, and the vocabulary consists of two unary functions, one interpreted as g_s, and the other one as successor.

Prove the following:

- A set of strings S is in NLIN iff there exists a sentence Φ of the form

$$\exists F_1 \ldots \exists F_m \; \forall x \; \varphi,$$

where F_i's are unary function symbols, and φ is quantifier-free, such that $S = \{s \mid \mathcal{M}_s \models \Phi\}$.
- Every set of strings in NLIN is definable by an $\exists$MSO sentence in the presence of a built-in addition relation.

Exercise 7.24. Using the fact that the MSO theory of finite trees is decidable (Rabin [202]), prove that the MSO theory of finite forests is decidable.

Exercise 7.25. Define $\mathrm{Th}^k_{\mathrm{MSO}}(\mathfrak{A})$ as the set of all MSO[k] sentences true in $\mathfrak{A}$. Notice that $\mathrm{Th}^k_{\mathrm{MSO}}(\mathfrak{A})$ is a finite object.

We call an m-ary operation F on structures of the same vocabulary MSO-*smooth* if $\mathrm{Th}^k_{\mathrm{MSO}}(F(\mathfrak{A}_1, \ldots, \mathfrak{A}_m))$ is uniquely determined by, and can be computed from, $\mathrm{Th}^k_{\mathrm{MSO}}(\mathfrak{A}_1), \ldots, \mathrm{Th}^k_{\mathrm{MSO}}(\mathfrak{A}_m)$, for every k. Prove that the disjoint union of structures, root joining of trees, and concatenation of words are MSO-smooth.

Exercise 7.26. A class $\mathcal{C}$ of structures is MSO-*inductive* if it is the smallest class of structures that can be constructed from a fixed finite set of structures using a fixed finite set of MSO-smooth operations. Such a construction naturally yields, for each structure $\mathfrak{A} \in \mathcal{C}$, its parse tree $T_{\mathfrak{A}}$.

Prove that the following are MSO-inductive classes of structures:

- words;
- forests;
- graphs of treewidth at most l, for a fixed l.

Also prove that for a fixed MSO sentence Φ, checking whether $\mathfrak{A} \models \Phi$ can be solved in time linear in the size of $T_{\mathfrak{A}}$, if $\mathfrak{A} \in \mathcal{C}$.

Exercise 7.27. Consider representation of graphs from Exercise 7.4. Prove that if $\mathcal{C}$ is a class of finite graphs, and its MSO theory in that representation is decidable, then $\mathcal{C}$ is of bounded treewidth.

Hint: you will have to use decidability of the MSO theory of graphs of bounded treewidth, undecidability of the MSO theory of grids (Cartesian products of successor relations), and the fact, due to Robertson and Seymour [204], that a class of graphs of unbounded treewidth has arbitrarily large grids as its minors.

Exercise 7.28.* Is every query in $(\mathrm{MSO}+<)_{\mathrm{inv}}$ definable in the expansion of MSO with unary generalized quantifiers (see the definition in the next chapter) $\mathcal{Q}_m x \; \varphi(x, \vec{y})$ such that $\mathfrak{A} \models \mathcal{Q}_m x \; \varphi(x, \vec{a})$ holds iff $|\varphi(\mathfrak{A}, \vec{a})| \bmod m = 0$?

8

Logics with Counting

We continue dealing with extensions of first-order logic. We have seen that the expressive power of FO on finite structures is limited in a number of ways: it cannot express counting properties, nor is it capable of expressing properties that require iterative algorithms, as those typically violate locality.

In this chapter we address FO's inability to count. As we saw earlier, nontrivial properties of cardinalities are not expressible in FO: for example, a sentence of quantifier rank n cannot distinguish any two linear orders of cardinality over 2^n. Comparisons of cardinalities, such as testing if $|A|>|B|$, are inexpressible too.

We first introduce two possible ways of extending FO that add counting power to it: one is to use counting quantifiers and two-sorted structures, the other is to use generalized unary quantifiers. We shall mostly concentrate on counting quantifiers, as unary quantifiers can be simulated with them. We shall see a very powerful counting logic, expressing *arbitrary* properties of cardinalities, and yet we show that this logic is local. We also address the problem of complexity of some of the counting extensions of FO.

8.1 Counting and Unary Quantifiers

Suppose we want to find an extension of FO capable of expressing the PARITY query: if U is a unary predicate in the vocabulary σ, and $\mathfrak{A} \in \text{STRUCT}[\sigma]$, is $|U^{\mathfrak{A}}|$ even? How can one do it?

One approach is to add enough expressiveness to the logic to find cardinalities of some sets: for example, sets definable by other formulae. Thus, if we have a formula $\varphi(x)$, we want to find the cardinality of $\varphi(\mathfrak{A}) = \{a \mid \mathfrak{A} \models \varphi(a)\}$. The problem is that $|\varphi(\mathfrak{A})|$ is a number, and hence the logic must be adequately equipped to deal with numbers. To be able to use $|\varphi(\mathfrak{A})|$, we introduce *counting quantifiers*:

$$\exists ix\ \varphi(x)$$

is a formula with a new free variable i, which states that there are at least i elements a of A such that $\varphi(a)$ holds.

The variable i must range over some numerical domain (which, as we shall see, is different for different counting logics). On that numerical domain, we should have some arithmetic operations available (e.g., addition and multiplication), as well as quantification over it, so that sentences in the logic could be formed.

Without yet giving a formal definition of the logic that extends FO with counting quantifiers, we show, as an example, how parity is definable in it:

$$\exists i \exists j \ \Big((i = j + j) \land \exists ix \varphi(x) \land \big(\forall k \ (k > i) \rightarrow \neg \exists kx \ \varphi(x)\big)\Big) .$$

This sentence says that we can find an even number i (since it is of the form $2j$) such that exactly i elements satisfy $\varphi(x)$: that is, at least i elements satisfy φ, and for every $k > i$, we cannot find k elements that satisfy φ.

Note that we really have two different kinds of variables: variables that range over the domain of $\mathfrak{A}$, and variables that range over some numerical domain. Such a logic is called two-sorted. Formally, a structure for such a logic has two universes: one is the non-numerical universe (we shall normally refer to it as *first-sort* universe) and the numerical, *second-sort* universe. We now give the formal definition of the logic FO(**Cnt**).

Definition 8.1 (FO with counting). *Given a vocabulary σ, a σ-structure for* FO *with counting,* FO(**Cnt**), *is a structure of the form*

$$\langle \{a_0, \ldots, a_{n-1}\}, \{0, \ldots, n-1\}, (R_i)^{\mathfrak{A}}, +, \times, \underline{\min}, \underline{\max} \rangle$$

where $\langle \{a_0, \ldots, a_{n-1}\}, (R_i)^{\mathfrak{A}} \rangle$ is a structure from STRUCT[σ] *(R_i ranges over the symbols in σ), $+$ and $\times$ are ternary relations $\{(i, j, k) \mid i + j = k\}$ and $\{(i, j, k) \mid i \cdot j = k\}$ on $\{0, \ldots, n-1\}$, $\underline{\min}$ denotes 0 and $\underline{\max}$ denotes $n-1$. We shall assume that the universes $\{a_0, \ldots, a_{n-1}\}$ and $\{0, \ldots, n-1\}$ are disjoint.*

Formulae of FO(**Cnt**) *can have free variables of two sorts, ranging over the two universes. We normally use $i, j, k, \vec{\imath}, \vec{\jmath}$ for second-sort variables.* FO(**Cnt**) *extends the definition of* FO *by the following rules:*

- *$\underline{\min}$, $\underline{\max}$ are terms of the second sort. Also, every second-sort variable i is a term of the second sort.*
- *If t_1, t_2, t_3 are terms of the second sort, then $+(t_1, t_2, t_3)$ and $\times(t_1, t_2, t_3)$ are formulae (which we shall normally write as $t_1 + t_2 = t_3$ and $t_1 \cdot t_2 = t_3$).*
- *If $\varphi(\vec{x}, \vec{\imath})$ is a formula, then $\exists i \ \varphi(\vec{x}, \vec{\imath})$ is a formula. The quantifier $\exists i$ binds the second-sort variable i.*
- *If $\varphi(y, \vec{x}, \vec{\imath})$ is a formula, then $\psi(\vec{x}, i, \vec{\imath}) \equiv \exists iy \varphi(y, \vec{x}, \vec{\imath})$ is a formula. The quantifier $\exists iy$ binds the first-sort variable y but not the second-sort variable i.*

For the semantics of this logic, only the last item needs explanation. Suppose we have a structure $\mathfrak{A}$, and we fix an interpretation $\vec{a}$ for $\vec{x}$ (from $\{a_0, \ldots, a_{n-1}\}$), $\vec{\imath}_0$ for $\vec{\imath}$, and i_0 for i (from $\{0, \ldots, n-1\}$). Then $\mathfrak{A} \models \psi(\vec{a}, i_0, \vec{\imath}_0)$ iff

$$|\{b \in \{a_0, \ldots, a_{n-1}\} \mid \mathfrak{A} \models \varphi(b, \vec{a}, \vec{\imath}_0)\}| \;\geq\; i_0.$$

If we have a σ-structure $\mathfrak{A}$, there is a two-sorted structure $\mathfrak{A}'$ naturally associated with $\mathfrak{A}$. Assuming $A = \{a_0, \ldots, a_{n-1}\}$, we let the numerical domain of $\mathfrak{A}'$ be $\{0, \ldots, n-1\}$, with $\underline{\min}$ and $\underline{\max}$ interpreted as 0 and $n-1$, and $+$ and $\times$ getting their usual interpretations. Hence, for $\mathfrak{A} \in \text{STRUCT}[\sigma]$, we shall write $\mathfrak{A} \models \varphi$ whenever φ is an FO(**Cnt**) formula, instead of the more formal $\mathfrak{A}' \models \varphi$.

Let us see a few examples of definability in FO(**Cnt**). First, the usual linear ordering on numbers is definable: $i \leq j$ iff $\exists k\ (i + k = j)$. Note that this does *not* imply definability of ordering on the first-sort universe; in fact we shall see that with such an ordering, FO(**Cnt**) is more powerful than FO(**Cnt**) on unordered first-sort structures (similarly to the case of FO, shown in Theorem 5.3, and MSO, shown in Corollary 7.13).

We can define a formula $\exists! ix\varphi(x, \cdots)$ saying that there are exactly i elements satisfying φ:

$$\exists! ix\varphi(x, \cdots) \;\equiv\; \exists ix\varphi(x, \cdots) \;\wedge\; \forall k\, \big((k > i) \rightarrow \neg\exists kx\varphi(x, \cdots)\big).$$

We can also compare cardinalities of two sets. Suppose we have two formulae $\varphi(x)$ and $\psi(x)$; to test if $|\varphi(\mathfrak{A})| > |\psi(\mathfrak{A})|$, one could write

$$\exists i\, \Big(\exists ix\varphi(x) \;\wedge\; \neg\exists ix\psi(x)\Big).$$

One can also write a formula for the majority predicate $\text{MAJ}(\varphi, \psi)$ testing if the set $\varphi(\mathfrak{A})$ contains at least half of the set $\psi(\mathfrak{A})$:

$$\exists i \exists j \Big((\exists! ix(\varphi(x) \wedge \psi(x))) \;\wedge\; (\exists! jx\psi(x)) \;\wedge\; (i + i \geq j) \Big).$$

Note that the definition of FO(**Cnt**) allows us to use formulae of the form $t_1(\vec{\imath})\ \{=, >, \geq\}\ t_2(\vec{\imath})$, where t_1 and t_2 are terms. For example, $(i + i \geq j)$ is $\exists k\ (k = i + i \;\wedge\; k \geq j)$.

We now present another way of adding counting power to FO that does not involve two-sorted structures. Suppose we want to state that $|\varphi(\mathfrak{A})|$ is even. We define a new *quantifier*, $\mathcal{Q}_{\text{EVEN}}$, that binds one variable, and write $\mathcal{Q}_{\text{EVEN}} x\ \varphi(x)$. In fact, more generally, for a formula with several free variables $\varphi(x, \vec{y})$, we can construct a new formula $\mathcal{Q}_{\text{EVEN}} x\ \varphi(x, \vec{y})$, with free variables $\vec{y}$. Its semantics is defined as follows. If $\vec{a}$ is the interpretation for $\vec{y}$, then

$$\mathfrak{A} \models \mathcal{Q}_{\text{EVEN}} x\ \varphi(x, \vec{a}) \;\Leftrightarrow\; |\{b \mid \mathfrak{A} \models \varphi(b, \vec{a})\}| \bmod 2 = 0.$$

Using the same approach, we can do cardinality comparisons. For example, let $\mathcal{Q}_H$ be a quantifier that binds two variables; then for two formulae $\varphi_1(x,\vec{y})$ and $\varphi_2(z,\vec{y})$, we have a new formula $\psi(\vec{y}) \equiv \mathcal{Q}_H x, z\ (\varphi_1(x,\vec{y}), \varphi_2(z,\vec{y}))$ such that

$$\mathfrak{A} \models \psi(\vec{a}) \;\Leftrightarrow\; |\{b \mid \mathfrak{A} \models \varphi_1(b,\vec{a})\}| \;=\; |\{b \mid \mathfrak{A} \models \varphi_2(b,\vec{a})\}|.$$

The quantifier $\mathcal{Q}_H$ is known as the Härtig, or equicardinality, quantifier. Another example is the Rescher quantifier $\mathcal{Q}_R$. The formation rule is the same as for the Härtig quantifier, and

$$\begin{aligned} &\mathfrak{A} \models \mathcal{Q}_R x, z\ (\varphi_1(x,\vec{a}), \varphi_2(z,\vec{a})) \\ \Leftrightarrow\ &|\{b \mid \mathfrak{A} \models \varphi_1(b,\vec{a})\}| \;>\; |\{b \mid \mathfrak{A} \models \varphi_2(b,\vec{a})\}|. \end{aligned}$$

What is common to these definitions? In all the cases, we construct sets of the form $\varphi(\mathfrak{A},\vec{a}) = \{b \in A \mid \mathfrak{A} \models \varphi(b,\vec{a})\} \subseteq A$, and then make some cardinality statements about those sets. This idea admits a nice generalization.

Definition 8.2 (Unary quantifiers). *Let σ_k^u be a vocabulary of k unary relation symbols $U_1,\ldots,U_k$, and let $\mathcal{K} \subseteq \mathrm{STRUCT}[\sigma_k^u]$ be a class of structures closed under isomorphisms. Then $\mathcal{Q}_\mathcal{K}$ is a* unary quantifier *and* $\mathrm{FO}(\mathcal{Q}_\mathcal{K})$ *extends the set of formulae of* FO *with the following additional rule:*

$$\begin{aligned} &\textit{if } \psi_1(x_1,\vec{y}_1),\ldots,\psi_k(x_k,\vec{y}_k) \textit{ are formulae,} \\ &\textit{then } \mathcal{Q}_\mathcal{K} x_1 \ldots x_k(\psi_1(x_1,\vec{y}_1),\ldots,\psi_k(x_k,\vec{y}_k)) \textit{ is a formula.} \end{aligned} \tag{8.1}$$

Here $\mathcal{Q}_\mathcal{K}$ binds x_i in the ith formula, for each $i = 1,\ldots,k$. A free occurrence of a variable y in $\psi_i(x_i,\vec{y}_i)$ remains free in this new formula unless $y = x_i$. The semantics of $\mathcal{Q}_\mathcal{K}$ is defined as follows:

$$\begin{aligned} &\mathfrak{A} \models Q_\mathcal{K} x_1 \ldots x_k(\psi_1(x_1,\vec{a}_1),\ldots,\psi_k(x_k,\vec{a}_k)) \\ \Leftrightarrow\ &\langle A, \psi_1(\mathfrak{A},\vec{a}_1),\ldots,\psi_k(\mathfrak{A},\vec{a}_k)\rangle \in \mathcal{K}. \end{aligned} \tag{8.2}$$

In this definition, $\vec{a}_i$ is a tuple of parameters that gives the interpretation for those free variables of $\psi_i(x_i,\vec{y}_i)$ which are not equal to x_i.

If $\mathbf{Q}$ is a set of unary quantifiers, then $\mathrm{FO}(\mathbf{Q})$ *is the extension of* FO *with the formation rule above for each $\mathcal{Q}_\mathcal{K} \in \mathbf{Q}$.*

The quantifier rank of formulae with unary quantifiers is defined by the additional rule:

$$\begin{aligned} &\mathsf{qr}(Q_\mathcal{K} x_1,\ldots,x_k(\psi_1(x_1,\vec{y}_1),\ldots,\psi_k(x_k,\vec{y}_k))) \\ &= \max\{\mathsf{qr}(\psi_i(x_i,\vec{y}_i)) \mid i \leq k\} + 1. \end{aligned} \tag{8.3}$$

The three examples seen earlier are all unary quantifiers: for $\mathcal{Q}_{\text{EVEN}}$, the class $\mathcal{K}$ consists of structures $\langle A, U\rangle$ such that $|U|$ is even; for $\mathcal{Q}_H$, it consists of structures $\langle A, U_1, U_2\rangle$ with $|U_1|=|U_2|$, and for $\mathcal{Q}_R$, it consists of structures $\langle A, U_1, U_2\rangle$ with $|U_1|>|U_2|$. Note that the usual quantifiers $\exists$ and $\forall$ are examples of unary quantifiers too: the classes of structures corresponding to them consist of $\langle A, U\rangle$ with $U \neq \emptyset$ and $U = A$, respectively.

We shall see that the two ways of adding counting power to a logic – by means of counting quantifiers, or unary quantifiers – are essentially equivalent in their expressiveness. Formulae with counting quantifiers tend to be easier to understand, but the logic becomes two-sorted. Unary quantifiers, on the other hand, let us keep the logic one-sorted, but then a new quantifier has to be introduced for each counting property we wish to express.

8.2 An Infinitary Counting Logic

The goal of this section is to introduce a very powerful counting logic: so powerful, in fact, that it can express arbitrary properties of cardinalities, even nonrecursive ones. Yet we shall see that this logic cannot address another limitation of FO, namely, expressing iterative computations. We shall later see another logic that expresses very powerful forms of iteration, and yet is unable to count. Both of these logics are based on the idea of expanding FO with infinitary connectives.

Definition 8.3 (Infinitary connectives and $\mathcal{L}_{\infty\omega}$). *The logic $\mathcal{L}_{\infty\omega}$ is defined as an extension of* FO *with* infinitary connectives $\bigvee$ *and* $\bigwedge$*: if φ_i's are formulae, for $i \in I$, where I is not necessarily finite, and the free variables of all the φ_i's are among $\vec{x}$, then*

$$\bigvee_{i\in I} \varphi_i \quad \text{and} \quad \bigwedge_{i\in I} \varphi_i$$

are formulae. Their free variables are those variables in $\vec{x}$ that occur freely in one of the φ's.

The semantics is defined as follows: $\mathfrak{A} \models \bigvee_{i\in I} \varphi_i(\vec{a})$ if for some $i \in I$, it is the case that $\mathfrak{A} \models \varphi_i(\vec{a})$, and $\mathfrak{A} \models \bigwedge_{i\in I} \varphi(\vec{a})$ if $\mathfrak{A} \models \varphi_i(\vec{a})$ for all $i \in I$.

This logic per se is too powerful to be of interest in finite model theory, in view of the following.

Proposition 8.4. *Let $\mathcal{C}$ be a class of finite structures closed under isomorphism. Then there is an $\mathcal{L}_{\infty\omega}$ sentence $\Phi_{\mathcal{C}}$ such that $\mathfrak{A} \in \mathcal{C}$ iff $\mathfrak{A} \models \Phi_{\mathcal{C}}$.*

Proof. Recall that by Lemma 3.4, for every finite $\mathfrak{B}$, there is a sentence $\Phi_{\mathfrak{B}}$ such that $\mathfrak{A} \models \Phi_{\mathfrak{B}}$ iff $\mathfrak{A} \cong \mathfrak{B}$. Hence we take $\Phi_{\mathcal{C}}$ to be

$$\bigvee_{\mathfrak{B}\in\mathcal{C}} \Phi_{\mathfrak{B}}.$$

Clearly, $\mathfrak{A} \models \Phi_{\mathcal{C}}$ iff $\mathfrak{A} \in \mathcal{C}$. ☐

However, we can make logics with infinitary connectives useful by putting some restrictions on them. Our goal now is to define a two-sorted counting logic $\mathcal{L}^*_{\infty\omega}(\mathbf{Cnt})$. We do it in two stages: first, we extend $\mathcal{L}_{\infty\omega}$ with some counting features, and second, we impose restrictions that make the logic suitable in the finite model theory context.

The structures for this logic are two-sorted, but the second sort is no longer interpreted as an initial segment of the natural numbers: now it is the whole set $\mathbb{N}$. Furthermore, there is a constant symbol for each $k \in \mathbb{N}$ (which we also denote by k). Hence, a structure is of the form

$$\langle \{a_1, \ldots, a_n\}, \mathbb{N}, (R_i^{\mathfrak{A}}), \ \{k\}_{k \in \mathbb{N}} \rangle, \tag{8.4}$$

where again $\langle \{a_1, \ldots, a_n\}, (R_i^{\mathfrak{A}}) \rangle$ is a finite σ-structure, and R_i's range over symbols in σ.

We now define $\mathcal{L}_{\infty\omega}(\mathbf{Cnt})$, an extremely powerful two-sorted logic, that extends infinitary logic $\mathcal{L}_{\infty\omega}$. Its structures are two-sorted structures (8.4), and the logic extends $\mathcal{L}_{\infty\omega}$ by the following rules:

- Each variable or constant of the second sort is a term of the second sort.
- If φ is a formula and $\vec{x}$ is a tuple of free first-sort variables in φ, then $\#\vec{x}.\varphi$ is a term of the second sort, and its free variables are those in φ except $\vec{x}$.

  The interpretation of this term is the number of tuples $\vec{a}$ over the finite first-sort universe that satisfy φ. That is, given a structure $\mathfrak{A}$ with the first-sort universe A, a formula $\varphi(\vec{x}, \vec{y}, \vec{\imath})$ and the interpretations $\vec{b}$ and $\vec{\imath}_0$ for $\vec{y}$ and $\vec{\imath}$, respectively, the value of the term $\#\vec{x}.\varphi(\vec{x}, \vec{b}, \vec{\imath}_0)$ is

$$|\{\vec{a} \in A^{|\vec{x}|} \ \mid \ \mathfrak{A} \models \varphi(\vec{a}, \vec{b}, \vec{\imath}_0)\}| \,.$$

- Counting quantifiers $\exists ix\varphi$, with the same semantics as before, except that i could be an arbitrary natural number.

The logic $\mathcal{L}_{\infty\omega}(\mathbf{Cnt})$ is enormously powerful: it can define not only every property of finite models (since it contains $\mathcal{L}_{\infty\omega}$), but also *every* predicate or function on $\mathbb{N}$. That is, $P \subseteq \mathbb{N}^k$ is definable by

$$\varphi_P(i_1, \ldots, i_k) \quad = \quad \bigvee_{(n_1, \ldots, n_k) \in P} \quad \big((i_1 = n_1) \wedge \ldots \wedge (i_k = n_k)\big). \tag{8.5}$$

Note that the definition is also redundant: for example, $\exists ix \ \varphi$ can be replaced by $\#x.\varphi \geq i$. However, we need counting quantifiers separately, as will become clear soon.

Next, we restrict the logic by defining the *rank* of a formula, $\mathsf{rk}(\varphi)$. Its definition is similar to that of quantifier rank, but there is one important difference. In a two-sorted logic, we may have quantification over two different universes. In the definition of the rank, we disregard quantification over $\mathbb{N}$. Thus, $\mathsf{rk}(\varphi)$ and $\mathsf{rk}(t)$, where t is a term, are defined inductively as follows:

- $\mathsf{rk}(t) = 0$ if t is a variable, or a term k for $k \in \mathbb{N}$.
- $\mathsf{rk}(\varphi) = 0$ if φ is an atomic formula of vocabulary σ (i.e., an atomic first-sort formula).
- $\mathsf{rk}(t_1 = t_2) = \max\{\mathsf{rk}(t_1), \mathsf{rk}(t_2)\}$, where t_1 and t_2 are terms.
- $\mathsf{rk}(\neg\varphi) = \mathsf{rk}(\varphi)$.
- $\mathsf{rk}(\#\vec{x}.\varphi) = \mathsf{rk}(\varphi) + |\vec{x}|$.
- $\mathsf{rk}(\bigvee \varphi_j) = \mathsf{rk}(\bigwedge \varphi_j) = \sup_j \mathsf{rk}(\varphi_j)$.
- $\mathsf{rk}(\forall x\ \varphi) = \mathsf{rk}(\exists x\ \varphi) = \mathsf{rk}(\exists ix\ \varphi) = \mathsf{rk}(\varphi) + 1$.
- $\mathsf{rk}(\forall i\ \varphi) = \mathsf{rk}(\exists i\ \varphi) = \mathsf{rk}(\varphi)$.

Note that if φ is an FO formula, then $\mathsf{rk}(\varphi) = \mathsf{qr}(\varphi)$.

Definition 8.5. *$\mathcal{L}^*_{\infty\omega}(\mathbf{Cnt})$ is defined as the restriction of $\mathcal{L}_{\infty\omega}(\mathbf{Cnt})$ to formulae and terms that have finite rank.* □

This logic is clearly closed under the Boolean connectives and both first- and second-sort quantification. It is *not* closed under infinitary connectives: for example, if Φ_i, $i > 0$, are $\mathcal{L}^*_{\infty\omega}(\mathbf{Cnt})$ sentences such that $\mathsf{rk}(\Phi_i) = i$, then $\bigvee_i \Phi_i$ is not an $\mathcal{L}^*_{\infty\omega}(\mathbf{Cnt})$ sentence.

Note also that (8.5) implies that every subset of $\mathbb{N}^k$, $k > 0$, is definable by an $\mathcal{L}^*_{\infty\omega}(\mathbf{Cnt})$ formula of rank 0. Thus, we assume that $+, \cdot, -, \leq$, and in fact *every* predicate on natural numbers is available. To give an example, we can express properties like: there is a node in the graph whose in-degree i and out-degree j satisfy $p_i^2 > p_j$ where p_i stands for the ith prime. This is done by $\exists x \exists i \exists j\ (i = \#y.E(y,x)) \wedge (j = \#y.E(x,y)) \wedge P(i,j)$, where P is the predicate on $\mathbb{N}$ for the property $p_i^2 > p_j$.

Known expansions of FO with counting properties are contained in $\mathcal{L}^*_{\infty\omega}(\mathbf{Cnt})$.

Proposition 8.6. *For every* FO, FO($\mathbf{Cnt}$), *or* FO($\mathbf{Q}$) *formula, where $\mathbf{Q}$ is a collection of unary quantifiers, there exists an equivalent $\mathcal{L}^*_{\infty\omega}(\mathbf{Cnt})$ formula of the same rank.*

Proof. The proof is trivial for FO and FO($\mathbf{Cnt}$). For FO($\mathbf{Q}$), assume we have a formula

$$\psi(\vec{y}_1, \ldots, \vec{y}_k) \ \equiv\ Q_{\mathcal{K}} x_1 \ldots x_k.(\psi_1(x_1, \vec{y}_1), \ldots, \psi_k(x_k, \vec{y}_k)), \qquad (8.6)$$

where $\mathcal{K}$ is a class of σ_k^u-structures $\mathfrak{A} = \langle A, U_1, \ldots, U_k \rangle$ closed under isomorphism. Let Π be the set of all 2^k mapping $\pi : \{1, \ldots, k\} \to \{0,1\}$, and for a structure $\mathfrak{A} \in \mathcal{K}$, let

$$\pi(\mathfrak{A}) \ =\ \Big| \bigcap_{i:\pi(i)=1} U_i^{\mathfrak{A}} \ \cap \bigcap_{j:\pi(j)=0} (A - U_j^{\mathfrak{A}}) \Big|.$$

With each structure $\mathfrak{A}$, we then associate a tuple $\Pi(\mathfrak{A}) = (\pi(\mathfrak{A}))_{\pi \in \Pi}$, with π's ordered lexicographically. Since $\mathcal{K}$ is a class of unary structures closed under isomorphism, $\mathfrak{A} \in \mathcal{K}$ and $\Pi(\mathfrak{A}) = \Pi(\mathfrak{B})$ imply $\mathfrak{B} \in \mathcal{K}$.

This provides a translation of (8.6) into $\mathcal{L}^*_{\infty\omega}(\mathbf{Cnt})$ as follows. Let $P_{\mathcal{K}}(n_0, \ldots, n_{2^k-1})$ be the predicate on $\mathbb{N}$ that holds iff $(n_0, \ldots, n_{2^k-1})$ is of the form $\Pi(\mathfrak{A})$ for some $\mathfrak{A} \in \mathcal{K}$. Then (8.6) translates into

$$P_{\mathcal{K}}(\#x.\psi_{\pi_0}(x, \vec{y}_1, \ldots, \vec{y}_k), \ldots, \#x.\psi_{\pi_{2^k-1}}(x, \vec{y}_1, \ldots, \vec{y}_k)), \tag{8.7}$$

where $\pi_0, \ldots, \pi_{2^k-1}$ is the enumeration of Π in the lexicographic ordering, and

$$\psi_\pi(x, \vec{y}_1, \ldots, \vec{y}_k) \;=\; \bigwedge_{i:\pi(i)=1} \psi_i(x, \vec{y}_i) \;\wedge\; \bigwedge_{j:\pi(j)=0} \neg\psi_i(x, \vec{y}_i).$$

Thus, if $\vec{b}_1, \ldots, \vec{b}_k$ interpret $\vec{y}_1, \ldots, \vec{y}_k$, respectively, in a structure $\mathfrak{B}$, then the value of $\#x.\psi_\pi(x, \vec{b}_1, \ldots, \vec{b}_k)$ in $\mathfrak{B}$ is precisely

$$\pi\big(\langle B, \psi_1(\mathfrak{B}, \vec{b}_1), \ldots, \psi_k(\mathfrak{B}, \vec{b}_k)\rangle\big).$$

Therefore, (8.7) holds for $\vec{b}_1, \ldots, \vec{b}_k$ in $\mathfrak{B}$ iff the σ^u_k-structure $\langle B, \psi_1(\mathfrak{B}, \vec{b}_1), \ldots, \psi_k(\mathfrak{B}, \vec{b}_k)\rangle$ is in $\mathcal{K}$. This proves the equivalence of (8.6) and (8.7). Finally, since $P_{\mathcal{K}}$ is a numerical predicate, it has rank 0, and hence the rank of (8.7) is $\max\{\mathsf{rk}(\psi_1), \ldots, \mathsf{rk}(\psi_k)\} + 1 = \mathsf{rk}(\psi)$, which proves the proposition. □

In general, $\mathcal{L}^*_{\infty\omega}(\mathbf{Cnt})$ can be viewed as an extremely powerful counting logic: we can define arbitrary cardinalities of sets of tuples over a structure, and on those, we can use arbitrary numerical predicates. Compared to $\mathcal{L}^*_{\infty\omega}(\mathbf{Cnt})$, a logic such as FO(**Cnt**) restricts us in what sort of cardinalities we can define (only those of sets given by formulae in one free variable), and what operations we can use on those cardinalities (those definable with addition and multiplication).

We now introduce what seems to be a drastic simplification of $\mathcal{L}^*_{\infty\omega}(\mathbf{Cnt})$.

Definition 8.7. *The logic $\mathcal{L}^\circ_{\infty\omega}(\mathbf{Cnt})$ is defined as $\mathcal{L}^*_{\infty\omega}(\mathbf{Cnt})$ where counting terms $\#\vec{x}.\varphi$ and quantification over $\mathbb{N}$ are not allowed.* □

On the surface, $\mathcal{L}^\circ_{\infty\omega}(\mathbf{Cnt})$ is a lot simpler than $\mathcal{L}^*_{\infty\omega}(\mathbf{Cnt})$, mainly because counting terms for vectors, $\#\vec{x}.\varphi$, are very convenient for defining complex counting properties. But it turns out that the power of $\mathcal{L}^\circ_{\infty\omega}(\mathbf{Cnt})$ and $\mathcal{L}^*_{\infty\omega}(\mathbf{Cnt})$ is identical.

Proposition 8.8. *There is a translation $\varphi \to \varphi^\circ$ of $\mathcal{L}^*_{\infty\omega}(\mathbf{Cnt})$ formulae into $\mathcal{L}^\circ_{\infty\omega}(\mathbf{Cnt})$ formulae such that φ and φ° are equivalent and $\mathsf{rk}(\varphi) = \mathsf{rk}(\varphi^\circ)$.*

Proof. It is easy to eliminate quantifiers over $\mathbb{N}$ without increasing the rank: $\exists i\ \varphi(i,\cdots)$ and $\forall i\ \varphi(i,\cdots)$ are equivalent to

$$\bigvee_{k\in\mathbb{N}} \varphi(k,\cdots) \quad \text{and} \quad \bigwedge_{k\in\mathbb{N}} \varphi(k,\cdots),$$

respectively. Thus, in the formulae below, we shall be using such quantifiers, assuming that they are eliminated in the last step of the translation from $\mathcal{L}^{*}_{\infty\omega}(\mathbf{Cnt})$ to $\mathcal{L}^{\circ}_{\infty\omega}(\mathbf{Cnt})$.

To eliminate counting terms, assume without loss of generality that every occurrence of $\#\vec{x}.\varphi$ is of the form $\#\vec{x}.\varphi = \#\vec{y}.\psi$ or $\#\vec{x}.\varphi = i$, where i is a variable or a constant (if $\#\vec{x}.\varphi$ occurs inside an arithmetic predicate P, we replace P by its explicit definition, using infinitary connectives). Since $\#\vec{x}.\varphi = \#\vec{y}.\psi$ is equivalent to $\exists i\ (\#\vec{x}.\varphi = i) \wedge (\#\vec{y}.\psi = i)$, whose rank is the same as the rank of $\#\vec{x}.\varphi = \#\vec{y}.\psi$, and $\#\vec{x}.\varphi = k$, for a constant k, is equivalent to $\exists i\ (\#\vec{x}.\varphi = i \wedge i = k)$, we may assume that all occurrences of $\#$-terms are of the form $\#\vec{x}.\varphi = i$, where i is a second-sort variable.

The proof is now by induction on the formula. The only nontrivial case is $\psi(\vec{y},\vec{\jmath}) \equiv (\#\vec{x}.\varphi(\vec{x},\vec{y},\vec{\jmath}) = i)$. Throughout this proof, we assume that i is in $\vec{\jmath}$.

By the hypothesis, there exists an $\mathcal{L}^{\circ}_{\infty\omega}(\mathbf{Cnt})$ formula φ° which is equivalent to φ and has the same rank. We must now produce an $\mathcal{L}^{\circ}_{\infty\omega}(\mathbf{Cnt})$ formula ψ° equivalent to ψ such that $\mathsf{rk}(\psi^{\circ}) = \mathsf{rk}(\varphi) + |\,\vec{x}\,|$. The existence of such a formula will follow from the lemma below.

Lemma 8.9. *Let $\varphi(\vec{x},\vec{y},\vec{\jmath})$ be an $\mathcal{L}^{\circ}_{\infty\omega}(\mathbf{Cnt})$ formula. Then there exists an $\mathcal{L}^{\circ}_{\infty\omega}(\mathbf{Cnt})$ formula $\gamma(\vec{y},\vec{\jmath})$ of rank $\mathsf{rk}(\varphi) + |\vec{x}|$ such that γ is equivalent to $\#\vec{x}.\varphi = i$.*

Proof of the lemma is by induction on $|\,\vec{x}\,|$. If $\vec{x}$ has a single component x, $\gamma(\vec{y},\vec{\jmath})$ is defined as

$$\exists l\ \left((l = i) \wedge (\exists! l x\ \varphi(x,\vec{y},\vec{\jmath}))\right),$$

which has rank $\mathsf{rk}(\varphi) + 1$. The quantifier $\exists l$ denotes an infinite disjunction, as explained earlier.

We next assume that $\vec{x} = \vec{z}x_0$. By the hypothesis, there is an $\mathcal{L}^{\circ}_{\infty\omega}(\mathbf{Cnt})$ formula $\alpha(x_0,\vec{y},\vec{\jmath},l)$ equivalent to $(l = \#\vec{z}.\varphi(\vec{z},x_0,\vec{y},\vec{\jmath}))$ such that $\mathsf{rk}(\alpha) = \mathsf{rk}(\varphi) + |\,\vec{z}\,|$. We define

$$\beta(\vec{y},\vec{\jmath},k,l) \ \equiv\ \exists! k x_0\ \alpha(x_0,\vec{y},\vec{\jmath},l).$$

Then $\mathsf{rk}(\beta) = \mathsf{rk}(\alpha) + 1 = \mathsf{rk}(\varphi) + |\,\vec{x}\,|$. The formula $\beta(\vec{y},\vec{\jmath},k,l)$ holds iff there exist exactly k elements x_0 such that the number of vectors $\vec{x}$ with x_0 in the last position that satisfy $\varphi(\vec{x},\cdots)$ is precisely l. Note that if $\beta(\vec{y},\vec{\jmath},k,l)$ and $\beta(\vec{y},\vec{\jmath},k',l)$ hold, then k' must equal k.

Thus, to check if $\#\vec{x}.\varphi = i$, one must check if

$$\sum_{\beta(\cdots,k,l) \text{ holds}} (k \cdot l) = i.$$

This is done as follows. Let $\gamma_p(\vec{y}, \vec{j})$ be defined as:

$$\exists i_1 \dots i_p \exists j_1 \dots j_p \left(\begin{array}{l} \bigwedge_{s=1}^{p} \beta(\vec{y}, \vec{j}, i_s, j_s) \\ \wedge \forall i, j \; \beta(\vec{y}, \vec{j}, i, j) \rightarrow \bigvee_{s=1}^{p} (i = i_s \wedge j = j_s) \\ \wedge \bigwedge_{s \neq s'} (\neg(i_s = i_{s'}) \vee \neg(j_s = j_{s'})) \\ \wedge\, i_1 \cdot j_1 + \ldots + i_p \cdot j_p = i \end{array} \right)$$

That is, γ_p says that there are precisely p pairs (i_s, j_s) that satisfy $\beta(\vec{y}, \vec{j}, k, l)$, and $\sum_{s=1}^{p} i_s \cdot j_s = i$. When $p = 0$, we define $\gamma_p(\vec{y}, \vec{j})$ as $(i = 0) \wedge \forall i', j' \; (\neg\beta(\vec{y}, \vec{j}, i', j'))$. We can see that $\mathsf{rk}(\gamma_p) = \mathsf{rk}(\beta)$. We finally define

$$\gamma(\vec{y}, \vec{j}) \;\equiv\; \bigvee_{p \in \mathbb{N}} \gamma_p(\vec{y}, \vec{j}).$$

It follows that γ is an $\mathcal{L}^{\circ}_{\infty\omega}(\mathbf{Cnt})$ formula of rank that is equal to $\mathsf{rk}(\beta)$, and hence to $\mathsf{rk}(\varphi) + |\,\vec{x}\,|$, and that γ is equivalent to $\#\vec{x}.\varphi = i$. This completes the proof of the lemma and the proposition. □

We next consider $\mathcal{L}^{*}_{\infty\omega}(\mathbf{Cnt})+<$; that is, $\mathcal{L}^{*}_{\infty\omega}(\mathbf{Cnt})$ over ordered structures. We shall see in the next section that, as for FO, there is a separation

$$\mathcal{L}^{*}_{\infty\omega}(\mathbf{Cnt}) \;\subsetneq\; (\mathcal{L}^{*}_{\infty\omega}(\mathbf{Cnt})+<)_{\text{inv}}.$$

As the first step, we show that $\mathcal{L}^{*}_{\infty\omega}(\mathbf{Cnt})+<$ defines every property of finite structures. Intuitively, with $<$, one can say that a given element of A is the first, second, etc., element of A. Then the unlimited counting power allows us to code finite structures with numbers.

Proposition 8.10. *Every property of finite ordered structures is definable in* $\mathcal{L}^{*}_{\infty\omega}(\mathbf{Cnt})$.

Proof. We show this for sentences in the language of graphs. Let $\mathcal{C}$ be a class of ordered graphs. We assume without loss of generality that the set of nodes of each such graph is a set of the form $\{0, \ldots, n\}$. Then the membership in $\mathcal{C}$ is tested by the following $\mathcal{L}^{*}_{\infty\omega}(\mathbf{Cnt})$ sentence of rank 3:

$$\bigvee_{G \in \mathcal{C}} \forall x \forall y \left(E(x,y) \leftrightarrow \bigvee_{(k,l) \in E^G} \left(\begin{array}{l} k = \#z.(z < x) \\ \wedge \;\; l = \#z.(z < y) \end{array} \right) \right),$$

where E^G stands for the set of edges of G. □

We finish this section by presenting a one-sorted version of $\mathcal{L}^*_{\infty\omega}(\mathbf{Cnt})$ that has the same expressiveness. This logic is obtained by adding infinitary connectives and unary quantifiers to FO.

Let $\mathfrak{Q}_{\mathsf{All}}$ be the collection of all unary quantifiers; that is, all quantifiers $\mathcal{Q}_{\mathcal{K}}$ where $\mathcal{K}$ ranges over all collections of unary structures closed under isomorphism. We define a logic $\mathcal{L}_{\infty\omega}(\mathfrak{Q}_{\mathsf{All}})$ by extending $\mathcal{L}_{\infty\omega}$ with the formation rules (8.1) for each $\mathcal{Q}_{\mathcal{K}} \in \mathfrak{Q}_{\mathsf{All}}$, with the semantics given by (8.2), and quantifier rank defined as in (8.3). We then define $\mathcal{L}^*_{\infty\omega}(\mathfrak{Q}_{\mathsf{All}})$ as the restriction of $\mathcal{L}_{\infty\omega}(\mathfrak{Q}_{\mathsf{All}})$ to formulae of finite quantifier rank. This logic turns out to express the same sentences as $\mathcal{L}^*_{\infty\omega}(\mathbf{Cnt})$. The proof of the proposition below is left as an exercise for the reader.

Proposition 8.11. *For every $\mathcal{L}^*_{\infty\omega}(\mathbf{Cnt})$ formula $\varphi(\vec{x})$ without free second-sort variables, there is an equivalent $\mathcal{L}^*_{\infty\omega}(\mathfrak{Q}_{\mathsf{All}})$ formula $\psi(\vec{x})$ such that $\mathsf{rk}(\varphi) = \mathsf{qr}(\psi)$, and conversely, for every $\mathcal{L}^*_{\infty\omega}(\mathfrak{Q}_{\mathsf{All}})$ formula $\psi(\vec{x})$, there is an equivalent $\mathcal{L}^*_{\infty\omega}(\mathbf{Cnt})$ formula $\varphi(\vec{x})$ with $\mathsf{rk}(\varphi) = \mathsf{qr}(\psi)$.* □

8.3 Games for $\mathcal{L}^*_{\infty\omega}(\mathbf{Cnt})$

We know that the expressive power of FO can be characterized via Ehrenfeucht-Fraïssé games. Is there a similar game characterization for $\mathcal{L}^*_{\infty\omega}(\mathbf{Cnt})$? We give a positive answer to this question, by showing that bijective games, introduced in Sect. 4.5, capture the expressiveness of $\mathcal{L}^*_{\infty\omega}(\mathbf{Cnt})$. We first review the definition of the game.

Definition 8.12 (Bijective games). *A* bijective Ehrenfeucht-Fraïssé game *is played by two players, the spoiler and the duplicator, on two structures $\mathfrak{A}, \mathfrak{B} \in \text{STRUCT}[\sigma]$. If $\mid A \mid \neq \mid B \mid$, the spoiler wins the game. If $\mid A \mid = \mid B \mid$, in each round $i = 1, \ldots, n$, the duplicator selects a bijection $f_i : A \to B$, and the spoiler selects a point $a_i \in A$. The duplicator responds by $b_i = f(a_i) \in B$. The duplicator wins the n-round game if the relation $\{(a_i, b_i) \mid 1 \leq i \leq n\}$ is a partial isomorphism between $\mathfrak{A}$ and $\mathfrak{B}$. If the duplicator has a winning strategy in the n-round bijective game on $\mathfrak{A}$ and $\mathfrak{B}$, we write $\mathfrak{A} \equiv^{bij}_n \mathfrak{B}$.*

Note that it is *harder* for the duplicator to win the bijective game. First, if $|A| \neq |B|$, the duplicator immediately loses the game. Even if $|A| = |B|$, in each round the duplicator must figure out what his response to each possible move by the spoiler is, before the move is made, and there must be a one-to-one correspondence between the spoiler's moves and the duplicator's responses. In particular, any strategy where the same element $b \in B$ could be used as a response to several moves by the spoiler is disallowed.

Theorem 8.13. *Given two structures $\mathfrak{A}, \mathfrak{B} \in \text{STRUCT}[\sigma]$, and $k \geq 0$, the following are equivalent:*

1. $\mathfrak{A} \equiv^{bij}_k \mathfrak{B}$*;*

2. $\mathfrak{A}$ and $\mathfrak{B}$ agree on all $\mathcal{L}^*_{\infty\omega}(\mathbf{Cnt})$ sentences of rank k.

Proof. Both implications $1 \to 2$ and $2 \to 1$ are proved by induction on k. We start with the easier implication $1 \to 2$. By Proposition 8.8, assume that there is no quantification over the numerical domain, and that all quantifiers are of the form $\exists ix$. For the base case $k = 0$, the proof is the same as in the case of Ehrenfeucht-Fraïssé games.

We now assume that the implication holds for k, and we prove it for $k+1$. Suppose $\mathfrak{A} \equiv^{bij}_{k+1} \mathfrak{B}$. First consider a sentence of the form $\Phi \equiv \exists nx\varphi(x)$ for a constant $n \in \mathbb{N}$. Suppose $\mathfrak{A} \models \Phi$, and let $c_1, \ldots, c_n$ be distinct elements of $\mathfrak{A}$ such that $\mathfrak{A} \models \varphi(c_i), i = 1, \ldots, n$. Since $\mathfrak{A} \equiv^{bij}_{k+1} \mathfrak{B}$, there is a bijection $f : A \to B$ such that $(\mathfrak{A}, a) \equiv^{bij}_k (\mathfrak{B}, f(a))$ for all $a \in A$; in particular, $(\mathfrak{A}, c_i) \equiv^{bij}_k (\mathfrak{B}, f(c_i))$ for all $i \leq n$. By the hypothesis, $(\mathfrak{A}, c_i)$ and $(\mathfrak{B}, f(c_i))$ agree on sentences of rank k; hence $\mathfrak{A} \models \varphi(c_i)$ implies $\mathfrak{B} \models \varphi(f(c_i))$. Since f is a bijection, all $f(c_i)$'s are distinct, and thus $\mathfrak{B} \models \exists nx\varphi(x)$. The converse, that $\mathfrak{B} \models \Phi$ implies $\mathfrak{A} \models \Phi$, is proved in exactly the same way, using the bijection f^{-1}.

Since every sentence of rank $k+1$ can be obtained from sentences of the form $\exists nx\varphi(x)$ by using the Boolean and infinitary connectives, we see that $\mathfrak{A} \models \Phi \Leftrightarrow \mathfrak{B} \models \Phi$ for any rank $k+1$ sentence Φ.

For the other direction, we use a proof similar to the proof of the Ehrenfeucht-Fraïssé theorem given in Exercise 3.11. We want to define explicitly formulae specifying rank-k types in $\mathcal{L}^*_{\infty\omega}(\mathbf{Cnt})$. The number of types can be infinite, but this is not a problem since we can use infinitary connectives, and rank-k types will be given by formulae of rank k.

We let $\varphi_i^{0,m}(\vec{x})$ be an enumeration of all the formulae that define distinct atomic types of $\vec{x}$ with $|\vec{x}| = m$; that is, all consistent conjunctions of the form

$$\alpha_1(\vec{x}) \wedge \ldots \wedge \alpha_M(\vec{x}),$$

where $\alpha_i(\vec{x})$ enumerate all (finitely many) atomic and negated atomic formulae in $\vec{x}$.

Next, inductively, let $\{\varphi_i^{k+1,m}(\vec{x}) \mid i \in \mathbb{N}\}$ be an enumeration of all the formulae of the form

$$\Big(\exists! l_1\, y\, \varphi_{i_1}^{k,m+1}(\vec{x}, y) \wedge \ldots \wedge \exists! l_p\, y\, \varphi_{i_p}^{k,m+1}(\vec{x}, y)\Big) \wedge \Big(\forall y \bigvee_{j=1}^{p} \varphi_{i_j}^{k,m+1}(\vec{x}, y)\Big), \quad (8.8)$$

as p ranges over $\mathbb{N}$ and $(l_1, \ldots, l_p)$ ranges over p-tuples of positive integers. Intuitively, each $\varphi_{i_j}^{k,m+1}(\vec{x}, y)$ defines the rank-k $m+1$-type of a tuple $(\vec{x}, y)$. Hence rank-$k+1$ types of the form (8.8) say that a given $\vec{x}$ can be extended to p different rank-k types in such a way that for each i_j, there are precisely l_j elements y such that $\varphi_{i_j}^{k,m+1}(\vec{x}, y)$ defines the i_jth rank-k of the tuple $(\vec{x}, y)$. Note that if the formula (8.8) is true in $(\mathfrak{A}, \vec{a})$, then $|A| = l_1 + \ldots + l_p$.

It follows immediately from the definition of formulae $\varphi_i^{k,m}$ that for every $\mathfrak{A}, \vec{a} \in A^m$, and every $k \geq 0$, there is exactly one $\varphi_i^{k,m}$ such that $\mathfrak{A} \models \varphi_i^{k,m}(\vec{a})$.

Next, we prove the following lemma by induction on k.

Lemma 8.14. *For every m, every two structures $\mathfrak{A}, \mathfrak{B}$, and every $\vec{a} \in A^m, \vec{b} \in B^m$, suppose there is a formula $\varphi_i^{k,m}(\vec{x})$ such that $\mathfrak{A} \models \varphi_i^{k,m}(\vec{a})$ and $\mathfrak{B} \models \varphi_i^{k,m}(\vec{b})$. Then $(\mathfrak{A}, \vec{a}) \equiv_k^{bij} (\mathfrak{B}, \vec{b})$.*

Proof of the lemma. The case $k = 0$ is the same as in the proof of the Ehrenfeucht-Fraïssé theorem. For the induction step, assume that the statement holds for k, and let $\varphi_i^{k+1,m}(\vec{x})$ be given by (8.8). If $\mathfrak{A} \models \varphi_i^{k+1,m}(\vec{a})$ and $\mathfrak{B} \models \varphi_i^{k+1,m}(\vec{b})$, then both A and B have exactly $l_1 + \ldots + l_p$ elements. Furthermore, for each $j \leq p$, let $A_j = \{a \in A \mid \mathfrak{A} \models \varphi_{i_j}^{k,m+1}(\vec{a}a)\}$ and $B_j = \{b \in B \mid \mathfrak{B} \models \varphi_{i_j}^{k,m+1}(\vec{b}b)\}$. Then $\mid A_j \mid = \mid B_j \mid = l_j$, and hence there exists a bijection $f : A \to B$ that maps each A_j to B_j. For any $a \in A$, if j is such that $\mathfrak{A} \models \varphi_{i_j}^{k,m+1}(\vec{a}a)$, then $\mathfrak{B} \models \varphi_{i_j}^{k,m+1}(\vec{b}f(a))$, and hence by the induction hypothesis, $(\mathfrak{A}, \vec{a}a) \equiv_k^{bij} (\mathfrak{B}, \vec{b}f(a))$. Thus, the bijection f proves that $(\mathfrak{A}, \vec{a}) \equiv_{k+1}^{bij} (\mathfrak{B}, \vec{b})$.

The implication $2 \to 1$ of Theorem 8.13 is now a special case of Lemma 8.14, since $\mathsf{rk}(\varphi_i^{k,m}) = k$. □

8.4 Counting and Locality

Theorem 8.13 and Corollary 4.21 stating that $(\mathfrak{A}, \vec{a}) \leftrightarrows_{(3^k-1)/2} (\mathfrak{B}, \vec{b})$ implies $(\mathfrak{A}, \vec{a}) \equiv_k^{bij} (\mathfrak{B}, \vec{b})$, immediately give us the following result.

Theorem 8.15. *Every $\mathcal{L}_{\infty\omega}^*(\mathbf{Cnt})$ formula $\varphi(\vec{x})$ without free second-sort variables is Hanf-local (and hence Gaifman-local, and has the BNDP).* □

Thus, despite its enormous counting power, $\mathcal{L}_{\infty\omega}^*(\mathbf{Cnt})$ remains local, and cannot express properties such as graph connectivity. Combining Theorem 8.15 and Proposition 8.6, we obtain the following.

Corollary 8.16. *If $\varphi(\vec{x})$ is an* FO(**Cnt**) *formula without free second-sort variables, or an* FO($\mathfrak{Q}$) *formula, where $\mathfrak{Q}$ is an arbitrary collection of unary quantifiers, then $\varphi(\vec{x})$ is Hanf-local (and hence Gaifman-local, and has the BNDP).*

Furthermore, we obtain the separation

$$\mathcal{L}_{\infty\omega}^*(\mathbf{Cnt}) \subsetneq (\mathcal{L}_{\infty\omega}^*(\mathbf{Cnt}) + <)_{\text{inv}}, \tag{8.9}$$

since $(\mathcal{L}_{\infty\omega}^*(\mathbf{Cnt}) + <)$ expresses every property of ordered structures (including nonlocal ones, such as graph connectivity), by Proposition 8.10.

Theorem 8.15 says nothing about formulae that may have free numerical variables. Next, we show how to extend the notions of Hanf- and Gaifman-locality to such formulae.

Definition 8.17. *An $\mathcal{L}^*_{\infty\omega}(\mathbf{Cnt})$ formula $\varphi(\vec{x},\vec{\imath})$ is* Hanf-local *if there exists $d \geq 0$ such that for all $\vec{\imath}_0 \in \mathbb{N}^{|\vec{\imath}|}$, any two structures $\mathfrak{A},\mathfrak{B}$, and $\vec{a} \in A^{|\vec{x}|}$, $\vec{b} \in B^{|\vec{x}|}$,*

$$(\mathfrak{A},\vec{a}) \leftrightarrows_d (\mathfrak{B},\vec{b}) \quad \textit{implies} \quad \Big(\mathfrak{A} \models \varphi(\vec{a},\vec{\imath}_0) \Leftrightarrow \mathfrak{B} \models \varphi(\vec{b},\vec{\imath}_0)\Big).$$

Furthermore, $\varphi(\vec{x},\vec{\imath})$ is Gaifman-local *if there is $d \geq 0$, such that for all $\vec{\imath}_0 \in \mathbb{N}^{|\vec{\imath}|}$, every structure $\mathfrak{A}$, and $\vec{a}_1,\vec{a}_2 \in A^{|\vec{x}|}$,*

$$\vec{a}_1 \approx^{\mathfrak{A}}_d \vec{a}_2 \quad \textit{implies} \quad \mathfrak{A} \models \varphi(\vec{a}_1,\vec{\imath}_0) \leftrightarrow \varphi(\vec{a}_2,\vec{\imath}_0).$$

The locality rank $\mathsf{lr}(\cdot)$ and the Hanf-locality rank $\mathsf{hlr}(\cdot)$ are defined as before: these are the smallest d that witnesses Gaifman-locality (Hanf-locality, respectively) of a formula.

In other words, the formula must be Hanf-local or Gaifman-local for any instantiation of its free second-sort variables, with the locality rank being uniformly bounded for all such instantiations.

A simple extension of Theorem 4.11 shows:

Proposition 8.18. *If an $\mathcal{L}^*_{\infty\omega}(\mathbf{Cnt})$ formula $\varphi(\vec{x},\vec{\imath})$ is Hanf-local, then it is Gaifman-local.* □

Furthermore, we can show Hanf-locality of all $\mathcal{L}^*_{\infty\omega}(\mathbf{Cnt})$ formulae (not just those without free numerical variables) by using essentially the same argument as in Theorem 4.12.

Theorem 8.19. *Every $\mathcal{L}^*_{\infty\omega}(\mathbf{Cnt})$ formula $\varphi(\vec{x},\vec{\imath})$ is Hanf-local, and hence Gaifman-local. Furthermore, $\mathsf{hlr}(\varphi) \leq (3^k-1)/2$, and $\mathsf{lr}(\varphi) \leq (3^{k+1}-1)/2$, where $k = \mathsf{rk}(\varphi)$.*

Proof. We give the proof for Hanf-locality; it is by induction on the structure of the formulae. For atomic formulae and Boolean connectives, it is the same as the proof of Theorem 4.12. For infinitary connectives, the argument is the same as for $\wedge$ and $\vee$. By Proposition 8.8, the only remaining case is that of counting quantifiers: $\varphi(\vec{x},\vec{\imath}) \equiv \exists j y\ \psi(y,\vec{x},\vec{\imath})$. We assume j is in $\vec{\imath}$. Let $\mathsf{rk}(\psi) = k$, so that $\mathsf{rk}(\varphi) = k+1$. Let $d = \mathsf{hlr}(\psi)$. It suffices to show that $\mathsf{hlr}(\varphi) \leq 3d+1$.

Fix an interpretation $\vec{\imath}_0$ for $\vec{\imath}$ (and j_0 for j). Assume $(\mathfrak{A},\vec{a}) \leftrightarrows_{3d+1} (\mathfrak{B},\vec{b})$. By Corollary 4.10, there is a bijection $f : A \to B$ such that $(\mathfrak{A},\vec{a}c) \leftrightarrows_d (\mathfrak{B},\vec{b}f(c))$ for every $c \in A$. Assume $\mathfrak{A} \models \varphi(\vec{a},\vec{\imath})$; then we can find $c_1,\ldots,c_{j_0}$ such that $\mathfrak{A} \models \psi(c_l,\vec{a},\vec{\imath})$, $l = 1,\ldots,j_0$. Since $\mathsf{hlr}(\psi) = d$, by the hypothesis, $(\mathfrak{A},\vec{a}c_l) \leftrightarrows_d (\mathfrak{B},\vec{b}f(c_l))$ implies $\mathfrak{B} \models \psi(f(c_l),\vec{b},\vec{\imath})$, $l = 1,\ldots,j_0$. Thus, $\mathfrak{B} \models \varphi(\vec{b},\vec{\imath})$, since f is a bijection. The converse, that $\mathfrak{B} \models \varphi(\vec{b},\vec{\imath})$ implies $\mathfrak{A} \models \varphi(\vec{a},\vec{\imath})$, is identical. This proves $\mathsf{hlr}(\varphi) \leq 3d+1$. □

8.5 Complexity of Counting Quantifiers

In this section we revisit the logic FO(**Cnt**), and give a circuit model that corresponds to it. This circuit model defines a complexity class that extends AC^0; the class is called TC^0, where TC stands for *threshold circuits.* There are different ways of defining the class TC^0; the one chosen here uses *majority circuits*, which have special gates for the majority function.

Definition 8.20. Majority circuits *are defined as the usual Boolean circuits except that they have additional majority gates. Such a gate has* $2k$ *inputs,* $x_1, \ldots, x_n$, $y_1, \ldots, y_n$, *for* $k > 0$. *The output of the gate is* 1 *if*

$$\sum_{i=1}^{n} x_i \;\geq\; \sum_{i=1}^{n} y_i,$$

and 0 *otherwise.*

A circuit family $\mathbf{C}$ *has one circuit* C_n *for each* n, *where* n *is the number of inputs. The size, the depth, and the language accepted by* $\mathbf{C}$, *are defined in exactly the same way as for Boolean circuits. The class* nonuniform TC^0 *is defined as the class of languages (subsets of* $\{0,1\}^*$*) accepted by polynomial-size constant-depth families of majority circuits.*

We now extend FO(**Cnt**) to a logic FO(**Cnt**)$_{\mathsf{All}}$. This logic, in addition to FO(**Cnt**), has the linear ordering $<$ on the non-numerical universe, and, furthermore, the restriction of *every* predicate $P \subseteq \mathbb{N}^k$ to the numerical universe $\{0, \ldots, n-1\}$; that is, $P \cap \{0, \ldots, n-1\}^k$.

Theorem 8.21. *The class of structures definable by an* FO(**Cnt**)$_{\mathsf{All}}$ *sentence is in nonuniform* TC^0. *Consequently, the data complexity of* FO(**Cnt**)$_{\mathsf{All}}$ *is nonuniform* TC^0.

Proof. As in the proof of Theorem 6.4, we code formulae by circuits. We first note that if a linear order is available on the non-numerical universe A, there is no need for the numerical universe $\{0, \ldots, n-1\}$, where $n = | A |$, since we can interpret $\underline{\min}$, $\underline{\max}$, $<$, and the arithmetic operations directly on A, associating the ith element of A in the ordering $<$ with $i \in \mathbb{N}$. Thus, counting quantifiers will be assumed to be of the form $\exists yx\varphi(x, \cdots)$, stating that there exist at least i elements x satisfying φ, where y is the ith element of A in the ordering $<$.

Recall that for each structure $\mathfrak{A}$ with $|A| = n$, its encoding enc($\mathfrak{A}$) starts with 0^n1 that represents the size of the universe. For each formula $\varphi(x_1, \ldots, x_m)$, and each tuple $\vec{b} = (b_1, \ldots, b_m)$ in A^m, we construct a circuit $\mathsf{C}^n_{\varphi(\vec{b})}$ with the input enc($\mathfrak{A}$) which outputs 1 iff $\mathfrak{A} \models \varphi(\vec{b})$.

If $\varphi(\vec{b})$ is an atomic formula of the form $S(\vec{b})$, where $S \in \sigma$, then we simply output the corresponding bit from enc($\mathfrak{A}$). If φ is a numerical formula, we output 1 or 0 depending on whether $\varphi(\vec{b})$ is true. For Boolean connectives, we simply use $\vee, \wedge$ or $\neg$ gates. Thus, it remains to show how to handle the case of counting quantifiers.

Let $\varphi(x_1, \ldots, x_m) \equiv \exists x_1\, y\, \psi(y, \vec{x})$. That is, there exist x_1 elements y satisfying φ (since structures are ordered, we associate an element x_1 with its ranking in the linear order).

Let $\vec{b} \in A^m$ be given, and let $a_0, \ldots, a_{n-1}$ enumerate all the elements of A. Let C_i be the circuit $\mathsf{C}^n_{\psi(a_i, \vec{b})}$. We then collect the n outputs of such circuits, and for each of the first n inputs (which are the first n zeros of enc($\mathfrak{A}$)), we produce 1 for the first a_1 zeros, and 0 for the remaining $n - a_1$ zeros. This can easily be done with small constant-depth circuits. We then feed all the $2n$ inputs to a majority gate as shown in Fig. 8.1.

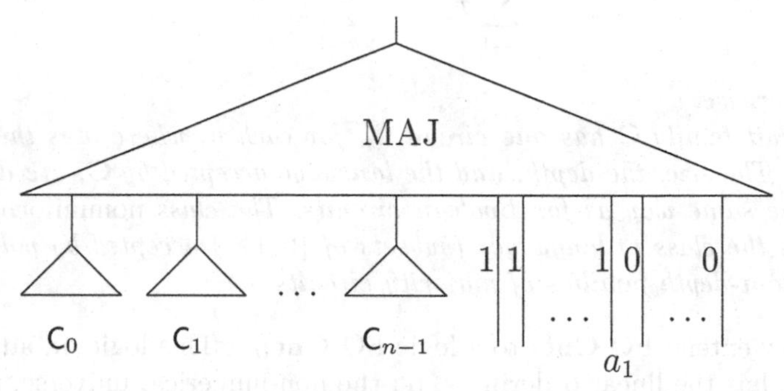

Fig. 8.1. Circuit for the proof of Theorem 8.21

It is clear from the construction that the family of circuits defined this way has a fixed constant depth (in fact, linear in the size of the formula), and polynomial size in terms of $\|\mathfrak{A}\|$. This completes the proof. □

As with nonuniform AC^0, the nonuniform version of TC^0 can define even noncomputable problems, since every predicate on $\mathbb{N}$ is available. The *uniform* version of TC^0 is defined as FO(**Cnt**)+ <: that is, FO(**Cnt**) with ordering available on the non-numerical universe. Thus, we restrict ourselves to addition and multiplication on natural numbers, and other functions and predicates definable with them (e.g., the BIT predicate).

Uniform TC^0 is a proper extension of uniform AC^0: for example, parity is in TC^0 but not in AC^0. It appears to be a rather modest extension: all we add is a simple form of counting. In particular, TC^0 is contained in PTIME, and in fact even in DLOG. Nevertheless, we still do not know if $\mathrm{TC}^0 \subsetneq \mathrm{NP}$.

We know, however, that FO(**Cnt**) is subsumed by $\mathcal{L}^*_{\infty\omega}$(**Cnt**), and that $\mathcal{L}^*_{\infty\omega}$(**Cnt**) is local – and hence it cannot express many PTIME problems such as graph connectivity, acyclicity, etc. Would not this give us the desired separation? Unfortunately, it would not, since we can only prove locality of FO(**Cnt**) but not FO(**Cnt**)+ <. We have seen that for FO, its extension with order, that is, (FO+ $<$)$_{\text{inv}}$, is local too. The same result, however, is *not* true for FO(**Cnt**). We now show a counterexample to locality of (FO(**Cnt**)+$<$)$_{\text{inv}}$.

Proposition 8.22. *There exist queries expressible in* (FO(**Cnt**)+$<$)$_{\text{inv}}$ *which are not Gaifman-local.*

Proof. The vocabulary σ contains a binary relation E and a unary relation P. We call a σ-structure *good* if three conditions are satisfied:

1. E has exactly one node of in-degree 0 and out-degree 1, exactly one node of out-degree 0 and in-degree 1, and all other nodes have both in-degree 1 and out-degree 1.
   That is, the relation E is a disjoint union of a chain $\{(a_0, a_1), (a_1, a_2), \ldots, (a_{k-1}, a_k)\}$ and zero or more cycles.
2. P contains a_0, does not contain a_k, and with each $a \in P$, except a_0, it contains its predecessor in E (the unique node b such that $(b, a) \in E$). Thus, P contains an initial segment of the successor part of E, and may contain some of the cycles in E.
3. $|P| \leq \log n$, where n is the size of the universe of the structure.

We claim that there is an FO(**Cnt**) sentence Φ_{good} that tests if a structure $\mathfrak{A} \in \text{STRUCT}[\sigma]$ is good. Clearly, conditions 1 and 2 can be verified by FO sentences. For condition 3, it suffices to check that the predicate $j \leq \log k$ is definable. Since $j \leq \log k$ iff $2^j \leq k$, and the predicate $i = 2^j$ is definable even in FO in the presence of addition and multiplication (see Sect. 6.4), we see that all three conditions can be defined in FO(**Cnt**).

We now consider the following binary query Q:

If $\mathfrak{A}$ is good, return the transitive closure of E restricted to P.

The result will follow from two claims. First, Q is definable in FO(**Cnt**)+$<$. Second, Q is not Gaifman-local. The latter is simple: assume, to the contrary, that Q is Gaifman-local and let $d = \text{lr}(Q)$. Let $k = 4d + 5$, and $n = 2^k$. Take E to be a successor (chain) of length n, with P interpreted as its initial k elements. Notice that this is a good structure. Then in P, we can find two elements a, b with isomorphic and disjoint d-neighborhoods. Hence, $(a, b) \approx_d (b, a)$, but the transitive closure query would distinguish (a, b) from (b, a).

It remains to show that Q is expressible in FO(**Cnt**)+$<$. First, we assume, without loss of generality, that in a given structure $\mathfrak{A}$, elements of P precede elements of $A - P$ in the ordering $<$. Indeed, if this is not true of $<$, we can always define, in FO, a new ordering $<_1$ which coincides with $<$ on P and on $A - P$, and, furthermore, $a <_1 b$ for all $a \in P$ and $b \notin P$.

Let $S \subseteq P$, with $S = \{s_1, \ldots, s_m\}$. Let each s_j be the i_jth element in the ordering $<$; that is, $\exists! i_j x \; (x \leq s_j)$ holds. Define a_S as the pth element of A in the ordering $<$, where $\text{BIT}(p, i_1), \ldots, \text{BIT}(p, i_m)$ are all true, and for every $i \notin \{i_1, \ldots, i_m\}$, the value of BIT$(p, i)$ is false. Since $|P| \leq \log n$, such an element a_S exists for every $S \subseteq P$. Moreover, since BIT is definable, there is a definable (in FO(**Cnt**)) predicate Code(u, v) which is true iff v is of the form a_S for a set S, and $u \in S$.

The query Q will now be definable by a formula $\exists z\ \psi(x, y, z)$, where ψ says that z codes the path from x to y. That is, it says the following:

- Code(x, z) and Code(y, z) hold.
- If x_0 is the predecessor of x and y_0 is the successor of y, then Code(x_0, z) and Code(y_0, z) do not hold.
- For every other element $u \neq x, y$ such that Code(u, z) holds, it is the case that Code(u_1, z) and Code(u_2, z) hold, where u_1 and u_2 are the predecessor and the successor of u.
- Code(a_0, z) holds iff $a_0 = x$, and Code(a_k, z) does not hold. Here a_0 and a_k are the elements of in-degree and out-degree 0, respectively.

Clearly, all these conditions can be expressed in FO(**Cnt**).

Given the special form of E, one can easily verify that this defines the transitive closure restricted to P. $\square$

As a corollary of Proposition 8.22, we get a separation

$$\text{FO}(\textbf{Cnt}) \subsetneq (\text{FO}(\textbf{Cnt})+<)_{\text{inv}},$$

since all FO(**Cnt**)-expressible queries are Gaifman-local, by Corollary 8.16.

8.6 Aggregate Operators

Aggregate operators occur in most practical database query languages. They allow one to apply functions for entire columns of relations. For example, if we have a ternary relation R whose tuples are (d, e, s), where d is the department name, e is the employee name, and s is his/her salary, a typical aggregate query would ask for the total salary for each department. Such a query would construct, for each department d, the set of all tuples $\{(e_1, s_1), \ldots, (e_n, s_n)\}$ such that $(d, e_i, s_i) \in R$ for $i = 1, \ldots, n$, and then output $(d, \sum_{i=1}^{n} s_i)$. We view this as applying the aggregate function `SUM` to the *multiset* $\{s_1, \ldots, s_n\}$ (it is a multiset since some of the s_i's can be the same, but we have to sum them all).

Logics with counting seen so far are not well suited for proving results about languages with aggregations, as they cannot talk about entire columns of relations. Nevertheless, we shall show here that aggregate operators can be simulated in $\mathcal{L}^*_{\infty\omega}(\textbf{Cnt})$, thereby giving us expressibility bounds for practical database query languages.

We first define the notion of an aggregate operator.

Definition 8.23. *An* aggregate operator *is a collection* $\mathcal{F} = \{f_0, f_1, f_2, \ldots, f_\omega\}$ *of functions, where each* f_n, $0 < n < \omega$, *takes an* n*-element multiset (bag) of natural numbers, and returns a number in* $\mathbb{N}$. *Furthermore,* f_0 *and* f_ω *are constants;* f_ω *is the fixed value associated to all infinite multisets.*

For example, the aggregate SUM will be represented as $\mathcal{F}_{\mathtt{SUM}} = \{f_0, f_1, f_2, \ldots, f_\omega\}$, where $f_0 = f_\omega = 0$, and

$$f_n(\{a_1, \ldots, a_n\}) = a_1 + \ldots + a_n.$$

Definition 8.24 (Aggregate logic). *The* aggregate logic $\mathcal{L}_{\text{aggr}}$ *is defined as the following extension of* $\mathcal{L}^*_{\infty\omega}(\mathbf{Cnt})$.

For every possible aggregate operator $\mathcal{F}$, *a numerical term* $t(\vec{x}, \vec{y})$ *and a formula* $\varphi(\vec{x}, \vec{y})$, *we have a new numerical term*

$$t'(\vec{x}) \quad = \quad \mathsf{Aggr}_{\mathcal{F}}\vec{y}\ \big(t(\vec{x}, \vec{y}), \varphi(\vec{x}, \vec{y})\big).$$

Variables $\vec{y}$ *become bound in* $\mathsf{Aggr}_{\mathcal{F}}\vec{y}\ \big(t(\vec{x}, \vec{y}), \varphi(\vec{x}, \vec{y})\big)$.

The value $t'(\vec{a})$ *is calculated as follows. If there are infinitely many* $\vec{b}$ *such that* $\varphi(\vec{a}, \vec{b})$ *holds, then* $t'(\vec{a}) = f_\omega$. *If there is no* $\vec{b}$ *such that* $\varphi(\vec{a}, \vec{b})$ *holds, then* $t'(\vec{a}) = f_0$. *Otherwise, let* $\vec{b}_1, \ldots, \vec{b}_m$ *enumerate all the* $\vec{b}$ *such that* $\varphi(\vec{a}, \vec{b})$ *holds. Then*

$$t'(\vec{a}) \quad = \quad f_m(\{t(\vec{a}, \vec{b}_1), \ldots, t(\vec{a}, \vec{b}_m)\}).$$

Note that the argument of f_m *is in general a multiset, since some of* $t(\vec{a}, \vec{b}_i)$ *may be the same. The rank of* t' *is defined as* $\max(\mathsf{rk}(t), \mathsf{rk}(\varphi)) + |\vec{y}|$.

For example, the query that computes the total salary for each department is given by the following $\mathcal{L}_{\text{aggr}}$ formula $\varphi(d, v)$:

$$\big(\exists e \exists s\ R(d, e, s)\big)\ \wedge\ \big(v = \mathsf{Aggr}_{\mathcal{F}_{\mathtt{SUM}}}(e, s)\big(s, R(d, e, s)\big)\big).$$

The above query assumes that some of the columns in a relation could be numerical. The results below are proved without this assumption, but it is easy to extend the proofs to relations with columns of different types (see Exercise 8.16).

It turns out that this seemingly powerful extension does not actually provide any additional power.

Theorem 8.25. *The expressive power of* $\mathcal{L}_{\text{aggr}}$ *and* $\mathcal{L}^*_{\infty\omega}(\mathbf{Cnt})$ *is the same.*

Proof. It suffices to show that for every formula $\varphi(\vec{x})$ of $\mathcal{L}_{\text{aggr}}$, there exists an equivalent formula $\varphi^\circ(\vec{x})$ of $\mathcal{L}^*_{\infty\omega}(\mathbf{Cnt})$ such that $\mathsf{rk}(\varphi^\circ) \leq \mathsf{rk}(\varphi)$. We prove this theorem by induction on the formulae and terms. We also produce, for each second-sort term $t(\vec{x})$ of $\mathcal{L}_{\text{aggr}}$, a formula $\psi_t(\vec{x}, z)$ of $\mathcal{L}^*_{\infty\omega}(\mathbf{Cnt})$, with z of the second sort, such that $\mathfrak{A} \models \psi_t(\vec{a}, n)$ iff the value of $t(\vec{a})$ on $\mathfrak{A}$ is n. Below we show how to produce such formulae ψ_t.

For a second-sort term t which is a variable i, we define $\psi_t(i,z)$ to be $(z=i)$. If t is a constant c, then $\psi_t(z) \equiv (z=c)$.

For a term

$$t'(\vec{x}) = \mathsf{Aggr}_{\mathcal{F}}\vec{y}\ (t(\vec{x},\vec{y}), \varphi(\vec{x},\vec{y})),$$

$\psi_{t'}(\vec{x},z)$ is defined as

$$(\varphi^\circ_\infty(\vec{x}) \wedge (z = f_\omega)) \ \vee \ (\neg\varphi^\circ_\infty(\vec{x}) \wedge \psi'(\vec{x},z)),$$

where $\varphi^\circ_\infty(\vec{x})$ tests if the number of $\vec{y}$ satisfying $\varphi(\vec{x},\vec{y})$ is infinite, and ψ' produces the value of the term in the case when the number of such $\vec{y}$ is finite.

The formula $\varphi^\circ_\infty(\vec{x})$ can be defined as

$$\bigvee_{i:y_i \text{ of 2nd sort}} \ \bigvee_{C\subseteq\mathbb{N},\ C \text{ infinite}} \ \bigwedge_{c\in C} \varphi^\circ_i(\vec{x},c)$$

where $\varphi^\circ_i(\vec{x},y_i) \equiv \exists y_1,\ldots,y_{i-1},y_{i+1},\ldots,y_m\ \varphi^\circ(\vec{x},\vec{y})$.

The formula $\psi'(\vec{x},z)$ is defined as the disjunction of $\neg\exists\vec{y}\varphi^\circ(\vec{x},\vec{y}) \wedge (z=f_0)$ and

$$\bigvee_{c,(c_1,n_1),\ldots,(c_l,n_l)} \left(\begin{array}{l} z = c \\ \wedge\ \exists! n_1 \vec{y}\ (\varphi^\circ(\vec{x},\vec{y}) \wedge \psi_t(\vec{x},\vec{y},c_1)) \\ \wedge \cdots \\ \wedge\ \exists! n_l \vec{y}\ (\varphi^\circ(\vec{x},\vec{y}) \wedge \psi_t(\vec{x},\vec{y},c_l)) \\ \wedge\ \forall\vec{y} \bigwedge_{a\in\mathbb{N}} (\varphi^\circ(\vec{x},\vec{y}) \wedge \psi_t(\vec{x},\vec{y},a) \rightarrow \bigvee_{i=1}^{l} (a = c_i)) \end{array} \right)$$

where the disjunction is taken over all tuples $(c_1,n_1),\ldots,(c_l,n_l)$, $l>0, n_i>0$, and values $c\in\mathbb{N}$ such that

$$\mathcal{F}(\{\underbrace{c_1,\ldots,c_1}_{n_1 \text{ times}},\ldots,\underbrace{c_l,\ldots,c_l}_{n_l \text{ times}}\}) \ = \ c.$$

Indeed, this formula asserts either that $\varphi(\vec{x},\cdot)$ does not hold and then $z = f_0$, or that $c_1,\ldots,c_l$ are exactly the values of the term $t(\vec{x},\vec{y})$ when $\varphi(\vec{x},\vec{y})$ holds, and that n_i's are the multiplicities of the c_i's.

A straightforward analysis of the produced formulae shows that $\mathsf{rk}(\psi_{t'}) \leq \max(\mathsf{rk}(\varphi^\circ),\mathsf{rk}(\psi_t))$ plus the number of first-sort variables in $\vec{y}$; that is, $\mathsf{rk}(\psi_{t'}) \leq \mathsf{rk}(t')$. This completes the proof of the theorem. $\square$

Corollary 8.26. *Every query expressible in $\mathcal{L}_{\text{aggr}}$ is Hanf-local and Gaifman-local.*

Thus, practical database query languages with aggregate functions still cannot express queries such as graph connectivity or transitive closure.

8.7 Bibliographic Notes

Extension of FO with counting quantifiers was proposed by Immerman and Lander [135]; the presentation here follows closely Etessami [68]. Generalized quantifiers are used extensively in logic, see Väänänen [237, 238].

The infinitary counting logic $\mathcal{L}^*_{\infty\omega}(\mathbf{Cnt})$ is from Libkin [166], although a closely related logic with unary quantifiers was studied in Hella [121]. Proposition 8.8 is a standard technique for eliminating counting terms over tuples, see, e.g., Kolaitis and Väänänen [149], and [166].

Bijective games were introduced by Hella [121], and the connection between bijective games and $\mathcal{L}^*_{\infty\omega}(\mathbf{Cnt})$ is essentially from that paper (it used a slightly different logic though). Locality of $\mathcal{L}^*_{\infty\omega}(\mathbf{Cnt})$ is from [166].

Connection between FO(**Cnt**) and TC^0 is from Barrington, Immerman, and Straubing [16]. The name TC^0 refers to threshold circuits that use threshold gates: such a gate has a threshold i, and it outputs 1 if at least i of its inputs are set to 1. The equivalence of threshold and majority gates is well known, see, e.g., Vollmer [247]. Proposition 8.22 is from Hella, Libkin, and Nurmonen [123]. Our treatment of aggregate operators follows Grädel and Gurevich [98]; the definition of the aggregate logic and Theorem 8.25 are from Hella et al. [124].

Sources for exercises:

Exercise 8.6:	Libkin [166]
Exercises 8.7 and 8.8:	Libkin [167]
Exercises 8.9 and 8.10:	Libkin and Wong [170]
Exercise 8.11:	Immerman and Lander [135]
Exercises 8.12 and 8.13:	Barrington, Immerman, and Straubing [16]
Exercises 8.14 and 8.15:	Nurmonen [189]
Exercise 8.16:	Hella et al. [124]

8.8 Exercises

Exercise 8.1. Show that none of the following is expressible in $\mathcal{L}^*_{\infty\omega}(\mathbf{Cnt})$: transitive closure of a graph, testing for planarity, acyclicity, 3-colorability.

Exercise 8.2. Prove Proposition 8.10 for arbitrary vocabularies.

Exercise 8.3. Prove Proposition 8.11.

Exercise 8.4. Prove Proposition 8.18.

Exercise 8.5. Prove Theorem 8.19 for Gaifman-locality.

Exercise 8.6. Extend Exercise 4.11 to counting logics. That is, define functions $\mathrm{Hanf\_rank}_{\mathcal{L}}, \mathrm{Gaifman\_rank}_{\mathcal{L}} : \mathbb{N} \to \mathbb{N}$, for a logic $\mathcal{L}$, as follows:

$$\mathrm{Hanf\_rank}_{\mathcal{L}}(n) \quad = \quad \max\{\mathsf{hlr}(\varphi) \mid \varphi \in \mathcal{L},\ \mathsf{rk}(\varphi) = n\},$$

$$\text{Gaifman\_rank}_{\mathcal{L}}(n) \quad = \quad \max\{\mathsf{lr}(\varphi) \mid \varphi \in \mathcal{L},\ \mathsf{rk}(\varphi) = n\}.$$

Assume that we deal with purely relational vocabularies. Prove that for every $n > 1$, $\text{Hanf\_rank}_{\mathcal{L}}(n) = 2^{n-1} - 1$ and $\text{Gaifman\_rank}_{\mathcal{L}}(n) = 2^n - 1$, when $\mathcal{L}$ is one of the following: FO(**Cnt**), FO($\mathfrak{Q}$) for any $\mathfrak{Q}$, $\mathcal{L}^*_{\infty\omega}(\mathbf{Cnt})$.

Exercise 8.7. Extend $\mathcal{L}^*_{\infty\omega}(\mathbf{Cnt})$ by additional atomic formulae $\iota_d(\vec{x}, \vec{y})$ (where $|\vec{x}| = |\vec{y}|$), such that $\mathfrak{A} \models \iota_d(\vec{a}, \vec{b})$ iff $\vec{a} \approx^{\mathfrak{A}}_d \vec{b}$. Let $\mathcal{L}^{*,r}_{\infty\omega}(\mathbf{Cnt})$ be the resulting logic, where every occurrence of ι_d satisfies $d \leq r$. Prove that $\mathcal{L}^{*,r}_{\infty\omega}(\mathbf{Cnt})$ is Hanf-local.

Exercise 8.8. Extend $\mathcal{L}^*_{\infty\omega}(\mathbf{Cnt})$ by adding local second-order quantification: that is, second-order quantification restricted to $N_d(\vec{a})$, where $\vec{a}$ is the interpretation of free first-order variables. Such an extension, like the one of Exercise 8.7, must have the radii of neighborhoods, over which local second-order quantification is done, uniformly bounded in infinitary formulae.

Complete the definition of this logic, and prove that it captures precisely all the Hanf-local queries.

Exercise 8.9. Let $\precsim^k$ be the class of preorders in which every equivalence class has size at most k. The equivalence associated with a preorder $\precsim$ is

$$x \sim y \ \Leftrightarrow\ (x \precsim y) \wedge (y \precsim x).$$

Prove that graph connectivity is not in $(\mathcal{L}^*_{\infty\omega}(\mathbf{Cnt}) + \precsim^k)_{\text{inv}}$.

Exercise 8.10. The goal of this exercise is to prove a statement much stronger than that of Exercise 8.9. Given a preorder $\precsim$, let $[x]$ be the equivalence class of x with respect to $\sim$. Let $g : \mathbb{N} \to \mathbb{N}$ be a nondecreasing function which is not bounded by a constant. Let $\precsim_g$ be the class of preorders $\precsim$ such that on an n-element set, for at most $g(n)$ elements we have $\|[x]\| = 2$, and for the remaining at least $n - g(n)$ elements, $\|[x]\| = 1$; furthermore, if $\|[x]\| = 2$ and $\|[y]\| = 1$, then $x \prec y$. In other words, such preorders are linear orders everywhere, except at most $g(n)$ initial elements.

Prove the following:

1. There are functions g for which $(\mathcal{L}^*_{\infty\omega}(\mathbf{Cnt}) + \precsim_g)_{\text{inv}}$ contains nonlocal queries.
2. For every g, every query in $(\mathcal{L}^*_{\infty\omega}(\mathbf{Cnt}) + \precsim_g)_{\text{inv}}$ has the BNDP.

Exercise 8.11. Define Ehrenfeucht-Fraïssé games for FO(**Cnt**), and prove their correctness.

Exercise 8.12. Consider the logic FO(MAJ) defined as follows. A universe of σ-structure is ordered, and is thus associated with $\{0, \ldots, n-1\}$. Furthermore, for each $k > 0$, and a formula $\varphi(\vec{x}, \vec{z})$, with $|\vec{x}| = k$, we have a new formula

$$\psi(\vec{z}) \quad \equiv \quad \text{MAJ}\ \vec{x}\ \varphi(\vec{x}, \vec{z}),$$

binding $\vec{x}$, such that $\mathfrak{A} \models \psi(\vec{c})$ iff $|\varphi(\mathfrak{A}, \vec{c})| \geq \frac{1}{2} \cdot |A|^k$. Recall that $\varphi(\mathfrak{A}, \vec{c})$ stands for $\{\vec{b} \mid \mathfrak{A} \models \varphi(\vec{b}, \vec{c})\}$.

Prove the following:

- Over ordered structures, the logics FO(MAJ) and FO(**Cnt**) express all the same queries.

- In the definition of FO(MAJ), it suffices to consider $k \leq 2$: that is, the majority quantifier MAJ $(x_1, x_2)\ \varphi(x_1, x_2, z)$.
- Over ordered structures with the BIT predicate, the fragment of FO(MAJ) in which $k = 1$ (i.e., only new formulae of the form MAJ $x\ \varphi(x, \vec{z})$ are allowed) is as expressive as FO(**Cnt**).

Exercise 8.13. Prove the converse of Theorem 8.21: that is, any class of structures in nonuniform TC^0 is definable in $\mathrm{FO}(\mathbf{Cnt})_{\mathsf{All}}$.

Exercise 8.14. Consider the generalized quantifier $\mathbf{D}_n$ defined as follows. If $\varphi(x, \vec{z})$ is a formula, then $\psi(\vec{z}) \equiv \mathbf{D}_n x\ \varphi(x, \vec{z})$ is a formula, such that $\mathfrak{A} \models \varphi(\vec{a})$ iff $|\,\varphi(\mathfrak{A}, \vec{a})\,|$ mod $n = 0$.

Next, consider strings over the alphabet $\{0, 1\}$ as finite structure (see Chap. 7), and prove that none of the following properties of strings $s_0 \dots s_{m-1}$ is expressible in FO($\mathbf{D}_n$):

- Majority: $\sum_{i=0}^{m-1} s_i \geq \frac{m}{2}$;
- m mod $p = 0$, for every prime p that does not divide n;
- $\left(\sum_{i=0}^{m-1} s_i\right)$ mod $p = 0$, again for every prime p that does not divide n.

Exercise 8.15. Consider the generalized quantifier $\mathbf{D}_n$ from Exercise 8.14. Consider ordered structures (in which we can associate elements with numbers), and define an additional predicate $y = nx$ over them. Prove that even in the presence of such an additional predicate, FO($\mathbf{D}_n$) cannot express the predicate $y = (n+1)x$.

Exercise 8.16. Aggregate operators in database query languages normally operate on rational numbers; for example, one of the standard aggregates is `AVG` $= \{f_0, f_1, f_2, \dots, f_\omega\}$, where $f_0 = f_\omega = 0$, and $f_n(\{a_1, \dots, a_n\}) = (a_1 + \dots + a_n)/n$.

Define $\mathcal{L}_{\mathrm{aggr}}^{\mathbb{Q}}$ as an extension of $\mathcal{L}_{\mathrm{aggr}}$ where the numerical domain is $\mathbb{Q}$, each $q \in \mathbb{Q}$ is a numerical term, and all aggregate operators $\mathcal{F}$ on $\mathbb{Q}$ are available.

Prove the following:

1. For every $\mathcal{L}_{\mathrm{aggr}}^{\mathbb{Q}}$ formula $\varphi(\vec{x})$ without free numerical variables, there exists an equivalent $\mathcal{L}_{\infty\omega}^{*}(\mathbf{Cnt})$ formula of the same rank.
2. Conclude that $\mathcal{L}_{\mathrm{aggr}}^{\mathbb{Q}}$ is Hanf-local and Gaifman-local.

Next, extend all the results to the case when different columns of σ-relations could be of different types: some of the universe of the first sort, and some numerical.

Exercise 8.17.* Prove that transitive closure is not expressible in FO(**Cnt**)$+<$.

9

Turing Machines and Finite Models

In this chapter we introduce the technique of coding Turing machines in various logics. It is precisely this technique that gave rise to numerous applications of finite model theory in computational complexity. We start by proving the earliest such result, Trakhtenbrot's theorem, stating that finite satisfiability is not decidable. For the proof of Trakhtenbrot's theorem, we code Turing machines with no inputs. By a refinement of this technique, we code nondeterministic polynomial time Turing machines in existential second-order logic ($\exists$SO), proving Fagin's theorem stating that $\exists$SO-definable properties of finite structures are precisely those whose complexity is NP.

9.1 Trakhtenbrot's Theorem and Failure of Completeness

Recall the completeness theorem for FO: a sentence Φ is valid (is true in all models) iff it is provable in some formal system. In particular, this implies that the set of all valid FO sentences is r.e. (recursively enumerable), since one can enumerate all the formal proofs of valid FO sentences. We now show that completeness fails over finite models.

What does it mean that Φ is valid? It means that all structures $\mathfrak{A}$, finite or infinite, are models of Φ: that is, $\mathfrak{A} \models \Phi$. Since we are interested in finite models only, we want to refine the notions of satisfiability and validity in the finite context.

Definition 9.1. *Given a vocabulary σ, a sentence Φ in that vocabulary is called* finitely satisfiable *if there is a finite structure $\mathfrak{A} \in$ STRUCT$[\sigma]$ such that $\mathfrak{A} \models \Phi$.*

The sentence Φ is called finitely valid *if $\mathfrak{A} \models \Phi$ holds for all finite structures $\mathfrak{A} \in$ STRUCT$[\sigma]$.*

Theorem 9.2 (Trakhtenbrot). *For every relational vocabulary σ with at least one binary relation symbol, it is undecidable whether a sentence Φ of vocabulary σ is finitely satisfiable.*

In the proof that we give, the vocabulary σ contains several binary relation symbols and a constant symbol. But it is easy to modify it to prove the result with just one binary relation symbol (this is done by coding several relations into one; see Exercise 9.1).

Before we prove Trakhtenbrot's theorem, we point out two corollaries. First, as we mentioned earlier, completeness fails in the finite.

Corollary 9.3. *For any vocabulary containing at least one binary relation symbol, the set of finitely valid sentences is not recursively enumerable.*

Proof. Notice that the set of finitely satisfiable sentences is recursively enumerable: one simply enumerates all pairs $(\mathfrak{A}, \Phi)$, where $\mathfrak{A}$ is finite, and outputs Φ whenever $\mathfrak{A} \models \Phi$. Assume that the set of finitely valid sentences is r.e. Since $\neg\Phi$ is finitely valid iff Φ is not finitely satisfiable, we conclude that the set of sentences which are not finitely satisfiable is r.e., too. However, if both a set X and its complement $\bar{X}$ are r.e., then X is recursive; hence, we conclude that the set of finitely satisfiable sentences is recursive, which contradicts Trakhtenbrot's theorem. □

Another corollary states that one cannot have an analog of the Löwenheim-Skolem theorem for finite models.

Corollary 9.4. *There is no recursive function f such that if Φ has a finite model, then it has a model of size at most $f(\Phi)$.*

Indeed, with such a recursive function one would be able to decide finite satisfiability.

We now prove Trakhtenbrot's theorem. The idea of the proof is to code Turing machines in FO: for every Turing machine M, we construct a sentence Φ_M of vocabulary σ such that Φ_M is finitely satisfiable iff M halts on the empty input. The latter is well known to be undecidable (this is an easy exercise in computability theory).

Let $M = (Q, \Sigma, \Delta, \delta, q_0, Q_a, Q_r)$ be a deterministic Turing machine with a one-way infinite tape. Here Q is the set of states, Σ is the input alphabet, Δ is the tape alphabet, q_0 is the initial state, Q_a (Q_r) is the set of accepting (rejecting) states, from which there are no transitions, and δ is the transition function. Since we are coding the problem of halting on the empty input, we can assume without loss of generality that $\Delta = \{0, 1\}$, with 0 playing the role of the blank symbol.

We define σ so that its structures represent computations of M. More precisely,

$$\sigma = \{<, \underline{\min}, T_0(\cdot,\cdot), T_1(\cdot,\cdot), (H_q(\cdot,\cdot))_{q\in Q}\},$$

where

- $<$ is a linear order and $\underline{\min}$ is a constant symbol for the minimal element with respect to $<$; hence the finite universe will be associated with an initial segment of natural numbers.
- T_0 and T_1 are *tape* predicates; $T_i(p, t)$ indicates that position p at time t contains i, for $i = 0, 1$.
- H_q's are *head* predicates; $H_q(p, t)$ indicates that at time t, the machine is in state q, and its head is in position p.

The sentence Φ_M states that $<$, $\underline{\min}$, T_i's, and H_q's are interpreted as indicated above, and that the machine eventually halts. Note that if the machine halts, then $H_q(p, t)$ holds for some p, t, and $q \in Q_a \cup Q_r$, and after that the configuration of the machine does not change. That is, all the configurations of the halting computation can be represented by a finite σ-structure.

We define Φ_M to be the conjunction of the following sentences:

- A sentence stating that $<$ is a linear order and $\underline{\min}$ is its minimal element.
- A sentence defining the initial configuration of M (it is in state q_0, the head is in the first position, and the tape contains only zeros):

$$H_{q_0}(\underline{\min}, \underline{\min}) \ \land \ \forall p \, T_0(p, \underline{\min}).$$

- A sentence stating that in every configuration of M, each cell of the tape contains exactly one element of Δ:

$$\forall p \forall t \, \big(T_0(p, t) \leftrightarrow \neg T_1(p, t)\big).$$

- A sentence imposing the basic consistency conditions on the predicates H_q's (at any time the machine is in exactly one state):

$$\forall t \exists ! p \, \Big(\bigvee_{q \in Q} H_q(p, t) \Big) \ \land \ \neg \exists p \exists t \, \Big(\bigvee_{q, q' \in Q, \, q \neq q'} H_q(p, t) \land H_{q'}(p, t) \Big).$$

- A set of sentences stating that T_i's and H_q's respect the transitions of M (with one sentence per transition). For example, assume that $\delta(q, 0) = (q', 1, \ell)$; that is, if M is in state q reading 0, then it writes 1, moves the head one position to the left and changes the state to q'. This transition is represented by the conjunction of

$$\forall p \forall t \begin{pmatrix} p \neq \underline{\min} \\ \land \ T_0(p, t) \\ \land \ H_q(p, t) \end{pmatrix} \rightarrow \begin{pmatrix} T_1(p, t+1) \\ \land \ H_{q'}(p - 1, t + 1) \\ \land \ \forall p' \, (p \neq p' \rightarrow \\ \quad \big(\bigwedge_{i=0,1} T_i(p', t+1) \leftrightarrow T_i(p', t) \big)) \end{pmatrix}$$

and

$$\forall p \forall t \begin{pmatrix} p = \underline{\min} \\ \wedge\, T_0(p,t) \\ \wedge\, H_q(p,t) \end{pmatrix} \rightarrow \begin{pmatrix} T_1(p,t+1) \\ \wedge\, H_{q'}(p,t+1) \\ \wedge\, \forall p'\, (p \neq p' \rightarrow \\ (\bigwedge_{i=0,1} T_i(p',t+1) \leftrightarrow T_i(p',t))) \end{pmatrix}.$$

We use abbreviations $p-1$ and $t+1$ for the predecessor of p and the successor of t in the ordering $<$; these are, of course, FO-definable. The first sentence above ensures that the tape content in position p changes from 0 to 1, the state changes from q to q', the rest of the tape remains the same, and the head moves to position $p-1$, assuming p is not the first position on the tape. The second sentence is very similar, and handles the case when p is the initial position: then the head does not move and stays in p.

- Finally, a sentence stating that at some point, M is in a halting state:

$$\exists p \exists t \bigvee_{q \in Q_a \cup Q_r} H_q(p,t).$$

If Φ_M has a finite model, then such a model represents a computation of M that starts with the tape containing all zeros, and ends in a halting state. If, on the other hand, M halts on the empty input, then the set of all configurations of the halting computations of M coded as relations $<, T_i$'s, and H_q's, is a model of Φ_M (necessarily finite). Thus, M halts on the empty input iff Φ_M has a finite model. Since testing if M halts on the empty model is undecidable, then so is finite satisfiability for Φ_M. □

9.2 Fagin's Theorem and NP

Fagin's theorem provides a purely logical characterization of the complexity class NP, by means of coding computations of nondeterministic polynomial time Turing machines in a fragment of second-order logic. Before stating the result, we give the following general definition. Recall that by properties, we mean Boolean queries, namely, collections of structures closed under isomorphism.

Definition 9.5. *Let $\mathcal{K}$ be a complexity class, $\mathcal{L}$ a logic, and $\mathcal{C}$ a class of finite structures. We say that $\mathcal{L}$* captures *$\mathcal{K}$ on $\mathcal{C}$ if the following hold:*

1. *The data complexity of $\mathcal{L}$ on $\mathcal{C}$ is $\mathcal{K}$; that is, for every $\mathcal{L}$-sentence Φ, testing if $\mathfrak{A} \models \Phi$ is in $\mathcal{K}$, provided $\mathfrak{A} \in \mathcal{C}$.*
2. *For every property $\mathcal{P}$ of structures from $\mathcal{C}$ that can be tested with complexity $\mathcal{K}$, there is a sentence $\Phi_{\mathcal{P}}$ of $\mathcal{L}$ such that $\mathfrak{A} \models \Phi_{\mathcal{P}}$ iff $\mathfrak{A}$ has the property $\mathcal{P}$, for every $\mathfrak{A} \in \mathcal{C}$.*

If $\mathcal{C}$ is the class of all finite structures, we say that $\mathcal{L}$ captures $\mathcal{K}$.

Theorem 9.6 (Fagin). $\exists$SO *captures* NP.

Before proving this theorem, we make several comments and point out some corollaries. Fagin's theorem is a very significant result as it was the first machine-independent characterization of a complexity class. Normally, we define complexity classes in terms of resources (time, space) that computations can use; here we use a purely logical formalism. Following Fagin's theorem, logical characterizations have been proven for many complexity classes (we already saw them for uniform AC^0 and TC^0, and later we shall see how to characterize NLOG, PTIME, and PSPACE over ordered structures).

The hardest open problems in complexity theory concern separation of complexity classes, with the "PTIME vs. NP" question being undoubtedly the most famous such problem. Logical characterizations of complexity classes show that such separation results can be formulated as inexpressibility results in logic. Suppose that we have two complexity classes $\mathcal{K}_1$ and $\mathcal{K}_2$, captured by logics $\mathcal{L}_1$ and $\mathcal{L}_2$. To prove that $\mathcal{K}_1 \neq \mathcal{K}_2$, it would then suffice to separate the logics $\mathcal{L}_1$ and $\mathcal{L}_2$; that is, to show that some problem definable in $\mathcal{L}_2$ is inexpressible in $\mathcal{L}_1$, or vice versa.

Since the class CONP consists of the problems whose complements are in NP, and the negation of an $\exists$SO sentence is an $\forall$SO sentence, we obtain:

Corollary 9.7. $\forall$SO *captures* CONP. □

Hence, to show that NP $\neq$ CONP, it would suffice to exhibit a property definable in $\forall$SO but not definable in $\exists$SO. While we still do not know if such a property exists, recall that we have a property definable in $\forall$MSO but not definable in $\exists$MSO: graph connectivity. In fact, for reasons obvious from Fagin's theorem and Corollary 9.7, $\exists$MSO is sometimes referred to as "monadic NP", and $\forall$MSO as "monadic CONP". Hence, Proposition 7.14 tells us that

$$\text{monadic NP} \neq \text{monadic coNP}.$$

Note that separating $\forall$SO from $\exists$SO would also resolve the "PTIME vs. NP" question:

$$\forall\mathrm{SO} \neq \exists\mathrm{SO} \quad \Rightarrow \quad \mathrm{NP} \neq \mathrm{coNP} \quad \Rightarrow \quad \mathrm{PTIME} \neq \mathrm{NP}$$

(if PTIME and NP were the same, NP would be closed under the complement, and hence NP and CONP would be the same).

As another remark, we point out that the above remark concerning the separation of $\exists$SO and $\forall$SO is specific to the finite case. Indeed, by Fagin's theorem, $\exists$SO $\neq$ $\forall$SO over finite structures iff NP $\neq$ CONP, but over some infinite structures (e.g., $\langle \mathbb{N}, +, \cdot \rangle$), the logics $\exists$SO and $\forall$SO are known to be different.

We now prove Fagin's theorem. First, we show that every ∃SO sentence Φ can be evaluated in NP. Suppose Φ is $\exists S_1 \dots \exists S_n\ \varphi$, where φ is FO. Given $\mathfrak{A}$, the nondeterministic machine first guesses $S_1, \dots, S_n$, and then checks if $\varphi(S_1, \dots, S_n)$ holds. The latter can be done in polynomial time in $\|\mathfrak{A}\|$ plus the size of $S_1, \dots, S_n$, and thus in polynomial time in $\|\mathfrak{A}\|$ (see Proposition 6.6). Hence, Φ can be evaluated in NP.

Next, we show that every NP property of finite structures can be expressed in ∃SO. The proof of this direction is very close to the proof of Trakhtenbrot's theorem, but there are two additional elements we have to take care of: time bounds, and the input.

Suppose we are given a property $\mathcal{P}$ of σ-structures that can be tested, on encodings of σ-structures, by a nondeterministic polynomial time Turing machine $M = (Q, \Sigma, \Delta, \delta, q_0, Q_a, Q_r)$ with a one-way infinite tape. Here $Q = \{q_0, \dots, q_{m-1}\}$ is the set of states, and we assume without loss of generality that $\Sigma = \{0, 1\}$ and Δ extends Σ with the blank symbol "␣". We assume that M runs in time n^k. Notice that n is the size of the encoding, so we always assume $n > 1$. We can also assume without loss of generality that M always visits the entire input; that is, that n^k always exceeds the size of the encodings of n-element structures (this is possible because the size of enc($\mathfrak{A}$), defined in Chap. 6, is polynomial in $\|\mathfrak{A}\|$).

The sentence describing acceptance by M on encodings of structures from STRUCT[σ] will be of the form

$$\exists L\ \exists T_0 \exists T_1 \exists T_2\ \exists H_{q_0} \dots \exists H_{q_{m-1}}\ \Psi, \tag{9.1}$$

where Ψ is a sentence of vocabulary $\sigma \cup \{L, T_0, T_1, T_2\} \cup \{H_q \mid q \in Q\}$. Here L is binary, and other symbols are of arity $2k$. The intended interpretation of these relational symbols is as follows:

- L is a linear order on the universe.

With L, one can define, in FO, the lexicographic linear order $\leq_k$ on k-tuples. Since M runs in time n^k and visits at most n^k cells, we can model both positions on the tape ($\vec{p}$) and time ($\vec{t}$) by k-tuples of the elements of the universe.

With this, the predicates T_i's and H_q's are defined similarly to the proof of Trakhtenbrot's theorem:

- T_0, T_1, and T_2 are *tape* predicates; $T_i(\vec{p}, \vec{t})$ indicates that position $\vec{p}$ at time $\vec{t}$ contains i, for $i = 0, 1$, and $T_2(\vec{p}, \vec{t})$ says that $\vec{p}$ at time $\vec{t}$ contains the blank symbol.
- H_q's are *head* predicates; $H_q(\vec{p}, \vec{t})$ indicates that at time $\vec{t}$, the machine is in state q, and its head is in position $\vec{p}$.

The sentence Ψ must now assert that when M starts on the encoding of $\mathfrak{A}$, the predicates T_i's and H_q's correspond to its computation, and eventually M reaches an accepting state. Note that the encoding of $\mathfrak{A}$ depends on a linear ordering of the universe of A. We may assume, without loss of generality, that this ordering is L. Indeed, since queries are closed under isomorphism, choosing one particular ordering to be used in the representation of enc($\mathfrak{A}$) does not affect the result.

We now define Ψ as the conjunction of the following sentences:

- The sentence stating that L defines a linear ordering.
- The sentence stating that
  - in every configuration of M, each cell of the tape contains exactly one element of Δ;
  - at any time the machine is in exactly one state;
  - eventually, M enters a state from Q_a.

  All these are expressed in exactly the same way as in the proof of Trakhtenbrot's theorem.
- Sentences stating that T_i's and H_q's respect the transitions of M. These are written almost as in the proof of Trakhtenbrot's theorem, but one has to take into account nondeterminism. For every $a \in \Delta$ and $q \in Q$, we have a sentence

$$\bigvee_{(q',b,move)\in\delta(q,a)} \alpha_{(q,a,q',b,move)},$$

  where $move \in \{\ell, r\}$ and $\alpha_{(q,a,q',b,move)}$ is the sentence describing the transition in which, upon reading a in state q, the machine writes b, makes the move *move*, and enters state q'. Such a sentence is written in exactly the same way as in the proof of Trakhtenbrot's theorem.
- The sentence defining the initial configuration of M. Suppose we have formulae $\iota(\vec{p})$ and $\xi(\vec{p})$ of vocabulary $\sigma \cup \{L\}$ such that $\mathfrak{A} \models \iota(\vec{p})$ iff the $\vec{p}$th position of enc($\mathfrak{A}$) is 1 (in the standard encoding of structures presented in Chap. 6), and $\mathfrak{A} \models \xi(\vec{p})$ iff $\vec{p}$ exceeds the length of enc($\mathfrak{A}$). Note that we need L in these formulae since the encoding refers to a linear order on the universe. With such formulae, we define the initial configuration by

$$\forall\vec{p}\,\forall\vec{t}\,\left(\neg\exists\vec{u}\;(\vec{u} <_k \vec{t}) \;\rightarrow\; \begin{bmatrix} (\iota(\vec{p}) \leftrightarrow T_1(\vec{t},\vec{p})) \\ \wedge\;(\xi(\vec{p}) \leftrightarrow T_2(\vec{t},\vec{p})) \end{bmatrix}\right).$$

  In other words, at time 0, the tape contains the encoding of the structure followed by blanks.

Just as in the proof of Trakhtenbrot's theorem, we conclude that (9.1) holds in $\mathfrak{A}$ iff M accepts enc($\mathfrak{A}$). It thus remains to show how to define the formulae $\iota(\vec{p})$ and $\xi(\vec{p})$.

We illustrate this for the case of $\sigma = \{E\}$, with E binary (to keep the notation simple; extension to arbitrary vocabularies is straightforward). Assume that the universe of the graph is $\{0, \ldots, n-1\}$, where $(i, j) \in L$ iff $i < j$. The graph is then encoded by the string $0^n 1 \cdot s$, where s is a string of length n^2, such that it has 1 in position $u \cdot n + v$, for $0 \leq u, v \leq n-1$, iff $(u, v) \in E$. That is, the actual encoding of E starts in position $(n+1)$. Already from here, one can see that in the presence of addition and multiplication (given as ternary relations), ι is definable. Indeed, $\vec{p} = (p_1, \ldots, p_k)$ represents the position $p_1 \cdot n^{k-1} + p_2 \cdot n^{k-2} + \ldots + p_{k-1} \cdot n + p_k$. Hence, $\iota(\vec{p})$ is equivalent to the disjunction of $\sum_{i=1}^{k} p_i \cdot n^{k-i} = n$ and

$$\exists u \leq (n-1) \exists v \leq (n-1) \left((n+1) + u \cdot n + v = \sum_{i=1}^{k} p_i \cdot n^{k-i} \ \wedge \ E(u, v) \right).$$

With addition and multiplication, this is a definable property, and addition and multiplication themselves can be introduced by means of additional existential second-order quantifiers (since one can state in FO that a given relation properly represents addition or multiplication with respect to the ordering L).

While this is certainly enough to conclude that ι is definable, we now sketch a proof of definability of ι without any additional arithmetic. Instead, we shall only refer to the linear ordering L, and we shall use the associated successor relation (i.e., we shall refer to $x+1$ or $x-1$). Assume $k = 3$. That is, a tuple $\vec{p}$ represents the position $p_1 n^2 + p_2 n + p_3$ on the tape. The first position where the encoding of E starts is $n+1$ (positions 0 to n represent the size of the universe) and the last one is $n^2 + n$. Hence, if $p_1 > 1$, then ι is *false*. Assume $p_1 = 0$. Then we are talking about the position $p_2 n + p_3$. Positions 0 to $n-1$ have zeros, so if $p_2 = 0$, then again ι is *false*. If $p_3 \neq 0$, then $(p_2 - 1)n + (p_3 - 1) + (n+1) = p_2 n + p_3$, and hence the position corresponds to $E(p_2 - 1, p_3 - 1)$. If $p_3 = 0$, then this position corresponds to $E(p_2 - 2, n-1)$. Hence, the formula $\iota(p_1, p_2, p_3)$ is of the form

$$\begin{aligned} &\left[\begin{pmatrix} (p_1 = 0) \\ \wedge (p_2 > 1) \end{pmatrix} \wedge \begin{pmatrix} (p_3 \neq 0) \wedge E(p_2 - 1, p_3 - 1) \\ \vee \ (p_3 = 0) \wedge E(p_2 - 2, n-1) \end{pmatrix} \right] \\ \vee \ &\Big[(p_1 = 0) \wedge (p_2 = 1) \wedge (p_3 = 0) \Big] \ \vee \ \Big[(p_1 = 1) \wedge \ldots \Big], \end{aligned}$$

where for the case of $p_1 = 1$ a similar case analysis is done. Clearly, with the linear order L, both 0 and $n-1$, and the predecessor function are definable, and hence ι is FO. (The details of writing down ι for arbitrary k are left as an exercise to the reader, see Exercise 9.4.) The formula $\xi(\vec{p})$ simply says that $\vec{p}$, considered as a number, exceeds $n^2 + n + 1$. This completes the proof of Fagin's theorem. ☐

We now show several more corollaries of Fagin's theorem. The first one is Cook's theorem stating that SAT, propositional satisfiability, is NP-complete.

Corollary 9.8 (Cook). SAT *is* NP*-complete.*

Proof. Let $\mathcal{P}$ be a problem (a class of σ-structures) in NP. By Fagin's theorem, there is an $\exists$SO sentence $\Phi \equiv \exists S_1 \ldots \exists S_n\ \varphi$ such that $\mathfrak{A}$ is in $\mathcal{P}$ iff $\mathfrak{A} \models \Phi$. Let $X = \{S_i(\vec{a}) \mid i = 1, \ldots, n,\ \vec{a} \in A^{\text{arity}(S_i)}\}$. We construct a propositional formula $\alpha_\varphi^{\mathfrak{A}}$ with variables from X such that $\mathfrak{A} \models \Phi$ iff $\alpha_\varphi^{\mathfrak{A}}$ is satisfiable.

The formula $\alpha_\varphi^{\mathfrak{A}}$ is obtained from φ by the following three transformations:

- replacing each $\exists x\ \psi(x, \cdot)$ by $\bigvee_{a \in A} \psi(a, \cdot)$;
- replacing each $\forall x\ \psi(x, \cdot)$ by $\bigwedge_{a \in A} \psi(a, \cdot)$; and
- replacing each $R(\vec{a})$, for $R \in \sigma$, by its truth value in $\mathfrak{A}$.

In the resulting formula, the variables are of the form $S_i(\vec{a})$; that is, they come from the set X. Clearly, $\mathfrak{A} \models \Phi$ iff $\alpha_\varphi^{\mathfrak{A}}$ is satisfiable, and $\alpha_\varphi^{\mathfrak{A}}$ can be constructed by a deterministic logarithmic space machine. This proves NP-completeness of SAT. ☐

The logics $\exists$SO and $\forall$SO characterize NP and coNP, the first level of the polynomial hierarchy PH. Recall that the levels of PH are defined inductively: $\Sigma_1^p = \text{NP}$, and $\Sigma_{k+1}^p = \text{NP}^{\Sigma_k^p}$. The level Π_k^p is defined as the set of complements of problems from Σ_k^p. Also recall that Σ_k^1 is the class of SO sentences of the form

$$(\exists \ldots \exists)(\forall \ldots \forall)(\exists \ldots \exists) \ldots \varphi,$$

with k quantifier blocks, and Π_k^1 is defined likewise but the first block of quantifiers is universal.

We now sketch an inductive argument showing that Σ_k^1 captures Σ_k^p, for every k. The base case is Fagin's theorem. Now consider a problem in Σ_{k+1}^p. By Fagin's theorem, there is an $\exists$SO sentence Φ (corresponding to the NP machine) with additional predicates expressing Σ_k^p properties. We know, by the hypothesis, that those properties are definable by Σ_k^1 formulae. Then pushing the second-order quantifier outwards, we convert Φ into a Σ_{k+1}^1 sentence. The extra quantifier alternation arises when these predicates for Σ_k^p properties are negated: suppose we have a formula $\exists \ldots \exists \varphi(P)$, where P is expressed by a formula $\exists \ldots \exists \psi$, with ψ being FO, and P may occur negatively. Then putting the resulting formula in the prenex form, we have a second-order quantifier prefix of the form $(\exists \ldots \exists)(\forall \ldots \forall)$. For example, $\exists \ldots \exists\ \neg(\exists \ldots \exists \psi)$ is equivalent to $\exists \ldots \exists \forall \ldots \forall\ \neg\psi$. Filling all the details of this inductive proof is left to the reader as an exercise (Exercise 9.5).

Thus, we have the the following result.

Corollary 9.9. *For each* $k \geq 1$,

- Σ_k^1 *captures* Σ_k^p, *and*
- Π_k^1 *captures* Π_k^p.

In particular, SO *captures the polynomial hierarchy.*

9.3 Bibliographic Notes

Trakhtenbrot's theorem, one of the earliest results in finite model theory, was published in 1950 [234].

Fagin's theorem was published in 1974 [70, 71]. His motivation came from the complementation problem for spectra. The spectrum of a sentence Φ is the set $\{n \in \mathbb{N} \mid \Phi \text{ has a finite model of size } n\}$. The complementation problem (Asser [14]) asks whether spectra are closed under complement; that is, where the complement of the spectrum of Φ is the spectrum of some sentence Ψ.

If $\sigma = \{R_1, \ldots, R_n\}$ is the vocabulary of Φ, then the spectrum of Φ can be alternatively viewed as finite models (of the empty vocabulary) of the $\exists$SO sentence $\exists R_1 \ldots \exists R_n\ \Phi$ (by associating a universe of size n with n). Fagin defined *generalized spectra* as finite models of $\exists$SO sentences (i.e., the vocabulary no longer needs to be empty). The complementation problem for generalized spectra is then the problem whether NP equals coNP.

The result that $\exists$SO and $\forall$SO are different on $\langle \mathbb{N}, +, \cdot \rangle$ is due to Kleene [146]. In fact, over $\langle \mathbb{N}, +, \cdot \rangle$, the intersection of $\exists$SO and $\forall$SO collapses to FO, while over finite structures it properly contains FO.

Cook's theorem is from [39] (and is presented in many texts of complexity and computability, e.g. [126, 195]).

The polynomial hierarchy and its connection with SO are from Stockmeyer [223].

Sources for exercises:

Exercises 9.6 and 9.7:	Grädel [97]
Exercise 9.8:	Jones and Selman [140]
Exercise 9.9:	Lautemann, Schwentick, and Thérien [162]
Exercise 9.10:	Eiter, Gottlob, and Gurevich [63]
Exercise 9.11:	Gottlob, Kolaitis, and Schwentick [95]
Exercise 9.12:	Makowsky and Pnueli [178]
Exercise 9.13:	(a) from Fagin [72]
	(b) from Ajtai [10]
	(see also Fagin [74])

9.4 Exercises

Exercise 9.1. Prove Trakhtenbrot's theorem for an arbitrary vocabulary with at least one binary relation symbol.

Hint: use the binary relation symbol to code several binary relations, used in our proof of Trakhtenbrot's theorem.

Exercise 9.2. Prove that Trakhtenbrot's theorem fails for unary vocabularies: that is, if all the symbols in σ are unary, then finite satisfiability is decidable.

Exercise 9.3. Use Trakhtenbrot's theorem to prove that order invariance for FO queries is undecidable.

Exercise 9.4. Give a general definition of the formula ι from the proof of Fagin's theorem (i.e., for arbitrary σ and k).

Exercise 9.5. Complete the proof of Corollary 9.9.

Exercise 9.6. Show that there is an encoding schema for finite σ-structures such that the formulae ι from the proof of Fagin's theorem can be assumed to be quantifier-free, if the successor relation and the minimal and maximal element with respect to it can be used in formulae.

Exercise 9.7. Use the encoding scheme of Exercise 9.6 to prove that every NP can be defined by an $\exists$SO sentence whose first-order part is universal (i.e., of the form $\forall \ldots \forall\ \psi$, where ψ is quantifier-free), under the assumption that we consider structures with explicitly given order and successor relations, as well as constants for the minimal and the maximal elements.

Prove that without these assumptions, universal first-order quantification in $\exists$SO formulae is not sufficient to capture all of NP. What kind of quantifier prefixes does one need in the general case?

Exercise 9.8. Prove that a set $X \subseteq \mathbb{N}$ is a spectrum iff it is in NEXPTIME. Explain why this does not contradict Fagin's theorem.

Exercise 9.9. Consider the vocabulary $\sigma_\Sigma = (<, (P_a)_{a\in\Sigma})$ used in Chap. 7 for coding strings as finite structures. Recall that a sentence Φ over such vocabulary defines a language (a subset of Σ^*) given by $\{s \in \Sigma^* \mid M_s \models \Phi\}$.

Consider a restriction $\exists\mathrm{SO}_{\mathrm{match}}$ of $\exists$SO in which existential second-order variables range over *matchings*: that is, binary relations of the form $\{(x_i, y_i) \mid i \leq k\}$ where all x_i's and y_i's are distinct.

Prove that a language is definable in $\exists\mathrm{SO}_{\mathrm{match}}$ iff it is context-free.

Exercise 9.10. Let $\mathcal{S}$ be a set of quantifier prefixes, and let $\exists\mathrm{SO}(\mathcal{S})$ be the fragment of $\exists$SO which consists of sentences of the form $\exists R_1 \ldots \exists R_n\ \varphi$, where φ is a prenex formula whose quantifier prefix is in $\mathcal{S}$. We call $\exists\mathrm{SO}(\mathcal{S})$ *regular* if over strings it only defines regular languages.

Prove the following:

- $\exists\mathrm{SO}(\forall^*\exists\forall^*)$ is regular;
- $\exists\mathrm{SO}(\exists^*\forall\forall)$ is regular;
- if $\exists\mathrm{SO}(\mathcal{S})$ is regular, then it is contained in the union of $\exists\mathrm{SO}(\forall^*\exists\forall^*)$ and $\exists\mathrm{SO}(\exists^*\forall\forall)$;
- if $\exists\mathrm{SO}(\mathcal{S})$ is not regular, then it defines some NP-complete language.

Exercise 9.11. We now consider $\exists\mathrm{SO}(\mathcal{S})$ and $\exists\mathrm{MSO}(\mathcal{S})$ over directed graphs. Prove the following:

- $\exists\mathrm{SO}(\exists^*\forall)$ only defines polynomial time properties of graphs;
- $\exists\mathrm{SO}(\forall\forall)$ and $\exists\mathrm{MSO}(\exists^*\forall\forall)$ in which at most one second-order quantifier is used only define polynomial time properties of graphs;
- each of the following defines some NP-complete problems on graphs:
  - $\exists\mathrm{SO}(\exists\forall\forall)$, where only one second-order quantifier over binary relations is used;

– $\exists$MSO($\forall\exists$) and $\exists$MSO($\forall\forall\forall$), where only one second-order quantifier is used;
– $\exists$MSO($\forall\forall$), where only two second-order quantifiers are used.

Exercise 9.12. Define SO(k,m) as the union of Σ^1_k and Π^1_k where all quantification is over relations of arity at most m. That is, SO(k,m) is the restriction of SO to at most $k-1$ alternations of quantifiers, and quantification is over relations of arity m. This is usually referred to as the *alternation-arity hierarchy.*

Prove that the alternation-arity hierarchy is strict: that is, there is a constant c such that

$$\mathrm{SO}(k,m) \subsetneqq \mathrm{SO}(k+c, m+c)$$

for all k, m.

Exercise 9.13. Define $\exists$SO(m) as the restriction of class of $\exists$SO to second-order quantification over relations of arity at most m. Prove the following:

(a) If $\exists\mathrm{SO}(m) = \exists\mathrm{SO}(m+1)$, then $\exists\mathrm{SO}(k) = \exists\mathrm{SO}(m)$ for every $k \geq m$.
(b) If σ contains an m-ary relation symbol P, then the class of structures in which P has an even number of tuples is not $\exists$SO$(m-1)$-definable.
(c) Conclude from (a) and (b) that, if σ contains an m-ary relation symbol P, then $\exists\mathrm{SO}(i) \subsetneqq \exists\mathrm{SO}(j)$ over σ-structures, for every $1 \leq i < j \leq m$.

Exercise 9.14.* Now consider just the *arity* hierarchy for SO: that is, SO(m) is defined as $\bigcup_{k\in\mathbb{N}} \mathrm{SO}(k,m)$. Is the arity hierarchy strict?

Exercise 9.15.* We call a sentence categorical if it has at most one model of each finite cardinality. Is it true that every spectrum is a spectrum of a categorical sentence?

10

Fixed Point Logics and Complexity Classes

Most logics we have seen so far are not well suited for expressing many tractable graph properties, such as graph connectivity, reachability, and so on. The limited expressiveness of FO and counting logics is due to the fact that they lack mechanisms for expressing fixed point computations. Other logics we have seen, such as MSO, ∃SO, and ∀SO, can express intractable graph properties.

Consider, for example, the transitive closure query. Given a binary relation R, we can express relations $R^0, R^1, R^2, R^3, \ldots$, where R^i contains pairs (a, b) such that there is a path from a to b of length at most i. To compute the transitive closure of R, we need the union of all those relations: that is,

$$R^\infty = \bigcup_{i=0}^{\infty} R^i.$$

How could one compute such a union? Since relation R is finite, starting with some n, the sequence $R^i, i \geq 0$, stabilizes: $R^n = R^{n+1} = R^{n+2} = \ldots$. Indeed, in this case n can be taken to be the number of elements of relation R. Hence, $R^\infty = R^n$; that is, R^n is the limit of the sequence $R^i, i > 0$. But we can also view R^n as a fixed point of an operator that sends each R^i to R^{i+1}.

In this chapter we study logics extended with operators for computing fixed points of various operators. We start by presenting the basics of fixed point theory (in a rather simplified way, adapted for finite structures). We then define various extensions of FO with fixed point operators, study their expressiveness, and show that on ordered structures these extensions capture complexity classes PTIME and PSPACE. Finally, we show how to extend FO with an operator for computing just the transitive closure, and prove that this extension captures NLOG on ordered structures.

10.1 Fixed Points of Operators on Sets

Typically the theory of fixed point operators is presented for complete lattices: that is, partially ordered sets $\langle U, \prec\rangle$ where every – finite or infinite – subset of U has a greatest lower bound and a least upper bound in the ordering $\prec$. However, here we deal only with finite sets, which somewhat simplifies the presentation.

Given a set U, let $\wp(U)$ be its powerset. An *operator* on U is a mapping $F : \wp(U) \to \wp(U)$. We say that an operator F is *monotone* if

$$X \subseteq Y \text{ implies } F(X) \subseteq F(Y),$$

and *inflationary* if

$$X \subseteq F(X)$$

for all $X \in \wp(U)$.

Definition 10.1. *Given an operator $F : \wp(U) \to \wp(U)$, a set $X \subseteq U$ is a* fixed point *of F if $F(X) = X$. A set $X \subseteq A$ is a* least fixed point *of F if it is a fixed point, and for every other fixed point Y of F we have $X \subseteq Y$. The least fixed point of F will be denoted by* $\mathbf{lfp}(F)$.

Let us now consider the following sequence:

$$X^0 = \emptyset, \quad X^{i+1} = F(X^i). \tag{10.1}$$

We call F *inductive* if the sequence (10.1) is increasing: $X^i \subseteq X^{i+1}$ for all i. Every monotone operator F is inductive, which is shown by a simple induction. Of course $X^0 \subseteq X^1$ since $X^0 = \emptyset$. If $X^i \subseteq X^{+1}$, then, by monotonicity, $F(X^i) \subseteq F(X^{i+1})$; that is, $X^{i+1} \subseteq X^{i+2}$. This shows that $X^i \subseteq X^{i+1}$ for all $i \in \mathbb{N}$.

If F is inductive, we define

$$X^\infty = \bigcup_{i=0}^{\infty} X^i. \tag{10.2}$$

Since U is assumed to be finite, the sequence (10.1) actually stabilizes after some finite number of steps, so there is a number n such that $X^\infty = X^n$.

To give an example, let R be a binary relation on a finite set A, and let $F : \wp(A^2) \to \wp(A^2)$ be the operator defined by $F(X) = R \cup (R \circ X)$. Here $\circ$ is the relational composition: $R \circ X = \{(a, b) \mid (a, c) \in R, (c, b) \in X, \text{ for some } c \in A\}$. Notice that this operator is monotone: if $X \subseteq Y$, then $R \circ X \subseteq R \circ Y$. Let us now define the sequence $X^i, i \geq 0$, as in (10.1). First, $X^0 = \emptyset$. Since $R \circ \emptyset = \emptyset$, we have $X^1 = R$. Then $X^2 = R \cup (R \circ R) = R \cup R^2$; that is, the set of pairs (a, b) such that there is a path of length at most 2 from a to b. Continuing, we see that $X^i = R \cup \ldots \cup R^i$, the set of pairs connected by paths of length at most i. This sequence reaches a fixed point X^∞, which is the transitive closure of R.

We now prove that every monotone operator has a least fixed point, which is the set X^∞ (10.2), defined as the union of the increasing sequence (10.1).

Theorem 10.2 (Tarski-Knaster). *Every monotone operator* $F : \wp(U) \to \wp(U)$ *has a least fixed point* $\mathbf{lfp}(F)$ *which can be defined as*

$$\mathbf{lfp}(F) \quad = \quad \bigcap\{Y \mid Y = F(Y)\}.$$

Furthermore, $\mathbf{lfp}(F) = X^\infty = \bigcup_i X^i$, *for the sequence* X^i *defined by (10.1).*

Proof. Let $\mathcal{W} = \{Y \mid F(Y) \subseteq Y\}$. Clearly, $\mathcal{W} \neq \emptyset$, since $U \in \mathcal{W}$. We first show that $S = \bigcap \mathcal{W}$ is a fixed point of F. Indeed, for every $Y \in \mathcal{W}$, we have $S \subseteq Y$ and hence $F(S) \subseteq F(Y) \subseteq Y$; therefore, $F(S) \subseteq \bigcap \mathcal{W} = S$. On the other hand, since $F(S) \subseteq S$, we have $F(F(S)) \subseteq F(S)$, and thus $F(S) \in \mathcal{W}$. Hence, $S = \bigcap \mathcal{W} \subseteq F(S)$, which proves $S = F(S)$.

Let $\mathcal{W}' = \{Y \mid F(Y) = Y\}$ and $S' = \bigcap \mathcal{W}'$. Then $S \in \mathcal{W}'$ and hence $S' \subseteq S$; on the other hand, $\mathcal{W}' \subseteq \mathcal{W}$, so $S = \bigcap \mathcal{W} \subseteq \bigcap \mathcal{W}' = S'$. Hence, $S = S'$. Thus, $S = \bigcap\{Y \mid Y = F(Y)\}$ is a fixed point of F. Since it is the intersection of all the fixed points of F, it is the least fixed point of F. This shows that

$$\mathbf{lfp}(F) \quad = \quad \bigcap\{Y \mid Y = F(Y)\} \quad = \quad \bigcap\{Y \mid F(Y) \subseteq Y\}.$$

To prove that $\mathbf{lfp}(F) = X^\infty$, note that the sequence X^i increases, and hence for some $n \in \mathbb{N}$, $X^n = X^{n+1} = \ldots = X^\infty$. Thus, $F(X^\infty) = X^\infty$ and X^∞ is a fixed point. To show that it is the least fixed point, it suffices to prove that $X^i \subseteq Y$ for every i and every $Y \in \mathcal{W}$. We prove this by induction on i. Clearly $X^0 \subseteq Y$ for all $Y \in \mathcal{W}$. Suppose we need to prove the statement for X^{i+1}. Let $Y \in \mathcal{W}$. We have $X^{i+1} = F(X^i)$. By the hypothesis, $X^i \subseteq Y$, and by monotonicity, $F(X^i) \subseteq F(Y) \subseteq Y$. Hence, $X^{i+1} \subseteq Y$. This shows that all the X^i's are contained in all the sets of $\mathcal{W}$, and completes the proof of the theorem. □

Not all the operators of interest are monotone. We now present two different constructions by means of which the fixed point of non-monotone operators can be defined.

Suppose F is inflationary: that is, $Y \subseteq F(Y)$ for all Y. Then F is inductive; that is, the sequence (10.1) is increasing, and hence it reaches a fixed point X^∞. Now suppose G is an arbitrary operator. With G, we associate an inflationary operator G_{infl} defined by $G_{\text{infl}}(Y) = Y \cup G(Y)$. Then X^∞ for G_{infl} is called the *inflationary fixed point* of G and is denoted by $\mathbf{ifp}(G)$. In other words, $\mathbf{ifp}(G)$ is the union of all sets X^i where $X^0 = \emptyset$ and $X^{i+1} = X^i \cup G(X^i)$.

Finally, we consider an arbitrary operator $F : \wp(U) \to \wp(U)$ and the sequence (10.1). This sequence need not be inductive, so there are two possibilities. The first is that this sequence reaches a fixed point; that is, for some

$n \in \mathbb{N}$ we have $X^n = X^{n+1}$, and thus for all $m > n$, $X^m = X^n$. If there is such an n, it must be the case that $n \leq 2^{|U|}$, since there are only $2^{|U|}$ subsets of U. The second possibility is that no such n exists.

We now define the *partial fixed point* of F as

$$\mathbf{pfp}(F) = \begin{cases} X^n & \text{if } X^n = X^{n+1} \\ \emptyset & \text{if } X^n \neq X^{n+1} \text{ for all } n \leq 2^{|U|}. \end{cases}$$

The definition is unambiguous: since $X^n = X^{n+1}$ implies that the sequence (10.1) stabilizes, then $X^n = X^{n+1}$ and $X^m = X^{m+1}$ imply that $X^n = X^m$.

We leave the following as an easy exercise to the reader.

Proposition 10.3. *If F is monotone, then* $\mathbf{lfp}(F) = \mathbf{ifp}(F) = \mathbf{pfp}(F)$. □

10.2 Fixed Point Logics

We now show how to add fixed point operators to FO. Suppose we have a relational vocabulary σ, and an additional relation symbol $R \notin \sigma$ of arity k. Let $\varphi(R, x_1, \ldots, x_k)$ be a formula of vocabulary $\sigma \cup \{R\}$. We put the symbol R explicitly as a parameter, since this formula will give rise to an operator on σ-structures.

For each $\mathfrak{A} \in \text{STRUCT}[\sigma]$, the formula $\varphi(R, \vec{x})$ gives rise to an operator $F_\varphi : \wp(A^k) \to \wp(A^k)$ defined as follows:

$$F_\varphi(X) = \{\vec{a} \mid \mathfrak{A} \models \varphi(X/R, \vec{a})\}. \tag{10.3}$$

Here the notation $\varphi(X/R, \vec{a})$ means that R is interpreted as X in φ; more precisely, if $\mathfrak{A}'$ is a $(\sigma \cup \{R\})$-structure expanding $\mathfrak{A}$, in which R is interpreted as X, then $\mathfrak{A}' \models \varphi(\vec{a})$.

The idea of fixed point logics is that we add formulae for computing fixed points of operators F_φ. This already gives us formal definitions of logics IFP and PFP.

Definition 10.4. *The logics* IFP *and* PFP *are defined as extensions of* FO *with the following formation rules:*

- *(For* IFP*): if $\varphi(R, \vec{x})$ is a formula, where R is k-ary, and $\vec{t}$ is a tuple of terms, where $|\vec{x}| = |\vec{t}| = k$, then*

  $$[\mathbf{ifp}_{R,\vec{x}}\varphi(R, \vec{x})](\vec{t})$$

  is a formula, whose free variables are those of $\vec{t}$.
- *(For* PFP*): if $\varphi(R, \vec{x})$ is a formula, where R is k-ary, and $\vec{t}$ is a tuple of terms, where $|\vec{x}| = |\vec{t}| = k$, then*

  $$[\mathbf{pfp}_{R,\vec{x}}\varphi(R, \vec{x})](\vec{t})$$

  is a formula, whose free variables are those of $\vec{t}$.

The semantics is defined as follows:

- *(For* IFP*):* $\mathfrak{A} \models [\mathbf{ifp}_{R,\vec{x}}\varphi(R,\vec{x})](\vec{a})$ *iff* $\vec{a} \in \mathbf{ifp}(F_\varphi)$.
- *(For* PFP*):* $\mathfrak{A} \models [\mathbf{pfp}_{R,\vec{x}}\varphi(R,\vec{x})](\vec{a})$ *iff* $\vec{a} \in \mathbf{pfp}(F_\varphi)$.

Why could we not define an extension with the least fixed point in exactly the same way? The reason is that least fixed points are guaranteed to exist only for monotone operators. However, monotonicity is not an easy property to deal with.

Lemma 10.5. *Testing if F_φ is monotone is undecidable for* FO *formulae φ.*

Proof. Let Φ be an arbitrary sentence, and $\varphi(S,x) \equiv (S(x) \to \Phi)$. Suppose Φ is valid. Then $\varphi(S,x)$ is always true and hence F_φ is monotone in every structure. Suppose now that $\mathfrak{A} \models \neg\Phi$ for some nonempty structure $\mathfrak{A}$. Then, over $\mathfrak{A}$, $\varphi(S,x)$ is equivalent to $\neg S(x)$, and hence F_φ is not monotone. Therefore, F_φ is monotone iff Φ is true in every nonempty structure, which is undecidable, by Trakhtenbrot's theorem. □

Thus, to ensure that least fixed points are only taken for monotone operators, we impose some syntactic restrictions. Given a formula φ that may contain a relation symbol R, we say that an occurrence of R is *negative* if it is under the scope of an odd number of negations, and is *positive*, if it is under the scope of an even number of negations. For example, in the formula $\exists x \neg R(x) \vee \neg\forall y \forall z \neg(R(y) \wedge \neg R(z))$, the first occurrence of R (i.e., $R(x)$) is negative, the second ($R(y)$) is positive (as it is under the scope of two negations), and the last one ($R(z)$) is negative again. We say that a formula is *positive in R* if there are no negative occurrences of R in it; in other words, either all occurrences of R are positive, or there are none at all.

Definition 10.6. *The logic* LFP *extends* FO *with the following formation rule:*

- *if $\varphi(R,\vec{x})$ is a formula positive in R, where R is k-ary and $\vec{t}$ is a tuple of terms, where $|\vec{x}|=|\vec{t}|=k$, then*

$$[\mathbf{lfp}_{R,\vec{x}}\varphi(R,\vec{x})](\vec{t})$$

is a formula, whose free variables are those of $\vec{t}$.

The semantics is defined as follows:

$$\mathfrak{A} \models [\mathbf{lfp}_{R,\vec{x}}\varphi(R,\vec{x})](\vec{a}) \quad \textit{iff} \quad \vec{a} \in \mathbf{lfp}(F_\varphi).$$

Of course, there is something to be proven here:

Lemma 10.7. *If $\varphi(R, \vec{x})$ is positive in R, then F_φ is monotone.* □

The proof is by an easy induction on the structure of the formula (which includes the cases of Boolean connectives, quantifiers, and **lfp** operators) and is left as an exercise to the reader.

We now give a few examples of queries definable in fixed point logics.

Transitive Closure and Acyclicity

Let E be a binary relation, and let $\varphi(R, x, y)$ be

$$E(x, y) \vee \exists z\ (E(x, z) \wedge R(z, y)).$$

Clearly, this is positive in R. Let $\psi(u, v)$ be $[\mathbf{lfp}_{R,x,y}\varphi(R, x, y)](u, v)$. What does this formula define?

To answer this, we must consider the operator F_φ. For a set X, we have $F_\varphi(X) = E \cup (E \circ X)$. We have seen this operator in the previous section, and know that its least fixed point is the transitive closure of E. Hence, $\psi(u, v)$ defines the transitive closure of E. This also implies that graph connectivity is LFP-definable by the sentence $\forall u \forall v\ \psi(u, v)$.

As the next example, we again consider graphs whose edge relation is E, and the formula $\alpha(S, x)$ given by

$$\forall y\ \big(E(y, x) \rightarrow S(y)\big).$$

This formula is again positive in S. The operator F_α associated with this formula takes a set X and returns the set of all nodes a such that all the nodes b from which there is an edge to a are in X. Let us now iterate this operator. Clearly, $F_\alpha(\emptyset)$ is the set of nodes of in-degree 0. Then $F_\alpha(F_\alpha(\emptyset))$ is the set of nodes a such that all nodes b with edges $(b, a) \in E$ have in-degree 0. Reformulating this, we can state that $F_\alpha(F_\alpha(\emptyset))$ is the set of nodes a such that all paths ending in a have length at most 1. Following this, at the ith stage of the iteration we get the set of nodes a such that all the paths ending in a have length at most i. When we reach the fixed point, we have nodes such that all the paths ending in them are finite. Hence, the formula

$$\forall u\ [\mathbf{lfp}_{S,x}\alpha(S, x)](u)$$

tests if a graph is acyclic.

Arithmetic on Successor Structures

As a third example, consider structures of vocabulary ($\underline{\text{min}}$, $\underline{\text{succ}}$), where $\underline{\text{succ}}$ is interpreted as a successor relation on the universe, and $\underline{\text{min}}$ is the minimal element with respect to $\underline{\text{succ}}$. That is, the structures will be of the form $\langle\{0, \ldots, n-1\},\ 0,\ \{(i, i+1) \mid i+1 \leq n-1\}\rangle$. We show how to define

$$+ = \{(i, j, k) \mid i + j = k\} \quad \text{and} \quad \times = \{(i, j, k) \mid i \cdot j = k\}$$

on such structures. For $+$, we use the recursive definition:

$$x + 0 = x$$
$$x + (y+1) = (x+y) + 1.$$

Let R be ternary and $\beta_+(R, x, y, z)$ be

$$(y = \underline{\min} \wedge z = x) \ \vee \ \exists u \exists v \ \big(R(x,u,v) \ \wedge \ \underline{\mathrm{succ}}(u,y) \ \wedge \ \underline{\mathrm{succ}}(v,z)\big).$$

Intuitively, it states the conditions for (x, y, z) to be in the graph of addition: either $y = 0$ and $x = z$, or, if we already know that $x + u = v$, and $y = u + 1, z = v + 1$, then we can infer $x + y = z$. This formula is positive in R, and the least fixed point computes the graph of addition:

$$\varphi_+(x,y,z) \ = \ [\mathbf{lfp}_{R,x,y,z}\beta_+(R,x,y,z)](x,y,z).$$

Using addition, we can define multiplication:

$$x \cdot 0 = 0$$
$$x \cdot (y+1) = x \cdot y + x.$$

Similarly to the case of addition, we define $\beta_\times(S, x, y, z)$ as

$$\big(y = \underline{\min} \wedge z = \underline{\min}\big) \ \vee \ \exists u \exists v \ \big(S(x,u,v) \ \wedge \ \underline{\mathrm{succ}}(u,y) \ \wedge \ \varphi_+(x,v,z)\big).$$

This formula is positive in S. Then

$$\varphi_\times(x,y,z) \ = \ [\mathbf{lfp}_{S,x,y,z}\beta_\times(S,x,y,z)](x,y,z)$$

defines the graph of multiplication. Since it uses φ_+ as a subformula, this gives us an example of *nested* least fixed point operators.

Combining this example with Theorem 6.12, we conclude that BIT is LFP-definable over successor structures.

A Game on Graphs

Consider the following game played on a graph $G = \langle V, E \rangle$ with a distinguished start node a. There are two players: player I and player II. At each round i, first player I selects a node b_i and then player II selects a node c_i, such that (a, b_1), as well as (b_i, c_i) and (c_i, b_{i+1}), are edges in E, for all i. The player who cannot make a legal move loses the game.

Let S be unary, and define $\alpha(S, x)$ as

$$\forall y \ \Big(E(x,y) \rightarrow \exists z \ \big(E(y,z) \wedge S(z)\big)\Big).$$

What is $F_\alpha(\emptyset)$? It is the set of nodes b of out-degree 0; that is, nodes in which player II wins, since player I does not have a single move. In general, $F_\alpha(X)$ is the set of nodes b such that no matter where player I moves from b, player

II will have a response from X. Thus, iterating F_α, we see that the ith stage consists of nodes from which player II has a winning strategy in at most $i-1$ rounds. Hence,

$$[\mathbf{lfp}_{S,x}\alpha(S,x)](a)$$

holds iff player II has a winning strategy from node a.

We conclude this section by a remark concerning free variables in fixed point formulae. So far, in the definition and all the examples we dealt with iterating formulae $\varphi(R,\vec{x})$ where $\vec{x}$ matched the arity of R. However, in general one can imagine that φ has additional free variables. For example, if we have a formula $\varphi(R,\vec{x},\vec{y})$ positive in R, we can, for each tuple $\vec{b}$, define an operator $F_\varphi^{\vec{b}}(X) = \{\vec{a} \mid \mathfrak{A} \models \varphi(X/R,\vec{a},\vec{b})\}$, and a formula $\psi(\vec{t},\vec{y}) \equiv [\mathbf{lfp}_{R,\vec{x}}\varphi(R,\vec{x},\vec{y})](\vec{t})$, with the semantics $\mathfrak{A} \models \psi(\vec{c},\vec{b})$ iff $\vec{c} \in \mathbf{lfp}(F_\varphi^{\vec{b}})$.

It turns out, however, that free variables in fixed point formulae can always be avoided, at the expense of relations of higher arity. Indeed, the formula $\psi(\vec{t},\vec{y})$ above is equivalent to $[\mathbf{lfp}_{R',\vec{x},\vec{y}}\varphi'(R',\vec{x},\vec{y})](\vec{t},\vec{y})$, where R' is of arity $\mid \vec{x} \mid + \mid \vec{y} \mid$, and φ' is obtained from φ by changing every occurrence of a subformula $R(\vec{z})$ to $R'(\vec{z},\vec{y})$. This is left as an exercise to the reader. Thus, we shall normally assume that no extra parameters are present in fixed point formulae.

10.3 Properties of LFP and IFP

In this section we study logics LFP and IFP. We start by introducing a very convenient tool of *simultaneous fixed points*, which allows one to iterate several formulae at once. We then analyze fixed point computations, and show how to define and compare their stages (that is, sets X^i as in (10.1)). From this analysis we shall derive two important conclusions. One is that LFP = IFP on finite structures. The other is a normal form for LFP, showing that nested occurrences of fixed point operators (which we saw in the multiplication example in the previous section) can be eliminated.

Let σ be a relational vocabulary, and $R_1,\ldots,R_n$ additional relation symbols, with R_i being of arity k_i. Let $\vec{x}_i$ be a tuple of variables of length k_i. Consider a sequence Φ of formulae

$$\begin{array}{c}\varphi_1(R_1,\ldots,R_n,\vec{x}_1),\\ \ldots,\\ \varphi_n(R_1,\ldots,R_n,\vec{x}_n)\end{array} \qquad (10.4)$$

of vocabulary $\sigma \cup \{R_1,\ldots,R_n\}$. Assume that all φ_i's are positive in all R_j's. Then, for a σ-structure $\mathfrak{A}$, each φ_i defines an operator

$$F_i : \wp(A^{k_1}) \times \ldots \times \wp(A^{k_n}) \to \wp(A^{k_i})$$

given by

$$F_i(X_1, \ldots, X_n) = \{\vec{a} \in A^{k_i} \mid \mathfrak{A} \models \varphi_i(X_1/R_1, \ldots, X_n/R_n, \vec{a})\}.$$

We can combine these operators F_i's into one operator

$$\vec{F} : \wp(A^{k_1}) \times \ldots \times \wp(A^{k_n}) \to \wp(A^{k_1}) \times \ldots \times \wp(A^{k_n})$$

given by

$$\vec{F}(X_1, \ldots, X_n) = (F_1(X_1, \ldots, X_n), \ldots, F_n(X_1, \ldots, X_n)).$$

A sequence of sets $(X_1, \ldots, X_n)$ is a *fixed point* of $\vec{F}$ if $\vec{F}(X_1, \ldots, X_n) = (X_1, \ldots, X_n)$. Furthermore, if for every fixed point $(Y_1, \ldots, Y_n)$ we have $X_1 \subseteq Y_1, \ldots, X_n \subseteq Y_n$, then we speak of the least fixed point of $\vec{F}$.

The product $\wp(A^{k_1}) \times \ldots \times \wp(A^{k_n})$ is partially ordered component-wise by $\subseteq$, and the operator $\vec{F}$ is component-wise monotone. Hence, it can be iterated in the same way as usual monotone operators on $\wp(U)$; that is,

$$\begin{aligned} \vec{X}^0 &= (\emptyset, \ldots, \emptyset) \\ \vec{X}^{i+1} &= \vec{F}(\vec{X}^i) \\ \vec{X}^\infty &= \bigcup_{i=1}^{\infty} \vec{X}^i = \Big(\bigcup_{i=1}^{\infty} X_1^i, \ldots, \bigcup_{i=1}^{\infty} X_n^i\Big). \end{aligned} \tag{10.5}$$

Just as for the case of the usual operators on sets, one can prove that $\vec{X}^\infty = \mathbf{lfp}(\vec{F})$. We then enrich the syntax of LFP with the rule that if Φ is a family of formulae (10.4), and $\vec{t}$ is a tuple of terms of length k_i, then

$$[\mathbf{lfp}_{R_i, \Phi}](\vec{t})$$

is a formula with the semantics $\mathfrak{A} \models [\mathbf{lfp}_{R_i, \Phi}](\vec{a})$ iff $\vec{a}$ belongs to the ith component of $\vec{X}^\infty$. The resulting logic will be denoted by $\text{LFP}^{\text{simult}}$.

As an example of a property expressible in $\text{LFP}^{\text{simult}}$, consider the following query Q on *undirected* graphs $G = \langle V, E \rangle$: it returns the set of nodes (a, b) such that there is a simple path of even length from a to b.

Let T be a ternary relation symbol, and R, S binary relation symbols. We consider the following system Φ of formulae:

$$\varphi_1(T, R, S, x, y, z) \equiv \begin{array}{l} (E(x,y) \wedge \neg(x = z) \wedge \neg(y = z)) \\ \vee\ \exists u\ \big(E(x,u) \wedge T(u,y,z) \wedge \neg(x = z)\big) \end{array}$$

$$\varphi_2(T, R, S, x, y) \equiv \begin{array}{l} E(x,y) \\ \vee\ \exists u\ \big(S(x,u) \wedge E(u,y) \wedge T(x,u,y)\big) \end{array}$$

$$\varphi_3(T, R, S, x, y) \equiv \exists u\ \big(R(x,u) \wedge R(u,y) \wedge T(x,u,y)\big).$$

Notice that these formulae are positive in R, S, T. We leave it to the reader to verify that the simultaneous least fixed point of this system Φ computes the following relations:

- $T^\infty(a, b, c)$ holds iff there is a simple path from a to b that does not pass through c;
- $R^\infty(a, b)$ holds iff there is a simple path from a to b of odd length; and
- $S^\infty(a, b)$ holds iff there is a simple path from a to b of even length.

Thus, $[\mathbf{lfp}_{S,\Phi}](x, y)$ expresses the query Q. (See Exercise 10.2.)

Simultaneous fixed points are often convenient for expressing complex properties, when several sets need to be defined at once. The question is then whether such fixed points enrich the expressiveness of the logic. The answer, as we are about to show, is negative.

Theorem 10.8. $\text{LFP}^{\text{simult}} = \text{LFP}$.

Proof. We give the proof for the case of a system Φ consisting of two formulae, $\varphi_1(R, S, \vec{x})$ and $\varphi_2(R, S, \vec{y})$. Extension to an arbitrary system is rather straightforward, and left as an exercise for the reader (Exercise 10.3). The idea is that we combine a simultaneous fixed point into two fixed point formulae, in which the **lfp** operators are nested.

We need an auxiliary result first. Assume we have two monotone operators

$$F_1 : \wp(U) \times \wp(V) \to \wp(U) \;\text{ and }\; F_2 : \wp(U) \times \wp(V) \to \wp(V).$$

Following (10.5), we define the stages of the operator (F_1, F_2) as $\vec{X}^0 = (X_1^0, X_2^0) = (\emptyset, \emptyset)$, $\vec{X}^{i+1} = (X_1^{i+1}, X_2^{i+1}) = (F_1(\vec{X}^i), F_2(\vec{X}^i))$, with the fixed point (X_1^∞, X_2^∞).

Fix a set $Y \subseteq U$, and define two operators:

$$F_2^Y : \wp(V) \to \wp(V), \quad F_2^Y(Z) = F_2(Y, Z);$$

$$G_1 : \wp(U) \to \wp(U), \quad G_1(Y) = F_1(Y, \mathbf{lfp}(F_2^Y)).$$

Clearly, F_2^Y is monotone, and hence $\mathbf{lfp}(F_2^Y)$ is well-defined. The operator G_1 is monotone as well (since for $Y \subseteq Y'$, it is the case that $\mathbf{lfp}(F_2^Y) \subseteq \mathbf{lfp}(F_2^{Y'})$), and hence it has a least fixed point.

To prove the theorem, we need the following lemma, which is sometimes referred to as the Bekic principle.

Lemma 10.9. $X_1^\infty = \mathbf{lfp}(G_1)$.

Before we prove the lemma, we show that the theorem follows from it. Since $X_1^\infty = \mathbf{lfp}(G_1)$, we have to express G_1 in **lfp**, which can be done, as G_1 is defined as the least fixed point of a certain operator. In fact, it follows from the definition of G_1 that $[\mathbf{lfp}_{R,\Phi}](\vec{t})$ is equivalent to

$$\Big[\mathbf{lfp}_{R,\vec{x}}\ \varphi_1\big(R,\ [\mathbf{lfp}_{S,\vec{y}}\varphi_2(R, S, \vec{y})]\ /\ S,\ \vec{x}\big)\Big](\vec{t}).$$

The roles of F_1 and F_2 can be reversed; that is, we can define $F_1^Y(Z) = F_1(Z,Y) : \wp(U) \to \wp(U)$ and $G_2 : \wp(V) \to \wp(V)$ by $G_2(Y) = F_2(\mathbf{lfp}(F_1^Y), Y)$, and prove, as in Lemma 10.9, that $X_2^\infty = \mathbf{lfp}(G_2)$. Therefore,

$$\Big[\mathbf{lfp}_{S,\vec{y}}\ \varphi_2\big([\mathbf{lfp}_{R,\vec{x}}\varphi_1(R,S,\vec{x})]\ /\ R,\ S,\ \vec{y}\big)\Big](\vec{t})$$

is equivalent to $[\mathbf{lfp}_{S,\Phi}](\vec{t})$.

It remains to prove Lemma 10.9. First, notice that $\mathbf{lfp}(F_2^{X_1^\infty}) \subseteq X_2^\infty$, because $F_2^{X_1^\infty}(X_2^\infty) = F_2(X_1^\infty, X_2^\infty) = X_2^\infty$. That is, X_2^∞ is a fixed point of $F_2^{X_1^\infty}$, and thus it must contain its least fixed point. Hence, $G_1(X_1^\infty) = F_1(X_1^\infty, \mathbf{lfp}(F_2^{X_1^\infty})) \subseteq F_1(X_1^\infty, X_2^\infty) = X_1^\infty$. Since $\mathbf{lfp}(G_1)$ is the intersection of all the set S such that $G_1(S) \subseteq S$, we conclude that $\mathbf{lfp}(G_1) \subseteq X_1^\infty$.

Next, we prove the reverse inclusion $X_1^\infty \subseteq \mathbf{lfp}(G_1)$. We use Z to denote $\mathbf{lfp}(G_1)$. We show inductively that for each i, $X_1^i \subseteq Z$ and $X_2^i \subseteq \mathbf{lfp}(F_2^Z)$.

This is clear for $i = 0$. To go from i to $i+1$, calculate

$$X_1^{i+1} = F_1(X_1^i, X_2^i) \subseteq F_1(Z, \mathbf{lfp}(F_2^Z)) = G_1(\mathbf{lfp}(G_1)) = \mathbf{lfp}(G_1) = Z,$$

and

$$X_2^{i+1} = F_2(X_1^i, X_2^i) \subseteq F_2(Z, \mathbf{lfp}(F_2^Z)) = F_2^Z(\mathbf{lfp}(F_2^Z)) = \mathbf{lfp}(F_2^Z).$$

Thus,

$$X_1^\infty = \bigcup_{i=0}^{\infty} X_1^i \subseteq \mathbf{lfp}(G_1).$$

This completes the proof of Lemma 10.9 and Theorem 10.8. ☐

One can similarly define logics $\text{IFP}^{\text{simult}}$ and $\text{PFP}^{\text{simult}}$, by allowing simultaneous inflationary and partial fixed points. It turns out that for IFP and PFP, simultaneous fixed points do not increase expressiveness either. The proof presented for LFP would not work, as it relies on the monotonicity of operators defined by formulae, which cannot be guaranteed for arbitrary formulae used in the definition of the logics IFP and PFP. Nevertheless, a different technique works for these logics. We explain it now by means of an example; details are left as an exercise for the reader.

Assume that the vocabulary σ has two constant symbols c_1 and c_2 interpreted as two distinct elements of σ-structure. This assumption is easy to get rid of, by existentially quantifying over two variables, u and w, and stating that $u \neq w$; however, formulae with constants will be easier to deal with. Furthermore, we can assume without loss of generality that structures have at least two elements, since the case of one-element structures can be dealt with explicitly by specifying the value of a fixed point operator on them.

Suppose we have two formulae, $\varphi_1(R_1, R_2, \vec{x})$ and $\varphi_2(R_1, R_2, \vec{x})$, where the arities of R_1 and R_2 are n, and the length of $\vec{x}$ is n. Let S be a relation symbol of arity $n+1$, and let $\psi(S, u, \vec{x})$ be the formula

$$\begin{aligned}&\Big((u = c_1) \ \wedge \ \varphi_1\big(S(c_1,\vec{z})/R_1(\vec{z}),\ S(c_2,\vec{z})/R_2(\vec{z}),\ \vec{x}\big)\Big)\\ \vee\ &\Big((u = c_2) \ \wedge \ \varphi_2\big(S(c_1,\vec{z})/R_1(\vec{z}),\ S(c_2,\vec{z})/R_2(\vec{z}),\ \vec{x}\big)\Big),\end{aligned}$$

where $S(c_i, \vec{z})/R_i(\vec{z})$ indicates that every occurrence of $R_i(\vec{z})$ is replaced by $S(c_i, \vec{z})$. Then the fixed point – inflationary or partial – of this formula ψ computes the simultaneous fixed point of the system $\{\varphi_1, \varphi_2\}$: the fixed point corresponding to R_i is the set of all n-tuples of the fixed point of ψ where the first coordinate is c_i.

This argument is generalized to arbitrary systems of formulae, thereby giving us the following result.

Theorem 10.10. $\mathrm{IFP}^{\mathrm{simult}} = \mathrm{IFP}$ *and* $\mathrm{PFP}^{\mathrm{simult}} = \mathrm{PFP}$.

We now come back to single fixed point definitions and analyze them in detail. Suppose we have a formula $\varphi(R, \vec{x})$. Assume for now that φ is positive in R. To construct the least fixed point of φ on a structure $\mathfrak{A}$, we inductively calculate $X^0 = \emptyset$, $X^{i+1} = F_\varphi(X^i)$, and then the fixed point is $X^\infty = \bigcup_i X^i$. We shall refer to X^i's as *stages* of the fixed point computation, with X^i being the ith stage.

First, we note that each stage is definable by an LFP formula, if φ is positive in R. Indeed, for each stage i, we have a formula $\varphi^i(\vec{x}_i)$, such that $\varphi^i(\mathfrak{A})$ is exactly X^i. These are defined inductively as follows:

$$\begin{aligned}\varphi^0(\vec{x}_0) &\equiv \neg(x = x) \qquad x \text{ is a variable in } \vec{x}_0\\ \varphi^{i+1}(\vec{x}_{i+1}) &\equiv \varphi(\varphi^i/R, \vec{x}_{i+1}).\end{aligned} \tag{10.6}$$

Here the notation $\varphi(\varphi^i/R, \vec{x}_{i+1})$ means that every occurrence $R(\vec{y})$ in φ is replaced by $\varphi^i(\vec{y})$ and, furthermore, all the bound variables in φ have been replaced by fresh ones. For example, consider the formula $\varphi(R, x, y) \equiv E(x, y) \vee \exists z\ (E(x, z) \wedge R(z, y))$. Following (10.6), we obtain the formulae

$$\begin{aligned}\varphi^0(x_0, y_0) &\equiv \neg(x_0 = x_0)\\ \varphi^1(x_1, y_1) &\equiv \quad E(x_1, y_1) \vee \exists z_1\ (E(x_1, z_1) \wedge \varphi^0(z_1, y_1))\\ &\leftrightarrow \quad E(x_1, y_1)\\ \varphi^1(x_2, y_2) &\equiv \quad E(x_2, y_2) \vee \exists z_2\ (E(x_2, z_2) \wedge \varphi^1(z_2, y_2))\\ &\leftrightarrow \quad E(x_2, y_2) \vee \exists z_2\ (E(x_2, z_2) \wedge E(z_2, y_2))\\ \ldots \quad & \ldots\end{aligned}$$

computing the stages of the transitive closure operator.

For an arbitrary φ, we can give formulae for computing stages of the inflationary fixed point computation. These are given by

$$\begin{aligned}\varphi^0(\vec{x}_0) &\equiv \neg(x = x)\\ \varphi^{i+1}(\vec{x}_{i+1}) &\equiv \varphi^i(\vec{x}_{i+1}) \ \vee \ \varphi(\varphi^i/R, \vec{x}_{i+1}).\end{aligned} \tag{10.7}$$

Thus, each stage of the inflationary fixed point computation is definable by an IFP formula.

What is more interesting is that we can write formulae that compare stages at which various tuples get into the sets X^i of fixed point computations. Suppose we are given a formula $\varphi(R, \vec{x})$ that gives rise to an inductive operator F_φ, where R is k-ary and $\vec{x}$ has k variables. For example, if we are interested in inflationary fixed point computation, we can always pass from $\varphi(R, \vec{x})$ to $R(\vec{x}) \vee \varphi(R, \vec{x})$, whose induced operator is inductive.

Given a structure $\mathfrak{A}$, we define $|\varphi|^{\mathfrak{A}}$ as the least n such that $X^n = X^\infty$. Furthermore, for a tuple $\vec{a} \in A^k$, we define $|\vec{a}|^{\mathfrak{A}}_\varphi$ as the least number i such that $\vec{a} \in X^i$ in the fixed point computation, and $|\varphi|^{\mathfrak{A}} + 1$ if no such i exists. Notice that if φ is positive in R, then the stages of the least and inflationary fixed point computation are the same.

We next define two relations $\prec^\varphi$ and $\preceq^\varphi$ on A^k as follows:

$$\begin{aligned} \vec{a} \prec^\varphi \vec{b} &\equiv |\vec{a}|^{\mathfrak{A}}_\varphi < |\vec{b}|^{\mathfrak{A}}_\varphi, \\ \vec{a} \preceq^\varphi \vec{b} &\equiv |\vec{a}|^{\mathfrak{A}}_\varphi \leq |\vec{b}|^{\mathfrak{A}}_\varphi \text{ and } |\vec{a}|^{\mathfrak{A}}_\varphi \leq |\varphi|^{\mathfrak{A}}. \end{aligned}$$

The theorem below shows that these can be defined with least fixed points of positive formulae.

Theorem 10.11 (Stage comparison). *If φ is in* LFP, *then the binary relations $\prec^\varphi$ and $\preceq^\varphi$ are* LFP*-definable.*

Proof. The idea of the proof is as follows. We want to define both $\prec^\varphi$ and $\preceq^\varphi$ as a simultaneous fixed point. This has to be done somehow from φ, but in φ we may have both positive and negative occurrences of R. So to find some relations to substitute for the negative occurrences of R, we explicitly introduce the complements of $\prec^\varphi$ and $\preceq^\varphi$:

$$\begin{aligned} \vec{a} \not\prec^\varphi \vec{b} &\equiv |\vec{a}|^{\mathfrak{A}}_\varphi \geq |\vec{b}|^{\mathfrak{A}}_\varphi, \\ \vec{a} \not\preceq^\varphi \vec{b} &\equiv |\vec{a}|^{\mathfrak{A}}_\varphi > |\vec{b}|^{\mathfrak{A}}_\varphi \text{ or } |\vec{a}|^{\mathfrak{A}}_\varphi = |\varphi|^{\mathfrak{A}} + 1. \end{aligned}$$

We shall be using formulae of the form

$$\varphi(\prec(\vec{y})/R, \vec{x}) \quad \text{and} \quad \varphi(\preceq(\vec{y})/R, \vec{x}).$$

This means that, for $\varphi(\prec(\vec{y})/R, \vec{x})$, every positive occurrence $R(\vec{z})$ of R is replaced by $\vec{z} \prec^\varphi \vec{y}$, and every negative occurrence of $R(\vec{z})$ of R is replaced by $\vec{z} \not\prec^\varphi \vec{y}$, and likewise for $\preceq^\varphi$. Note that all the occurrences of the four relations $\prec^\varphi, \preceq^\varphi, \not\prec^\varphi, \not\preceq^\varphi$ become positive. Also, we shall write

$$\varphi(\neg\prec(\vec{y})/R, \vec{x}),$$

meaning that every positive occurrence $R(\vec{z})$ of R is replaced by $\neg(\vec{z} \prec^\varphi \vec{y})$, and every negative occurrence of $R(\vec{z})$ of R is replaced by $\neg(\vec{z} \not\prec^\varphi \vec{y})$. These will be used in subformulae $\neg\varphi(\neg \prec (\vec{y})/R, \vec{x})$, again ensuring that all the occurrences of $\prec^\varphi, \preceq^\varphi, \not\prec^\varphi, \not\preceq^\varphi$ are positive.

These four relations will be defined by a simultaneous fixed point. For technical reasons, we shall add one more relation:

$$\vec{a} \lhd^\varphi \vec{b} \equiv |\vec{a}|^{\mathfrak{A}}_\varphi + 1 = |\vec{b}|^{\mathfrak{A}}_\varphi,$$

and show how to define $(\prec, \preceq, \lhd, \not\prec, \not\preceq)$ by a simultaneous fixed point. For readability only, we may omit the superscript φ. We define the system Ψ of five formulae $\psi_i(\prec, \preceq, \lhd, \not\prec, \not\preceq, x, y)$, $i = 1, \ldots, 5$, as follows:

$$\begin{aligned}
\psi_1 &\equiv \exists\vec{z}\ (\vec{x} \preceq \vec{z} \wedge \vec{z} \lhd \vec{y}),\\
\psi_2 &\equiv \varphi(\prec(\vec{y})/R,\ \vec{x}),\\
\psi_3 &\equiv \varphi(\prec(\vec{x})/R,\ \vec{x}) \wedge \neg\varphi(\not\prec(\vec{x})/R,\ \vec{y}) \qquad (10.8)\\
&\quad \wedge \Big(\varphi(\preceq(\vec{x})/R,\ \vec{y}) \vee \forall\vec{z}\ \big(\neg\varphi(\neg \not\preceq(\vec{x})/R,\ \vec{z}) \vee \varphi(\prec(\vec{x})/R,\ \vec{z})\big)\Big),\\
\psi_4 &\equiv \exists\vec{z}\ (\vec{x} \not\preceq \vec{z} \wedge \vec{z} \lhd \vec{y})\ \vee\ \varphi(\emptyset/R,\ \vec{y})\ \vee\ \forall\vec{z}\neg\varphi(\emptyset/R,\ \vec{z}),\\
\psi_5 &\equiv \neg\varphi(\neg \not\prec(\vec{y})/R,\ \vec{x})
\end{aligned}$$

where $\varphi(\emptyset/R, \cdot)$ means that all occurrences of R are eliminated and replaced by *false*.

Note that all the occurrences of $\prec, \preceq, \lhd, \not\prec, \not\preceq$ in Ψ are positive. We next claim that the simultaneous least fixed point of Ψ indeed defines $\prec^\varphi$, $\preceq^\varphi$, $\lhd^\varphi, \not\prec^\varphi, \not\preceq^\varphi$.

To prove the result, we have to show that $(\prec^\varphi, \preceq^\varphi, \lhd^\varphi, \not\prec^\varphi, \not\preceq^\varphi)$ satisfy (10.8), and that for each $* \in \{\prec^\varphi, \preceq^\varphi, \lhd^\varphi, \not\prec^\varphi, \not\preceq^\varphi\}$, if $\vec{a} * \vec{b}$ holds, then $(\vec{a}, \vec{b})$ is in the corresponding fixed point of Ψ (10.8). This will be proved by induction on $|\vec{b}|^{\mathfrak{A}}_\varphi$.

Below, we prove a few cases for both directions. The remaining cases are very similar, and are left as an exercise for the reader.

First, we prove that $\lhd^\varphi$ satisfies (10.8). Consider a tuple $(\vec{a}, \vec{b})$ in this relation. The result is immediate if $|\vec{a}|^{\mathfrak{A}}_\varphi = |\varphi|^{\mathfrak{A}} + 1$. If $|\vec{a}|^{\mathfrak{A}}_\varphi < |\varphi|^{\mathfrak{A}}$, then the third conjunct in $\psi_3(\vec{a}, \vec{b})$ is equivalent to $\varphi(\preceq^\varphi(\vec{a})/R,\ \vec{b})$ and, therefore, $\psi_3(\vec{a}, \vec{b})$ holds iff $|\vec{b}|^{\mathfrak{A}}_\varphi = |\vec{a}|^{\mathfrak{A}}_\varphi + 1$ iff $\vec{a} \lhd^\varphi \vec{b}$. Finally, if $|\vec{a}|^{\mathfrak{A}}_\varphi = |\varphi|^{\mathfrak{A}}$, then the third conjunct in ψ_3 is equivalent to the formula $\forall\vec{z}\ (\neg\varphi(\neg \not\preceq^\varphi(\vec{a})/R,\ \vec{z}) \vee \varphi(\prec^\varphi(\vec{a})/R,\ \vec{z}))$ and, thus, $\psi_3(\vec{a}, \vec{b})$ holds iff $\vec{b}$ is not in the fixed point of ψ_3 iff $|\vec{b}|^{\mathfrak{A}}_\varphi = |\varphi|^{\mathfrak{A}} + 1 = |\vec{a}|^{\mathfrak{A}}_\varphi + 1$.

Second, we prove by induction on $|\vec{b}|^{\mathfrak{A}}_\varphi$ that, for every $\vec{a}$, if $\vec{a} \lhd^\varphi \vec{b}$ or $\vec{a} \not\prec^\varphi \vec{b}$, then $(\vec{a}, \vec{b})$ is in the corresponding fixed point of Ψ.

Induction Basis: $|\vec{b}|^{\mathfrak{A}}_\varphi = 1$.

- *The case for* $\lhd^\varphi$. This is the simplest case, since $|\vec{b}|^{\mathfrak{A}}_\varphi = 1$ implies that $\vec{a} \lhd^\varphi \vec{b}$ holds for no $\vec{a}$.

- *The case for* $\not\prec^\varphi$. Since $|\vec{b}|^{\mathfrak{A}}_\varphi = 1$, we conclude that $\varphi(\emptyset/R,\ \vec{b})$ holds. We have $\vec{a} \not\prec^\varphi \vec{b}$ for all $\vec{a}$, and since $\varphi(\emptyset/R,\ \vec{b})$ is true, $(\vec{a}, \vec{b})$ is in the fixed point of ψ_4 for every $\vec{a}$.

Induction Step: Assume that $|\vec{b}|^{\mathfrak{A}}_\varphi = k+1$ and that the property holds for all $\vec{c}$ such that $|\vec{c}|^{\mathfrak{A}}_\varphi \le k$.

- *The case for* $\lhd^\varphi$. Suppose that $\vec{a} \lhd^\varphi \vec{b}$. Then $|\vec{a}|^{\mathfrak{A}}_\varphi \le k$. We show that the three conjuncts in ψ_3 hold for $(\vec{a}, \vec{b})$ and, thus, we conclude that $(\vec{a}, \vec{b})$ is in the fixed point of ψ_3.
  Since $|\vec{a}|^{\mathfrak{A}}_\varphi < |\vec{b}|^{\mathfrak{A}}_\varphi$, we have $|\vec{a}|^{\mathfrak{A}}_\varphi \le |\varphi|^{\mathfrak{A}}$ and, therefore, $\varphi(\prec^\varphi(\vec{a})/R,\ \vec{a})$ holds. By the induction hypothesis, $\prec^\varphi(\vec{a}) = \prec(\vec{a})$, so $\varphi(\prec(\vec{a})/R,\ \vec{a})$ holds.
  Since $|\vec{a}|^{\mathfrak{A}}_\varphi < |\vec{b}|^{\mathfrak{A}}_\varphi$, $\neg\varphi(\neg \not\prec^\varphi(\vec{a})/R,\ \vec{b})$ holds. By the induction hypothesis, $\not\prec^\varphi(\vec{a}) = \not\prec(\vec{a})$ and, hence, $\neg\varphi(\neg \not\prec(\vec{a})/R,\ \vec{b})$ holds.
  To prove that the third conjunct in ψ_3 holds, we consider two cases. If $|\vec{b}|^{\mathfrak{A}}_\varphi \le |\varphi|^{\mathfrak{A}}$, then $\varphi(\preceq^\varphi(\vec{a})/R,\ \vec{b})$ holds. By the hypothesis, $\preceq^\varphi(\vec{a}) = \preceq(\vec{a})$ and, therefore, $\varphi(\preceq(\vec{a})/R,\ \vec{b})$ holds. Otherwise $|\vec{b}|^{\mathfrak{A}}_\varphi = |\varphi|^{\mathfrak{A}} + 1$ and $|\vec{a}|^{\mathfrak{A}}_\varphi = |\varphi|^{\mathfrak{A}}$. In this case all the elements generated at stage $|\vec{a}|^{\mathfrak{A}}_\varphi + 1$ are already in stage $|\vec{a}|^{\mathfrak{A}}_\varphi$ and, therefore, the formula $\forall \vec{z}\ (\neg\varphi(\neg \not\preceq^\varphi(\vec{a})/R,\ \vec{z}) \vee \varphi(\prec^\varphi(\vec{a})/R,\ \vec{z}))$ holds. As in the previous cases, by the induction hypothesis we conclude that $\forall \vec{z}\ (\neg\varphi(\neg \not\preceq(\vec{a})/R,\ \vec{z}) \vee \varphi(\prec(\vec{a})/R,\ \vec{z}))$ holds.
- *The case for* $\not\prec^\varphi$. Suppose that $\vec{a} \not\prec^\varphi \vec{b}$, and that the second and third disjuncts in ψ_4 do not hold. Then we show that the first disjunct in ψ_4 holds and conclude that $(\vec{a}, \vec{b})$ is in the fixed point of ψ_4.
  Since $\varphi(\emptyset/R,\ \vec{b})$ and $\forall\vec{z}\neg\varphi(\emptyset/R,\ \vec{z})$ do not hold, we have $|\vec{b}|^{\mathfrak{A}}_\varphi > 1$ and the fixed point of ψ_4 contains at least one element. Thus, there exists $\vec{c}$ such that $\vec{c} \lhd^\varphi \vec{b}$.
  Given that $\vec{a} \not\prec^\varphi \vec{b}$, we have $\vec{a} \not\preceq^\varphi \vec{c}$ and $|\vec{c}|^{\mathfrak{A}}_\varphi \le k$. Therefore, we have a tuple $\vec{c}$ with $|\vec{c}|^{\mathfrak{A}}_\varphi \le k$ such that both $\vec{a} \not\preceq^\varphi \vec{c}$ and $\vec{c} \lhd^\varphi \vec{b}$ hold. Now using the equivalence from the previous case for $\vec{c} \lhd^\varphi \vec{b}$, and applying the induction hypothesis to $\vec{a} \not\preceq^\varphi \vec{c}$, we conclude that $(\vec{a}, \vec{b})$ satisfies $\exists \vec{z}\ (\vec{a} \not\preceq \vec{z} \wedge \vec{z} \lhd \vec{b})$, which finishes the proof. ☐

Corollary 10.12 (Gurevich-Shelah). IFP = LFP.

Proof. The inclusion LFP $\subseteq$ IFP is immediate. For the converse, proceed by induction on the formulae. The only case to consider is $\mathbf{ifp}_{R,\vec{x}}\varphi(R, \vec{x})$. We can assume, without loss of generality, that φ defines an inductive operator (if not, consider $R(\vec{x}) \vee \varphi$). Then $[\mathbf{ifp}_{R,\vec{x}}\varphi(R, \vec{x})](\vec{t})$ is equivalent to

$$\varphi(\prec^\varphi(\vec{t})/R,\ \vec{t}),$$

which, by the stage comparison theorem, is an LFP formula. ☐

As another corollary of stage comparison, we establish a normal form for LFP formulae. Define a logic LFP_0 which extends FO with the following. If Φ is a system of FO formulae $\varphi_i(R_1, \ldots, R_n, \vec{x})$ positive in all the R_i's, then $[\mathbf{lfp}_{R_i,\Phi}](\vec{x})$ is an LFP_0 formula. Note the difference between this and general LFP: we only allow fixed points to be applicable to FO formulae, and we do not close those fixed points under the Boolean connectives and quantification. In other words, every formula of LFP_0 is either FO, or of the form $[\mathbf{lfp}_{R_i,\Phi}](\vec{x})$, where Φ consists of FO formulae.

Corollary 10.13. $\mathrm{LFP} = \mathrm{LFP}_0$.

Proof. We first show that LFP_0 is closed under $\vee, \wedge$, and $\neg$. For $\vee$ and $\wedge$ this is easy: just introduce an extra relation to hold the union or intersection of two fixed points. For example, given $\varphi_1(R_1, \vec{x})$ and $\varphi_2(R_2, \vec{x})$, we define a system Φ that consists of formulae $\varphi_1(R_1, R_2, S, \vec{x})$, $\varphi_2(R_1, R_2, S, \vec{x})$, and $\varphi_3(R_1, R_2, S, \vec{x}) \equiv (R_1(\vec{x}) \vee R_2(\vec{x}))$. Then $\mathbf{lfp}_{S,\Phi}$ is the union of fixed points of φ_1 and φ_2.

The closure under negation follows from the stage comparison: $\neg[\mathbf{lfp}_{R,\vec{x}}\varphi](\vec{t})$ is equivalent to $\vec{t} \not\preceq^{\varphi} \vec{t}$.

The closure of LFP_0 under fixed point operators is immediate (one simply adds an extra formula to the system). Thus, $\mathrm{LFP}_0 = \mathrm{LFP}$. □

10.4 LFP, PFP, and Polynomial Time and Space

The goal of this section is to show that the fixed point logics we introduced capture familiar complexity classes over ordered structures. A structure is *ordered* if one of the symbols of its vocabulary σ is $<$, interpreted as a linear order on the universe. Recall that we used a linear order for defining an encoding of a structure: indeed, a string on the tape of a Turing machine is naturally ordered from left to right. For capturing NP and the polynomial hierarchy, we did not need the assumption that the structures are ordered, since we could guess an order by second-order quantifiers. However, fixed point logics are not sufficiently expressive for guessing a linear order (in fact, this will be proved formally).

Theorem 10.14 (Immerman-Vardi). *Both* LFP *and* IFP *capture* PTIME *over the class of ordered structures. That is,*

$$\mathrm{LFP}{+}{<} \;=\; \mathrm{IFP}{+}{<} \;=\; \mathrm{PTIME}.$$

Proof. By the Gurevich-Shelah theorem (Corollary 10.12), we can use IFP and LFP interchangeably. First, we show that LFP formulae can be evaluated in polynomial time. The proof is by induction on the formulae. The cases of the Boolean connectives and quantifiers are handled in exactly the same way as for FO (see, e.g., Proposition 6.6). For formulae of the form $\mathbf{lfp}_{R,\vec{x}}\varphi$, it suffices to observe the following: if $F : \wp(U) \to \wp(U)$ is a PTIME-computable monotone operator, then $\mathbf{lfp}(F)$ can be computed in polynomial time in $|U|$. Indeed, we know that the fixed point computation stops after at most $|U|$ iterations, and each iteration is PTIME-computable. Hence, every LFP formula can be evaluated in polynomial time.

For the converse, we use the same technique as in the proofs of Trakhtenbrot's and Fagin's theorems. Suppose we are given a property $\mathcal{P}$ of σ-structures which can be tested, on encodings of σ-structures, by a deterministic polynomial time Turing machine $M = (Q, \Sigma, \Delta, \delta, q_0, Q_a, Q_r)$ with a one-way infinite tape. We assume, without loss of generality, that there is only one accepting state, q_a, that $\Sigma = \{0, 1\}$, and that Δ extends Σ with the blank symbol. Let M run in time n^k. As before, we assume that n^k exceeds the size of the encodings of n-element structures.

With the linear order $<$, we can again define the lexicographic linear order $\leq_k$ on k-tuples, and use the ordered k-tuples to model both positions of M and time. We shall define, by means of fixed point formulae, the $2k$-ary predicates $T_0, T_1, T_2, (H_q)_{q\in Q}$, where $T_i(\vec{p}, \vec{t})$ indicates that position $\vec{p}$ at time $\vec{t}$ contains i, for $i = 0, 1$, and blank, for $i = 2$, and $H_q(\vec{p}, \vec{t})$ indicates that at time $\vec{t}$, the machine is in state q, and its head is in position $\vec{p}$. We shall provide a system Ψ of formulae whose simultaneous inflationary fixed point is exactly $(T_0, T_1, T_2, (H_q)_{q\in Q})$. Once we have such a system, the sentence testing $\mathcal{P}$ will be given by

$$\exists \vec{p}\, \exists \vec{t}\, [\mathbf{ifp}_{H_{q_a},\Psi}](\vec{p}, \vec{t}). \tag{10.9}$$

Since $\text{IFP}^{\text{simult}} = \text{IFP}$ and $\text{IFP} = \text{LFP}$, the formula (10.9) can be expressed in LFP.

The system Ψ contains formulae $\psi_i(\vec{p}, \vec{t}, T_0, T_1, T_2, (H_q)_{q\in Q}), i = 0, 1, 2$, defining T_i's, and $\psi_q(\vec{p}, \vec{t}, T_0, T_1, T_2, (H_q)_{q\in Q}), q \in Q$, defining H_q's. It has the property that the jth iteration for each of the relations it defines, R^j, contains $\{(\vec{p}, \vec{t}) \mid R(\vec{p}, \vec{t}) \text{ and } \vec{t} < j\}$, where $\vec{t} < j$ means that $\vec{t}$ is among the first $j-1$ k-tuples in the lexicographic ordering $<_k$. That is, we build the relations T_i's and H_q's in stages, where the jth stage represents the configuration at times up to $j-1$.

The formulae ψ_i are straightforward to write, and we only sketch a few of them. The formula ψ_0 is of the form

$$\Big(\vec{t} = 0 \wedge \neg\iota(\vec{p}) \wedge \neg\xi(\vec{p})\Big) \vee \Big(\neg(\vec{t} = 0) \wedge \alpha_0(\vec{t} - 1, \vec{p}, T_0, T_1, (H_q)_{q\in Q})\Big).$$

Here ι and ξ are formula from the proof of Fagin's theorem (ι holds iff the $\vec{p}$th position of the encoding of the input is 1, and ξ holds iff $\vec{p}$ is past the last position of the encoding of the input on the tape). Thus the first disjunct says that at time 0, the tape of M contains the encoding of the input structure. The formula $\alpha_0(\vec{t} - 1, \vec{p}, T_0, T_1, (H_q)_{q\in Q})$ lists conditions under which at the following time instant, $\vec{t}$, the position $\vec{p}$ will contain zero. It is similar to the formulae we used for modeling M's transitions in the proof of Fagin's theorem. The formula ψ_1 is similar to ψ_0.

The formula ψ_{q_0} is of the form

$$\Big(\vec{t} = 0 \land \vec{p} = 0\Big) \lor \Big(\neg(\vec{t} = 0) \land \alpha_{q_0}(\vec{t} - 1, \vec{p}, T_0, T_1, (H_q)_{q \in Q})\Big),$$

and other ψ_q's are of the form $(\vec{t} > 0) \land \alpha_q(\vec{t} - 1, \vec{p}, T_0, T_1, (H_q)_{q \in Q})$, where α_q again lists conditions under which at the next time instant, M will enter state q while having the head pointing at $\vec{p}$. The first disjunct in ψ_{q_0} states that at time 0, M is in state q_0 with its head in position 0.

We leave it as a routine exercise to the reader to write the α_i's and α_q's, based on M's transitions, and verify that that jth stage of the fixed point computation for the system Ψ indeed computes the configuration of M for times not exceeding $j - 1$. Hence, the fixed point formula (10.9) checks membership in $\mathcal{P}$, which completes the proof. □

Note that using inflationary fixed points instead of least fixed points in the proof of Theorem 10.14 gives us extra freedom in writing down formulae of the system Ψ: we do not have to ensure that these are positive in T_i's and H_q's. However, one can write those formulae carefully so that they would be positive in all those relation symbols. In that case, one can replace **ifp** with **lfp** in (10.9). Hence, the proof of Theorem 10.14 then shows that every LFP-definable property over ordered structures can be defined by a formula of the form

$$\exists \vec{x}\ [\mathbf{lfp}_{R_i, \Psi}](\vec{x}),$$

where Ψ is a system of FO formulae positive in relation symbols $R_1, \ldots, R_n$. This, of course, would follow from Corollary 10.13, stating that LFP = LFP$_0$, but notice that for ordered structures, we obtained the normal form result without using the stage comparison theorem.

We have seen that for several logics, adding an order increases their expressiveness; that is, $\mathcal{L} \subsetneq (\mathcal{L}+<)_{\text{inv}}$ for $\mathcal{L}$ being FO, or one of its counting extensions, or MSO. The same is true for LFP, IFP, and PFP; the proof of this will be given in the next chapter when we describe additional tools such as finite variable logics and pebble games. At this point we only say that the query that separates these logics on ordered and unordered structures is EVEN: it is not expressible in any of the fixed point logics without a linear order, but is obviously already in LFP+<, since it is PTIME-computable.

We conclude this section by considering the partial fixed point logic, PFP. Over ordered structures, it corresponds to another well-known complexity class.

Theorem 10.15. *Over ordered structures,* PFP *captures* PSPACE.

The proof, of course, follows the proofs of Trakhtenbrot's, Fagin's, and Immerman-Vardi's theorems. We only explain why PFP formulae can be evaluated in PSPACE. Consider $\mathbf{pfp}_{R,\vec{x}}\varphi(R,\vec{x})$, where R is k-ary, and let X^i's be the stages of the partial fixed point computation on $\mathfrak{A}$ with $|A|= n$. There are two possibilities. Either $X^{m+1} = X^m$ for some m, in which case a fixed point is reached. Otherwise, for some $0 \leq i, j \leq 2^{n^k}$, $i+1 < j$, we have $X^i = X^j$, and in this case the formula $[\mathbf{pfp}_{R,\vec{x}}\varphi(R,\vec{x})](\vec{t})$ would evaluate to *false*, since the partial fixed point is the empty set. Hence, one has to check which of these cases is true. For that, it suffices to enumerate all the subsets of A^k, one by one (which can be done in PSPACE), and proceed with computing the sequence X^i, checking whether a fixed point is reached. Since only 2^{n^k} steps need to be made, the entire computation is in PSPACE.

To show that PSPACE $\subseteq$ PFP$+<$, one modifies the proof of the Immerman-Vardi theorem, to simulate the accepting condition of a Turing machine by means of a partial fixed point formula. We leave the details to the reader (Exercise 10.9).

10.5 DATALOG and LFP

In this section we review a database query language DATALOG, and relate it to fixed point logics.

Recall that FO is used as the basic relational query language (it is known under the name *relational calculus* in the database literature). Conjunctive queries, seen in Sect. 6.7, constitute an important subclass of FO queries. They can be defined in the fragment of FO that only includes conjunction $\wedge$ and existential quantification $\exists$. There is another convenient form for writing conjunctive queries that in fact is used most often in the literature. Instead of $\psi(\vec{x}) \equiv \exists \vec{y} \bigwedge_i \alpha_i(\vec{x},\vec{y})$, one omits the existential quantifiers and replaces the $\wedge$'s with commas:

$$R_\psi(\vec{x}) \text{ :- } \alpha_1(\vec{x},\vec{y}), \alpha_2(\vec{x},\vec{y}), \ldots, \alpha_m(\vec{x},\vec{y}). \tag{10.10}$$

Here R_ψ is a new relation symbol; the meaning of (10.10) is that, for a given structure $\mathfrak{A}$, this new relation contains the set of all tuples $\vec{a}$ such that $\mathfrak{A} \models \psi(\vec{a})$.

Expressions of the form (10.10) are called *rules*; the part of the rule that appears on the left of the :- (in this case, $R_\psi(\vec{x})$) is called its *head*, and the part of the rule on the right of the :- is called its *body*. A rule is converted into a conjunctive query by replacing commas with conjunctions, and existentially quantifying all the variables that appear in the body but not in the head.

For example, the rule

$$q(x,y) \text{ :- } E(x,z), E(z,v), E(v,y)$$

is translated into $\exists z \exists v \; \big(E(x,z) \wedge E(z,v) \wedge E(v,y)\big)$.

DATALOG programs contain several rules some of which may be *recursive*: that is, the same predicate symbol may appear in both the head and the body of a rule. A typical DATALOG program would be of the following form:

$$\begin{array}{lcl} trcl(x,y) & \text{:-} & E(x,y) \\ trcl(x,y) & \text{:-} & E(x,z), trcl(z,y) \end{array} \tag{10.11}$$

This program computes the transitive closure of E: it says that (x, y) is in the transitive closure if there is an edge (x, y), or there is an edge (x, z) such that (z, y) is in the transitive closure. As with the fixed point definition of the transitive closure, to evaluate this program we iterate this definition, starting with the empty set, until a fixed point is reached.

Definition 10.16. *A* DATALOG *program over vocabulary σ is a pair (Π, Q), where Π is a set of rules of the form*

$$P(\vec{x}) \;\; \text{:-} \;\; \alpha_1(\vec{x},\vec{y}), \ldots, \alpha_m(\vec{x},\vec{y}). \tag{10.12}$$

Here the relation symbol P in the head of rule (10.12) does not occur in σ, and each α_i is an atomic formula of the form $R(\vec{x},\vec{y})$, for $R \in \sigma$, or $P'(\vec{x},\vec{y})$, for P' that occurs as a head of one of the rules of Π. Furthermore, Q is the head of one of the rules of Π.

By $\text{DATALOG}_\neg$ *we mean the extension of* DATALOG *where negated atomic formulae of the form $\neg R(\cdot)$, for $R \in \sigma$, can appear in the bodies of rules (10.12).*

For example, the transitive closure program consists of the rules (10.11), and *trcl* is the output predicate Q.

In the standard DATALOG terminology, relation symbols from σ are called *extensional predicates*, and symbols not in σ that appear as heads of rules are called *intensional predicates.* These are the predicates computed by the program, and Q is its output.

To define the semantics of a DATALOG (or $\text{DATALOG}_\neg$) program (Π, Q), we introduce the *immediate consequence* operator F_Π. Let $P_1, \ldots, P_k$ list all the intensional predicates (with Q being one of them). Let n_i be the arity of P_i, $i = 1, \ldots, k$. Let

$$\begin{array}{lcl} P_i(\vec{x}) & \text{:-} & \gamma_1^1(\vec{x},\vec{y}_1), \;\ldots,\; \gamma_{m_1}^1(\vec{x},\vec{y}_1) \\ \ldots & & \ldots \quad\;\; \ldots \;\; \ldots \\ P_i(\vec{x}) & \text{:-} & \gamma_1^l(\vec{x},\vec{y}_l), \;\ldots,\; \gamma_{m_l}^l(\vec{x},\vec{y}_l) \end{array} \tag{10.13}$$

enumerate all the rules in Π with P_i as the head.

Given a structure $\mathfrak{A}$ and a tuple of sets $\vec{Y} = (Y_1, \ldots, Y_k)$, $Y_i \subseteq A^{n_i}$, $i = 1, \ldots, k$, we define $F_\Pi(\vec{Y}) = (Z_1, \ldots, Z_k)$, where

$$Z_i \;=\; \left\{ \vec{a} \in A^{n_i} \;\middle|\; (\mathfrak{A}, Y_1, \ldots, Y_k) \models \bigvee_{j=1}^{l} \exists \vec{y}_j \left(\gamma_1^j(\vec{a},\vec{y}_j) \wedge \ldots \wedge \gamma_{m_j}^j(\vec{a},\vec{y}_j) \right) \right\},$$

where formulae γ_l^j are the formulae from the rules (10.13) for the intensional predicate P_i. In other words, $\vec{a} \in Z_i$ can be derived by applying one of the rules of Π whose head is P_i, using $\vec{Y}$ as the interpretation for the intensional predicates.

Since the formula above is positive in all the intensional predicates (even for a DATALOG¬ program), the operator F_Π is monotone. Hence, starting with $(\emptyset, \ldots, \emptyset)$ and iterating this operator, we reach the least fixed point $\mathbf{lfp}(F_\Pi) = (P_1^\infty, \ldots, P_k^\infty)$. The output of (Π, Q) on $\mathfrak{A}$ is defined as Q^∞ (recall that Q is one of the P_i's).

Returning to the transitive closure example, the stages of the fixed point computation of the immediate consequence operator are exactly the same as the stages of computing the least fixed point of $E(x,y) \vee \exists z\, (E(x,z) \wedge R(z,y))$, and hence, on an arbitrary finite graph, the program (10.11) computes its transitive closure.

Analyzing the semantics of a DATALOG program (Π, Q), we can see that it is simply a simultaneous least fixed point of a system Ψ of formulae

$$\psi_i(\vec{x}, P_1, \ldots, P_k) \equiv \bigvee_j \exists \vec{y}_j \left(\gamma_1^j(\vec{a}, \vec{y}_j) \wedge \ldots \wedge \gamma_{m_j}^j(\vec{a}, \vec{y}_j) \right). \tag{10.14}$$

That is, the answer to (Π, Q) on $\mathfrak{A}$ is

$$\{ \vec{a} \mid \mathfrak{A} \models [\mathbf{lfp}_{Q,\Psi}](\vec{a}) \}.$$

Hence, each DATALOG or DATALOG¬ program can be expressed in LFP^{simult}, and thus in LFP.

What fragment of LFP does DATALOG¬ correspond to? The special form of formulae ψ_i (10.14) indicates that there are some syntactic restrictions on LFP formulae into which DATALOG¬ is translated. We can capture these syntactic restrictions by a notion of existential least fixed point logic.

Definition 10.17. *The* existential least fixed point logic, ∃LFP, *over vocabulary* σ, *is defined as a restriction of* LFP *over* σ, *where:*

- *negation can only be applied to atomic formulae of vocabulary* σ *(i.e., formulae* $R(\cdot)$, *where* $R \in \sigma$*), and*
- *universal quantification is not allowed.*

Theorem 10.18. ∃LFP = DATALOG¬.

Proof. We have seen one direction already, since every DATALOG¬ query can be translated into one simultaneous fixed point of a system of FO formulae ψ_i (10.14), in which no universal quantifiers are used, and negation only applies to atomic σ-formulae. Elimination of the simultaneous fixed point introduces no negation and no universal quantification, and hence DATALOG¬ ⊆ ∃LFP.

For the converse, we translate each $\exists$LFP formula $\varphi(x_1, \ldots, x_k)$ into an equivalent DATALOG$_\neg$ program (Π_φ, Q_φ), which, on any structure $\mathfrak{A}$, computes $Q_\varphi^\infty = \varphi(\mathfrak{A})$. Moreover, the translation ensures that no relation symbol that appears positively in φ is negated in Π_φ. The translation proceeds by induction on the structure of the formulae as follows:

- If $\varphi(\vec{x})$ is an atomic or negated atomic formula (i.e., $R(\vec{x})$ or $\neg R(\vec{x})$), then Π_φ contains one rule $Q_\varphi(\vec{x})$:- $\varphi(\vec{x})$.
- If $\varphi \equiv \alpha \wedge \beta$, then
$$\Pi_\varphi \;=\; \Pi_\alpha \,\cup\, \Pi_\beta \,\cup\, \{Q_\varphi(\vec{x}) \text{ :- } Q_\alpha(\vec{x}), Q_\beta(\vec{x})\}.$$
- If $\varphi \equiv \alpha \vee \beta$, then
$$\Pi_\varphi \;=\; \Pi_\alpha \,\cup\, \Pi_\beta \,\cup\, \{Q_\varphi(\vec{x}) \text{ :- } Q_\alpha(\vec{x}),\; Q_\varphi(\vec{x}) \text{ :- } Q_\beta(\vec{x})\}.$$
- If $\varphi(\vec{x}) \equiv \exists y \alpha(y, \vec{x})$, then
$$\Pi_\varphi \;=\; \Pi_\alpha \,\cup\, \{Q_\varphi(\vec{x}) \text{ :- } Q_\alpha(y, \vec{x})\}.$$
- Let $\varphi(\vec{x}) \equiv [\mathbf{lfp}_{R,\vec{y}}\alpha(R, \vec{y})](\vec{x})$. By the induction hypothesis, we have a program (Π_α, Q_α) for α; notice that R appears positively in α, and thus does not appear negated in Π_α. Hence, we can define the following program, in which R is an intensional predicate:
$$\Pi_\varphi \;=\; \Pi_\alpha \,\cup\, \{R(\vec{y}) \text{ :- } Q_\alpha(\vec{y}),\; Q_\varphi(\vec{x}) \text{ :- } R(\vec{x})\},$$
and which computes the least fixed point of α. ☐

Thus, DATALOG and DATALOG$_\neg$ correspond to syntactic restrictions of LFP. But could they still be sufficient for capturing PTIME?

Let us first look at a DATALOG program (Π, Q), and suppose we have two σ-structures, $\mathfrak{A}_1$ and $\mathfrak{A}_2$, on the same universe A, such that for every symbol $R \in \sigma$, we have $R^{\mathfrak{A}_1} \subseteq R^{\mathfrak{A}_2}$. Then a straightforward induction on the stages of the immediate consequence operator shows that $(\Pi, Q)[\mathfrak{A}_1] \subseteq (\Pi, Q)[\mathfrak{A}_2]$, where by $(\Pi, Q)[\mathfrak{A}]$ we denote the result of (Π, Q) on $\mathfrak{A}$. Hence, DATALOG only expresses monotone properties, and thus cannot capture PTIME (exercise: exhibit a non-monotone PTIME property).

Queries expressible in DATALOG$_\neg$ satisfy a slightly different monotonicity property. Suppose $\mathfrak{A}$ is a substructure of $\mathfrak{B}$; that is, $A \subseteq B$, and for each $R \in \sigma$, $R^{\mathfrak{A}}$ is the restriction of $R^{\mathfrak{B}}$ to A. Then $(\Pi, Q)[\mathfrak{A}] \subseteq (\Pi, Q)[\mathfrak{B}]$, where (Π, Q) is a DATALOG$_\neg$ program. Indeed, when you look at the formulae (10.14), it is clear that if a witness $\vec{a}$ is found in $\mathfrak{A}$, it will be a witness for the existential quantifiers in $\mathfrak{B}$. Since it is again not hard to find a PTIME property that fails this notion of monotonicity, DATALOG$_\neg$ fails to capture PTIME. Furthermore, even adding order preserves monotonicity, and hence DATALOG$_\neg$ fails to capture PTIME even over ordered structures.

But now assume that on all the structures, we have a successor relation $\underline{\text{succ}}$ available, as well as constants $\underline{\text{min}}, \underline{\text{max}}$ for the minimal and maximal element with respect to the successor relation. It is impossible for $\langle A, \underline{\text{succ}}^{\mathfrak{A}}, \underline{\text{min}}^{\mathfrak{A}}, \underline{\text{max}}^{\mathfrak{A}}, \ldots\rangle$ to be a substructure of $\langle B, \underline{\text{succ}}^{\mathfrak{B}}, \underline{\text{min}}^{\mathfrak{B}}, \underline{\text{max}}^{\mathfrak{B}}, \ldots\rangle$, and hence the previous monotonicity argument does not work. In fact, the following theorem can be shown.

Theorem 10.19. *Over structures with successor relation and constants for the minimal and maximal elements,* DATALOG$_\neg$ *captures* PTIME. □

The proof mimics the proofs of Fagin's and Immerman-Vardi's theorems, by directly coding deterministic polynomial time Turing machines in DATALOG$_\neg$, and is left to the reader as an exercise.

10.6 Transitive Closure Logic

One of the standard examples of queries expressible in LFP is the transitive closure. In this section, we study a logic based on the transitive closure operator, rather than the least or inflationary fixed point, and prove that it corresponds to a well-known complexity class.

Definition 10.20. *The* transitive closure logic TRCL *is defined as an extension of* FO *with the following formation rule: if $\varphi(\vec{x}, \vec{y}, \vec{z})$ is a formula, where $|\vec{x}| = |\vec{y}| = k$, and $\vec{t}_1, \vec{t}_2$ are tuples of terms of length k, then*

$$[\mathbf{trcl}_{\vec{x},\vec{y}}\varphi(\vec{x}, \vec{y}, \vec{z})](\vec{t}_1, \vec{t}_2)$$

is a formula whose free variables are $\vec{z}$ plus the free variables of $\vec{t}_1, \vec{t}_2$.

The semantics is defined as follows. Given a structure $\mathfrak{A}$, values $\vec{a}$ for $\vec{z}$ and $\vec{a}_i$ for $\vec{t}_i$, $i = 1, 2$, construct the graph G on A^k with the set of edges

$$\{(\vec{b}_1, \vec{b}_2) \mid \mathfrak{A} \models \varphi(\vec{b}_1, \vec{b}_2, \vec{a})\}.$$

Then

$$\mathfrak{A} \models [\mathbf{trcl}_{\vec{x},\vec{y}}\varphi(\vec{x}, \vec{y}, \vec{a})](\vec{a}_1, \vec{a}_2)$$

iff $(\vec{a}_1, \vec{a}_2)$ is in the transitive closure of G.

For example, connectivity of directed graphs can be expressed by the TRCL formula $\forall u \forall v\ [\mathbf{trcl}_{x,y}(E(x,y) \vee E(y,x))](u,v)$.

We now state the main result of this section.

Theorem 10.21. *Over ordered structures,* TRCL *captures* NLOG.

Having seen a number of results of this type, one might be tempted to think that the proof is by a simple modification of the proofs of Trakhtenbrot's, Fagin's, and Immerman-Vardi's theorems. However, in this case we are running into problems, and the problems arise in the "easy" part of the proof: TRCL $\subseteq$ NLOG.

It is well known that the transitive closure of a graph can be computed by a nondeterministic logspace machine. Hence, trying to show the inclusion TRCL $\subseteq$ NLOG by induction on the structure of the formulae, we have no problems with the transitive closure operator. The problematic operation is *negation*. Since NLOG is a nondeterministic class, acceptance means that *some* computation ends in an accepting state. The negation of this statement is that *all* computations end in rejecting states, and it is not clear whether this can be reformulated as an existential statement. Our strategy for proving Theorem 10.21 is to split it into two statements. First, we define a logic POSTRCL in which all occurrences of the transitive closure operator are *positive* (i.e., occur under the scope of an even number of negations). In fact, one can always convert such a formula into an equivalent formula in which no **trcl** operator would be contained in the scope of any negation symbol. We then prove two results.

Proposition 10.22. *Over ordered structures,* POSTRCL *captures* NLOG.

Proposition 10.23. *Over ordered structures,* POSTRCL = TRCL.

Clearly, Theorem 10.21 will follow from these. Furthermore, they yield the following corollary.

Corollary 10.24 (Immerman–Szelepcsényi). NLOG *is closed under complementation.* □

This is in sharp contrast to other nondeterministic classes such as NP or the levels Σ_i^p of the polynomial hierarchy, where closure under complementation remains a major unsolved problem. In particular, for NP this is the problem of whether NP = CONP.

We start by showing how to prove Proposition 10.22. With negation gone, this proof becomes very similar to the other capture proofs seen in this and the previous chapters. Indeed, the inclusion POSTRCL $\subseteq$ NLOG is proved by straightforward induction (since negation is only applied to FO formulae). For the converse, suppose we have a nondeterministic logspace machine M. In such a machine, we have one read-only tape that stores the input, enc($\mathfrak{A}$), and one work tape, whose size is bounded by $c \log n$ for some constant c (where $n = |A|$). Let Q be the set of states. To model a configuration of M, we need to model both tapes. The input tape can be described by a tuple of variables $\vec{p}$, where $\vec{p}$ indicates a position on the tape, just as in the proof of Fagin's and Immerman-Vardi's theorems.

For the work tape, we need to describe its content, and the position of the head, together with the state. The latter (position and the state) can be described with $|Q|$ variables (assuming $c \log n$ is shorter than the encoding of structures with an n-element universe). If the alphabet of the work tape is $\{0, 1\}$, there are $2^{c \log n} = n^c$ possible configurations, which can be described with c variables. Hence, the entire configuration can be described by tuples $\vec{s}$ of length at most $c(\sigma) + |Q| + c$, where $c(\sigma)$ is a constant depending on σ that gives an upper bound on the size of tuples $\vec{p}$ describing positions in the input.

Then the class of structures accepted by M is definable by the formula

$$\exists \vec{s}_0 \exists \vec{s}_1 \left(\varphi_{\text{init}}(\vec{s}_0) \wedge \varphi_{\text{final}}(\vec{s}_1) \wedge [\mathbf{trcl}_{\vec{x},\vec{y}} \varphi_{\text{next}}(\vec{x}, \vec{y})](\vec{s}_0, \vec{s}_1) \right). \tag{10.15}$$

Here $\varphi_{\text{init}}(\vec{s}_0)$ says that $\vec{s}_0$ is the initial configuration, with the input tape head pointing at the first position in the initial state, and the work tape containing all zeros; $\varphi_{\text{final}}(\vec{s}_1)$ says that $\vec{s}_1$ is an accepting configuration (it is in an accepting state), and $\varphi_{\text{next}}(\vec{x}, \vec{y})$ says that the configuration $\vec{y}$ is obtained from the configuration $\vec{x}$ in one move. It is a straightforward (but somewhat tedious) task to write these three formulae in FO, and it is done similarly to the proofs of other capture theorems. This proves Proposition 10.22.

Before we prove Proposition 10.23, we re-examine (10.15). Let $\underline{\min}$ and $\underline{\max}$, as before, stand for the constants for the minimal and the maximal element with respect to the ordering, and let $\underline{\mathbf{min}}$ and $\underline{\mathbf{max}}$ stand for the tuples of these constants, of the same length as the configuration description. Suppose instead of $\varphi_{\text{next}}(\vec{x}, \vec{y})$ we use the formula φ'_{next}:

$$\varphi_{\text{next}}(\vec{x}, \vec{y}) \ \vee \ (\vec{x} = \underline{\mathbf{min}} \wedge \varphi_{\text{init}}(\vec{y})) \ \vee \ (\varphi_{\text{final}}(\vec{x}) \wedge \vec{y} = \underline{\mathbf{max}}),$$

allowing jumps from $\underline{\mathbf{min}}$ to the initial configuration, and from any final configuration to $\underline{\mathbf{max}}$. Then (10.15) is equivalent to

$$[\mathbf{trcl}_{\vec{x},\vec{y}} \varphi'_{\text{next}}(\vec{x}, \vec{y})](\underline{\mathbf{min}}, \underline{\mathbf{max}}). \tag{10.16}$$

Thus, every posTrCl formula over ordered structures defines an NLog property, which can be expressed by (10.15), and hence by (10.16). We therefore obtained the following.

Corollary 10.25. *Over ordered structures, every* posTrCl *formula is equivalent to a formula of the form* $[\mathbf{trcl}_{\vec{x},\vec{y}} \varphi](\underline{\mathbf{min}}, \underline{\mathbf{max}})$, *where* φ *is* FO.

We now prove Proposition 10.23. The proof is by induction on the structure of TrCl formulae, and the only nontrivial case is that of negation. By Corollary 10.25, we may assume that negation is applied to a formula of the form (10.16); that is, we have to show that

$$\neg \ [\mathbf{trcl}_{\vec{x},\vec{y}} \varphi(\vec{x}, \vec{y})](\underline{\mathbf{min}}, \underline{\mathbf{max}}), \tag{10.17}$$

where φ is FO, is equivalent to a posTrCl formula.

Assume $\vec{x} = k$. For an arbitrary formula $\alpha(\vec{x}, \vec{y})$ with $|\vec{y}| = k$, and a structure $\mathfrak{A}$, let $d_\alpha^\mathfrak{A}(\vec{a}, \vec{b})$ be the shortest distance between $\vec{a}$ and $\vec{b}$ in $\alpha(\mathfrak{A})$ (viewed as a graph on A^k). If no path between $\vec{a}$ and $\vec{b}$ exists, we assume $d_\alpha^\mathfrak{A}(\vec{a}, \vec{b}) = \infty$. We define

$$\text{Reach}_\alpha^\mathfrak{A}(\vec{a}) = \{\vec{b} \in A^k \mid d_\alpha^\mathfrak{A}(\vec{a}, \vec{b}) \neq \infty\}.$$

Thus, (10.17) holds in $\mathfrak{A}$ iff

$$|\text{Reach}_\varphi^\mathfrak{A}(\underline{\textbf{min}})| = |\text{Reach}_{\varphi(\vec{x},\vec{y}) \wedge \neg(\vec{y} = \underline{\textbf{max}})}^\mathfrak{A}(\underline{\textbf{min}})|. \tag{10.18}$$

Notice that the maximal finite value of $d_\alpha^\mathfrak{A}(\vec{a}, \vec{b})$ is $|A|^k$. Since structures are ordered, we can count up to $|A|^k$ using $(k+1)$-tuples of variables: associating the universe A with $\{0, \ldots, n-1\}$, we let a $(k+1)$-tuple $(c_1, \ldots, c_{k+1})$ represent

$$c_1 \cdot n^k + c_2 \cdot n^{k-1} + \ldots + c_k \cdot n + c_{k+1}. \tag{10.19}$$

As it will not cause any confusion, we shall use the notation $\vec{c}$ for both the tuple and the number (10.19) it represents. Note also that constants $0 = \underline{\text{min}}$ and 1, as well as successor and predecessor $\vec{c} + 1$ and $\vec{c} - 1$, are FO-definable in the presence of order, so we shall use them in formulae. Also notice that the maximum value of $d_\alpha^\mathfrak{A}(\vec{a}, \vec{b})$, $|A|^k$, is represented by $1\vec{0} = (1, 0, \ldots, 0)$.

One useful property of posTrCl is that over ordered structures it can count: for a formula $\beta(\vec{x})$ of posTrCl, one can construct another posTrCl formula $\text{count}_\beta(\vec{y})$ such that $\mathfrak{A} \models \text{count}_\beta(\vec{c})$ if there are at least $\vec{c}$ tuples $\vec{a}$ in $\beta(\mathfrak{A})$. Indeed, we can enumerate all the tuples $\vec{a}$, and go over all of them, checking if $\beta(\vec{a})$ holds. Since β can be checked in NLog, the whole algorithm has NLog complexity, and thus is definable in posTrCl. One can also express this counting directly: if $\psi(\vec{x}_1\vec{v}_1, \vec{x}_2\vec{v}_2)$ is $\big((\vec{x}_2 = \vec{x}_1 + 1) \wedge (\vec{v}_2 = \vec{v}_1)\big) \vee \big((\vec{x}_2 = \vec{x}_1 + 1) \wedge \beta(\vec{x}_2) \wedge (\vec{v}_2 = \vec{v}_1 + 1)\big)$, then

$$\exists \vec{z} \left(\begin{array}{l} \big[\textbf{trcl}_{\vec{x}_1\vec{v}_1, \vec{x}_2\vec{v}_2} \psi(\vec{x}_1\vec{v}_1, \vec{x}_2\vec{v}_2)\big] (\underline{\textbf{min}}, \underline{\textbf{min}}, \underline{\textbf{max}}, \vec{z}) \\ \wedge \Big((\vec{y} = \vec{z}) \vee \big(\beta(\underline{\textbf{min}}) \wedge (\vec{y} = \vec{z} + 1)\big)\Big) \end{array} \right)$$

expresses $\text{count}_\beta(\vec{y})$ (exercise: explain why).

Our next goal is to prove the following lemma.

Lemma 10.26. *For every* FO *formula $\alpha(\vec{x}, \vec{y})$, there exists a* posTrCl *formula $\rho_\alpha(\vec{x}, \vec{z})$ such that for every $\mathfrak{A}$,*

$$\mathfrak{A} \models \rho_\alpha(\vec{a}, \vec{c}) \quad \textit{iff} \quad |\text{Reach}_\alpha^\mathfrak{A}(\vec{a})| = \vec{c}.$$

Before proving this, notice that Lemma 10.26 immediately implies Proposition 10.23, since by (10.18), (10.17) is equivalent to

$$\exists \vec{z} \left(\rho_{\varphi}(\underline{\mathbf{min}}, \vec{z}) \ \wedge \ \rho_{\varphi(\vec{x},\vec{y}) \wedge \neg(\vec{y} = \underline{\mathbf{max}})}(\underline{\mathbf{min}}, \vec{z}) \right),$$

which is a POSTRCL formula.

Let $r^{\mathfrak{A}}_{\alpha}(\vec{a}, \vec{c})$ denote the cardinality of $\{\vec{b} \mid d^{\mathfrak{A}}_{\alpha}(\vec{a}, \vec{b}) \leq \vec{c}\}$, so that the cardinality of the set $\text{Reach}^{\mathfrak{A}}_{\alpha}(\vec{a})$ is $r^{\mathfrak{A}}_{\alpha}(\vec{a}, \vec{10})$.

Assume that there is a formula $\gamma_{\alpha}(\vec{x}, \vec{v}, \vec{z}_1, \vec{z}_2)$ such that $\mathfrak{A} \models \gamma_{\alpha}(\vec{a}, \vec{e}, \vec{c}_1, \vec{c}_2)$ means that if $r^{\mathfrak{A}}_{\alpha}(\vec{a}, \vec{e}) = \vec{c}_1$, then $r^{\mathfrak{A}}_{\alpha}(\vec{a}, \vec{e} + 1) = \vec{c}_2$. With such a formula γ_{α}, $\rho_{\alpha}(\vec{x}, \vec{z})$ is definable by

$$[\mathbf{trcl}_{\vec{v}_1 \vec{z}_1, \vec{v}_2 \vec{z}_2} \left((\vec{v}_2 = \vec{v}_1 + 1) \wedge \gamma_{\alpha}(\vec{x}, \vec{v}_1, \vec{z}_1, \vec{z}_2) \right)](\underline{\mathbf{min}}, \underline{\mathbf{min}}, \vec{10}, \vec{z}),$$

since the above formula says that $r_{\alpha}(\vec{x}, \vec{10}) = \vec{z}$. Thus, it remains to show how to define γ_{α}.

In preparation for writing down the formula γ_{α}, notice that there is a POSTRCL formula $d_{\alpha}(\vec{x}, \vec{y}, \vec{z})$ such that $\mathfrak{A} \models d_{\alpha}(\vec{a}, \vec{b}, \vec{c})$ iff $d^{\mathfrak{A}}_{\alpha}(\vec{a}, \vec{b}) \leq \vec{c}$. Indeed, it is given by

$$[\mathbf{trcl}_{\vec{x}_1 \vec{z}_1, \vec{x}_2 \vec{z}_2} \left(\alpha(\vec{x}_1, \vec{x}_2) \wedge (\vec{z}_1 < \vec{z}_2) \right)](\vec{x}, \underline{\mathbf{min}}, \vec{y}, \vec{z}).$$

Coming back to γ_{α}, notice that $r^{\mathfrak{A}}_{\alpha}(\vec{a}, \vec{e} + 1) = \vec{c}_2$ iff

$$\vec{c}_2 + |\{\vec{b} \mid d^{\mathfrak{A}}_{\alpha}(\vec{a}, \vec{b}) > \vec{e} + 1\}| = \vec{10} \ \ (= n^k).$$

Hence, if we could write a POSTRCL formula expressing this condition, we would be able to express γ_{α} in POSTRCL.

Suppose we can express $d^{\mathfrak{A}}_{\alpha}(\vec{a}, \vec{b}) > \vec{e} + 1$ in POSTRCL. Then γ_{α} is straightforward to write, since we already saw how to count: we start with $\vec{c}_2$ and increment the count every time $\vec{b}$ with $d^{\mathfrak{A}}_{\alpha}(\vec{a}, \vec{b}) > \vec{e} + 1$ is found; then **trcl** is applied to see if $\vec{10}$ is reached (we leave the details of this formula to the reader).

Thus, our last task is to express the condition $d^{\mathfrak{A}}_{\alpha}(\vec{a}, \vec{b}) > \vec{e} + 1$ in POSTRCL. Even though we have a formula $d_{\alpha}(\vec{x}, \vec{y}, \vec{z})$ in POSTRCL (meaning $d_{\alpha}(\vec{x}, \vec{y}) \leq \vec{z}$), what we need now is the *negation* of such a formula, which is not in POSTRCL. However, it is possible to express $d^{\mathfrak{A}}_{\alpha}(\vec{a}, \vec{b}) > \vec{e} + 1$ in POSTRCL under the condition $r^{\mathfrak{A}}_{\alpha}(\vec{a}, \vec{e}) = \vec{c}_1$ (which is all that we need anyway, by the definition of γ_{α}).

If $\vec{e} = \underline{\mathbf{min}}$, then $d^{\mathfrak{A}}_{\alpha}(\vec{a}, \vec{b}) > 1$ is equivalent to $\neg\alpha(\vec{a}, \vec{b})$. Otherwise, $d^{\mathfrak{A}}_{\alpha}(\vec{a}, \vec{b}) > \vec{e} + 1$ iff one can find $\vec{c}$ tuples $\vec{f}$ different from $\vec{b}$ such that $d^{\mathfrak{A}}_{\alpha}(\vec{a}, \vec{f}) \leq \vec{e}$ and $\neg\alpha(\vec{f}, \vec{b})$ for all such $\vec{f}$. Now the distance formula (which itself is a POSTRCL formula) occurs positively, and to express $d^{\mathfrak{A}}_{\alpha}(\vec{a}, \vec{b}) > \vec{e} + 1$, we simply count the number of $\vec{f}$ satisfying the conditions above, and compare that number with $\vec{c}$. As we have seen earlier, such counting of $\vec{f}$'s can be done by a POSTRCL formula. Thus, γ_{α} is expressible in POSTRCL, which completes the proof of Lemma 10.26 and Theorem 10.21. □

10.7 A Logic for PTIME?

We have seen that LFP and IFP capture PTIME on the class of ordered structures. On the other hand, for classes such as NP and CONP we have logics that capture them over all structures. The question that immediately arises is whether there is a logic that captures PTIME, without the additional restriction to ordered structures. If there were such a logic, answering the "PTIME vs. NP" question would become a purely logical problem: one would have to separate two logics over the class of all finite structures.

However, all attempts to produce a logic that captures PTIME have failed so far. In fact, it is even conjectured that no such logic exists:

Conjecture (Gurevich) *There is no logic that captures* PTIME *over the class of all finite structures.*

This is a very strong conjecture: since there is a logic for NP, by Fagin's theorem, it would imply that PTIME $\neq$ NP! The conjecture described precisely what a logic is. We shall not go into technical details, but the main idea is to rule out the possibility of taking an arbitrary collection of properties and stating that they constitute a logic. For example, is the collection of all PTIME properties a logic? If we want the conjecture to hold, clearly the answer ought to be *no*.

In this short section, we shall present a few attempts to refute Gurevich's conjecture and find a logic for PTIME – and show how they all failed. The results here will be presented without proofs, and the interested reader should consult the bibliographic notes section for the references.

What are examples of properties not expressible in LFP or IFP over unordered structures? Although we have not proved this yet, we mentioned one example: the query EVEN. We shall see later, in Chap. 11, that in general IFP cannot express nontrivial counting properties over unordered structures. Hence, one might try to add counting to IFP (it is better to use IFP, so that positiveness would not constrain us), and hope that such an extension captures PTIME.

This extension of IFP, denoted by IFP(**Cnt**), can be defined in the same way as we defined FO(**Cnt**) from FO: one introduces the additional universe $\{0, \ldots, n-1\}$, where n is the cardinality of the universe of a σ-structure $\mathfrak{A}$, and extends the logic with counting quantifiers $\exists ix$. However, this extension still falls short of PTIME, and the separating example is very complicated.

Theorem 10.27. *There are* PTIME *properties which are not definable in* IFP(**Cnt**). □

Another attempt to expand IFP is to introduce *generalized quantifiers*, already seen in Chap. 8. There, we only dealt with unary generalized quantifiers; here we present a general definition, but for notational simplicity deal with the case of one additional relation per quantifier.

Let R be a relation symbol of arity k, $R \notin \sigma$. Let $\mathcal{C} \subseteq \text{STRUCT}[\{R\}]$ be a class of structures closed under isomorphism. This gives rise to a generalized quantifier $\mathcal{Q}_{\mathcal{C}}$ and the extension of IFP with $\mathcal{Q}_{\mathcal{C}}$, denoted by IFP($\mathcal{Q}_{\mathcal{C}}$), which is defined as follows. If $\varphi(\vec{x}, \vec{y})$ is an IFP($\mathcal{Q}_{\mathcal{C}}$) formula of vocabulary σ, and $|\vec{x}| = k$, then

$$\psi(\vec{y}) \quad \equiv \quad \mathcal{Q}_{\mathcal{C}}\vec{x}\ \varphi(\vec{x}, \vec{y}) \tag{10.20}$$

is an IFP($\mathcal{Q}_{\mathcal{C}}$) formula. The other formation rules are exactly the same as for IFP. The semantics of (10.20) is as follows:

$$\mathfrak{A} \models \psi(\vec{b}) \quad \Leftrightarrow \quad \langle A, \{\vec{a} \mid \mathfrak{A} \models \varphi(\vec{a}, \vec{b})\}\rangle \in \mathcal{C}.$$

For example, if $\mathcal{C}$ is the class of connected graphs, then the sentence $\mathcal{Q}_{\mathcal{C}}x, y\ E(x, y)$ simply tests if the input graph is connected.

If $\mathfrak{Q}$ is a set of generalized quantifiers, then by IFP($\mathfrak{Q}$) we mean the extension of IFP with the formulae (10.20) for all the generalized quantifiers in $\mathfrak{Q}$.

There is a "simple" way of getting a logic that captures PTIME: it is IFP($\mathfrak{Q}_p$), where $\mathfrak{Q}_p$ is the collection of all PTIME properties. However, this is cheating: we define the logic in terms of itself. But perhaps there is a nicely behaving set $\mathfrak{Q}$ of generalized quantifiers such that IFP($\mathfrak{Q}$) captures PTIME.

The first result, showing that such a class – if it exists – will be hard to find, says the following.

Proposition 10.28. *Let $\mathfrak{Q}_n$ be a collection of generalized quantifiers of arity at most n. There there exists a vocabulary σ_n such that over σ_n-structures,* IFP($\mathfrak{Q}_n$) *fails to capture* PTIME. □

The reason this result is not completely satisfactory is that the arity of relations in σ_n depends on n. For example, Proposition 10.28 says nothing about the impossibility of capturing PTIME over graphs. And in fact there is a collection $\mathfrak{Q}_{\text{gr}}$ of generalized binary quantifiers (i.e., of arity 2) such that IFP($\mathfrak{Q}_{\text{gr}}$) expresses all the PTIME properties of graphs (why?). In fact, one can even show that there is a single *ternary* generalized quantifier $\mathcal{Q}_3$ such that IFP($\mathcal{Q}_3$) expresses all the PTIME properties of graphs (intuitively, it is possible to code $\mathfrak{Q}_{\text{gr}}$ with one ternary generalized quantifier), but $\mathcal{Q}_3$ itself is not PTIME-computable, and hence IFP($\mathcal{Q}_3$) fails to capture PTIME on graphs.

The existence of a generalized quantifier $\mathcal{Q}_3$ raises an intriguing possibility that for some finite collection $\mathfrak{Q}_{\text{fin}}$ of PTIME-computable generalized quantifiers, IFP($\mathfrak{Q}_{\text{fin}}$) captures PTIME on unordered graphs. However, this attempt to refute Gurevich's conjecture does not work either.

Theorem 10.29. *There is no finite collection $\mathfrak{Q}_{\text{fin}}$ of* PTIME*-computable generalized quantifiers such that* IFP($\mathfrak{Q}_{\text{fin}}$) *captures* PTIME *on unordered graphs.*

Thus, given all that we know today, Gurevich's conjecture may well be true, as it has withstood a number of attempts to produce a logic for PTIME over unordered structures.

10.8 Bibliographic Notes

Inductive operators and fixed point logics are studied extensively in Moschovakis [185] in the context of arbitrary models. The systematic study of fixed point logics in finite model theory originated with Chandra and Harel [33], who introduced the least fixed point operator in the context of database query languages to overcome well-known limitations of FO. The subject is treated in detail in Ebbinghaus and Flum [60], Immerman [133], Grohe [106]; see also a recent survey by Dawar and Gurevich [51]. All of these references present the Tarski-Knaster theorem, least and inflationary fixed point logics, and simultaneous fixed points.

The "even simple path" example is taken from Kolaitis [148], where it is attributed to Yannakakis. See also Exercise 10.2.

The stage comparison theorem was proved in Moschovakis [185], and specialized for the finite case in Immerman [130] and Gurevich and Shelah [119]; the proof presented here follows Leivant [165]. Corollary 10.12 is from Gurevich and Shelah [119], and Corollary 10.13 from [130].

The connection between fixed point logics and polynomial time was discovered by several people in the early 1980s. Sazonov [212] showed in 1980 that a certain least fixed point construction – of recursive-theoretic flavor – captures PTIME. Then, in 1982, Immerman [129], Vardi [244], and Livchak [172] proved what is now known as the Immerman-Vardi theorem. Both Immerman's and Vardi's papers appeared in the proceedings of the STOC 1982 conference; Livchak's paper was published in Russian and became known much later; hence Theorem 10.14 is usually referred to as the Immerman-Vardi theorem. In 1986, Immerman published a full version of his 1982 paper (see [130]). Theorem 10.15 is from Vardi [244].

DATALOG has been studied extensively in the database literature, see, e.g., Abiteboul, Hull, and Vianu [3] for many additional results and references. Theorem 10.19 is from Papadimitriou [194].

Theorem 10.21 is from Immerman [130, 132]: the first of these papers showed that posTRCL captures NLOG, and the other paper proved closure under complementation (see also Szelepcsényi [226]).

A number of references discuss Gurevich's conjecture in detail (e.g., Otto [191], Kolaitis [147], as well as [60]); they also discuss the notion of a "logic" suitable for capturing PTIME. Theorem 10.27 is from Cai, Fürer, and Immerman [30] (see also Otto [191], as well as Gire and Hoang [91] for extensions). Theorem 10.29 is from Dawar and Hella [52].

Sources for exercises:

Exercise 10.10:	Ajtai and Gurevich [13]
Exercise 10.11:	Immerman [130]
Exercises 10.12 and 10.13:	Grädel [97]
Exercise 10.14:	Immerman [131]
Exercise 10.15:	Grädel and McColm [101]

Exercise 10.16:	Abiteboul and Vianu [5]
Exercises 10.17 and 10.18:	Afrati, Cosmadakis, and Yannakakis [8]
Exercise 10.19:	Grädel and Otto [102]
Exercises 10.20 and 10.21:	Grohe [107]
Exercise 10.22:	Shmueli [220] and Cosmadakis et al. [43]
Exercise 10.23:	Marcinkowski [179]
Exercise 10.24:	Gottlob and Koch [94]
Exercise 10.25:	Gurevich, Immerman, and Shelah [118]
Exercise 10.26:	Dawar and Hella [52]
Exercise 10.27:	Dawar, Lindell, and Weinstein [54]

10.9 Exercises

Exercise 10.1. Prove Proposition 10.3.

Exercise 10.2. Prove that the simultaneous fixed point shown before Theorem 10.8 defines pairs of nodes connected by a simple path of even length.
Hint: use Menger's theorem in graph theory.
Also show that this does not generalize to directed graphs.

Exercise 10.3. Prove Theorem 10.8 for a system involving an arbitrary number of formulae.

Exercise 10.4. Prove Theorem 10.10.

Exercise 10.5. Prove Theorem 10.15.

Exercise 10.6. Prove Theorem 10.19.

Exercise 10.7. Prove that the combined complexity of LFP is EXPTIME-complete.

Exercise 10.8. Consider an alternative semantics for DATALOG programs. Given a set of rules Π and a structure $\mathfrak{A}$, an instantiation $\vec{P}$ of all the intensional predicates is called a *model* of Π on $\mathfrak{A}$ if every rule of Π is satisfied. Show that for any Π, there exists a minimal, with respect to inclusion, model $\vec{P}_{\min}$. The *minimal model semantics* of DATALOG defines the answer to (Π, Q) on $\mathfrak{A}$ as the interpretation of Q in $\vec{P}_{\min}$.

Prove that the fixed point and the minimal model semantics of DATALOG coincide.

Exercise 10.9. Write down the formulae ψ_i and ψ_q from the proof of the Immerman-Vardi theorem, and show that their simultaneous least fixed point computes the relations T_i and H_q.

Exercise 10.10. Show that over finite structures, monotone and positive are two different concepts (they are known to be the same over infinite structures, see Lyndon [175]). That is, give an example of an FO formula $\varphi(P, \cdot)$ which is monotone in P, but not equivalent to any FO formula positive in P.

Exercise 10.11. Assume that the vocabulary σ contains at least two distinct constants. Prove a stronger normal form result for LFP: every LFP formula is equivalent to a formula of the form $[\mathbf{lfp}_{R,\vec{x}}\varphi(R,\vec{x})](\vec{t})$, where φ is an FO formula.

Hint: use two constants to eliminate nested fixed points.

Exercise 10.12. Consider a restriction of SO that consists of formulae of the form

$$\mathbf{Q}R_1 \ldots \mathbf{Q}R_n \forall \vec{x} \bigwedge_l \alpha_l,$$

where each $\mathbf{Q}$ is either $\exists$ or $\forall$, and each α_l is *Horn with respect to* $R_1, \ldots, R_n$. That is, it is of the form

$$\gamma_1 \wedge \ldots \wedge \gamma_m \to \beta,$$

where each γ_j either does not mention R_i's, or is of the form $R_i(\vec{u})$, and β is either of the form $R_i(\vec{u})$, or *false*. We denote such restriction by SO-Horn. If all the quantifiers $\mathbf{Q}$ are existential, we speak of $\exists$SO-Horn.

Prove that over ordered structures, SO-Horn and $\exists$SO-Horn capture Ptime.

Exercise 10.13. The class SO-Krom is defined similarly to SO-Horn, except that each α_l is a disjunction of at most two atoms of the form $R_i(\vec{u})$ or $\neg R_j(\vec{u})$, and a formula that does not mention the R_i's. $\exists$SO-Krom is defined as the restriction where all second-order quantifiers are existential.

Prove that both SO-Krom and $\exists$SO-Krom capture NLog over ordered structures.

Exercise 10.14. Define a variant of the transitive closure logic, denoted by DetTrCl, where the transitive closure operator **trcl** is replaced by the *deterministic* transitive closure. When applied to a graph $\langle V, E\rangle$, it finds pairs (a, b) which are connected by a deterministic path: on such a path, every node except b must be of out-degree 1.

Prove that DetTrCl captures DLog over ordered structures.

Exercise 10.15. Prove that over unordered structures, DetTrCl $\subsetneq$ TrCl $\subsetneq$ LFP.

Exercise 10.16. Consider the following language that computes queries over STRUCT[σ]. Given an input structure $\mathfrak{A}$, its programs compute sequences of relations, and are defined inductively as follows:

- $\emptyset$ is a program that computes no relation.
- If $\Pi(R_1, \ldots, R_n)$ is a program that computes relations $R_1, \ldots, R_n$, where $R_1, \ldots, R_n \notin \sigma$, then

$$\Pi(R_1, \ldots, R_n); \;\; R(\vec{x}) \;\; \text{:-} \;\; \varphi(\vec{x});$$

  where $R \notin \sigma \cup \{R_1, \ldots, R_n\}$, and φ is an FO formula in the vocabulary of σ expanded with $R_1, \ldots, R_n$, is a program that computes relations $R_1, \ldots, R_n, R$, with R obtained by evaluating φ on the expansion of $\mathfrak{A}$ with $R_1, \ldots, R_n, R$.
- If $\Pi(R_1, \ldots, R_n)$ is a program that computes relations $R_1, \ldots, R_n$, and $\Pi'(T_1, \ldots, T_k)$ is a program over STRUCT[$\sigma \cup \{R_1, \ldots, R_n\} \cup \{S_1, \ldots, S_k\}$], where the arity of each S_i matches the arity of T_i, then

$$\Pi(R_1, \ldots, R_n); \;\; \textbf{while change do} \;\; \Pi'(T_1, \ldots, T_k) \;\; \textbf{end};$$

is a program that computes $(R_1, \ldots, R_n, T_1, \ldots, T_k)$ over σ-structures. The meaning of the last statement is that starting with $(\emptyset, \ldots, \emptyset)$ as the interpretation of the S_i's, one iterates Π'; it computes the T_i's, which are then reused as S_i's, and so on. This is done as long as it changes one relation among the S_i's. If this program terminates, the values of the relations $(T_1, \ldots, T_k)$ in that state become the output.

For example, the while loop

$$\textbf{while change do } T(x,y) :- E(x,y) \vee \exists z\ (E(x,z) \wedge S(z,y)) \textbf{ end;}$$

computes the transitive closure of E.

Prove that over ordered structures, such **while** programs compute precisely the PSPACE queries.

Exercise 10.17. Let monotone PTIME be the class of all monotone PTIME properties. Show that DATALOG, even in the presence of a successor relation, fails to capture monotone PTIME.

Hint: Let $\sigma = \{R, S\}$, where R is ternary, and S is unary. The separating query is defined as follows: Q is true in $\mathfrak{A}$ iff the system of linear equations

$$\{x_1 + x_2 + x_3 = 1 \mid (x_1, x_2, x_3) \in R^{\mathfrak{A}}\} \cup \{x = 0 \mid x \in S^{\mathfrak{A}}\}$$

does not have a non-negative solution.

Exercise 10.18. Prove that without the successor relation, DATALOG$_\neg$ fails to capture PTIME on ordered structures, even if one allows atoms $\neg(x = y)$.

Hint: The separating query takes a graph, and outputs pairs of nodes (a, b) such that there is a path from a to b whose length is a perfect square.

Exercise 10.19. Show how to expand DATALOG with counting, and prove that the resulting language is equivalent to the expansion of IFP with counting.

Exercise 10.20. Prove that the expansion of IFP with counting captures PTIME on the class of planar graphs.

Exercise 10.21. Prove that the class of planar graphs is definable in IFP.

Exercise 10.22. You may recall that containment of conjunctive queries is NP-complete (Exercise 6.19). Prove that containment of arbitrary DATALOG queries is undecidable, but becomes decidable if all intensional predicates are unary.

Exercise 10.23. We say that a DATALOG program Π is *uniformly bounded* if there is a number n such that on every structure $\mathfrak{A}$, the fixed point of F_Π is reached after at most n steps.

Prove that uniform boundedness is undecidable for DATALOG, even for programs that consist of a single rule.

Exercise 10.24. Consider trees represented as in Chap. 7, i.e., structures with two successor predicates, labeling predicates, and, furthermore, assume that we have unary predicates Leaf and Root interpreted as the set of leaves, and the singleton set containing the root.

Define *monadic* DATALOG as the restriction of DATALOG where all intensional predicates are unary.

Prove that over trees, Boolean and unary queries definable in monadic DATALOG and in MSO are precisely the same. In particular, a tree language is definable in monadic DATALOG iff it is regular.

Exercise 10.25. Prove that there exists a class $\mathcal{C}$ of graphs which admits fixed points of unbounded depth (i.e., for every n there is an inductive operator that reaches its fixed point on some graph from $\mathcal{C}$ in at least n iterations), and yet LFP = FO on $\mathcal{C}$.

Remark: this exercise says that it is possible for LFP and FO to coincide on a class of graphs which admits fixed points of unbounded depth. The negation of this was known as McColm's conjecture; hence the goal of this exercise is to disprove McColm's conjecture. McColm [181] made two conjectures relating boundedness of fixed points and collapse of logics; the second conjecture that talks about FO and the finite variable logic is known to be true (see Exercise 11.19).

For the next three exercises, consider the following statement, known as the *ordered conjecture* (see Kolaitis and Vardi [153]):

If $\mathcal{C}$ is an infinite class of finite ordered structures, then FO $\subsetneq$ LFP *on* $\mathcal{C}$.

Exercise 10.26. Prove that if the ordered conjecture does not hold, then PTIME $\neq$ PSPACE.

Exercise 10.27. Prove that if the ordered conjecture holds, then LINH $\neq$ ETIME.

Here LINH is the linear time hierarchy: the class of languages computed in linear time by alternating Turing machines, with a constant number of alternations, and ETIME is the class of languages computed by deterministic Turing machines in time $2^{O(n)}$.

Exercise 10.28.* Does the ordered conjecture hold?

11

Finite Variable Logics

In this chapter, we introduce finite variable logics: a unifying tool for studying fixed point logics. These logics use infinitary connectives already seen in Chap. 8, but here we impose a different restriction: each formula can use only finitely many variables. We show that fixed point logics LFP, IFP, and PFP can be embedded in such a finite variable logic. Furthermore, the finite variable logic is easier to study: it can be characterized by games, and this gives us bounds on the expressive power of fixed point logics; in particular, we show that without a linear ordering, they fail to capture complexity classes. We then study definability and ordering of types in finite variable logics, and use these techniques to relate separating complexity classes to separating some fixed point logics over unordered structures.

11.1 Logics with Finitely Many Variables

Let us revisit the example of the transitive closure of a relation. Suppose E is a binary relation. We know how to write FO formulae $\varphi_n(x, y)$ stating that there is a path from x to y of length n (that is, formulae defining the stages of the fixed point computation of the transitive closure). One can express $\varphi_n(x, y)$, $n > 1$, as $\exists x_1 \ldots \exists x_{n-1} \left(E(x, x_1) \wedge \ldots \wedge E(x_{n-1}, y)\right)$, and $\varphi_1(x, y)$ as $E(x, y)$. If we could use infinitary disjunctions (i.e., the logic $\mathcal{L}_{\infty\omega}$ of Chap. 8), we could express the transitive closure query by

$$\bigvee_{n \geq 1} \varphi_n(x, y). \tag{11.1}$$

One could even define $\varphi_n(x, y)$ by induction, as we did in Chap. 10:

$$\varphi_1(x, y) \equiv E(x, y), \quad \varphi_{n+1}(x, y) \equiv \exists z_n \left(E(x, z_n) \wedge \varphi_n(z_n, y)\right), \tag{11.2}$$

where z_n is a fresh variable. The problem with either definition of the φ_n's together with (11.1) is that the logic $\mathcal{L}_{\infty\omega}$ is useless in the context of finite model theory: as we saw in Chap. 8, it defines *every* property of finite structures (Proposition 8.4).

However, if we look carefully at the definition of the φ_n's given in (11.2), we can see that there is no need to introduce a fresh variable z_n for each new formula. In fact, we can define formulae φ_n as follows:

$$\begin{aligned} \varphi_1(x,y) &\equiv E(x,y) \\ &\cdots\cdots\cdot \\ \varphi_{n+1}(x,y) &\equiv \exists z \Big(E(x,z) \wedge \exists x \ (z = x \wedge \varphi_n(x,y))\Big). \end{aligned} \tag{11.3}$$

In definition (11.3), each formula φ_n uses only three variables, x, y, and z, by carefully reusing them. To define $\varphi_n(x,y)$, we need to say that there is a z such that $E(x,z)$ holds, and $\varphi_n(z,y)$ holds. But with three variables, we only know how to say that $\varphi_n(x,y)$ holds. So once z is used in $E(x,z)$, it is no longer needed, and we replace it by x: that is, we say that there is an x such that x happens to be equal to z, and $\varphi_n(x,y)$ holds: and we know that the latter is definable with three variables.

With these formulae (11.3), we can still define the transitive closure by (11.1). What makes the difference now is the fact that the resulting formula only uses three variables. If one checks the proof of Proposition 8.4, one discovers that, to define arbitrary classes of finite structures in $\mathcal{L}_{\infty\omega}$, one needs, in general, infinitely many variables. So perhaps an infinitary logic in which the number of variables is finite could be useful after all?

The answer to this question is a resounding *yes*: we shall see that all fixed point logics can be coded in a way very similar to (11.3), and that the resulting infinitary logic can be analyzed by the same techniques we have seen in previous chapters.

Definition 11.1 (Finite variable logics). *The class of* FO *formulae that use at most k distinct variables will be denoted by* FO^k. *The class of* $\mathcal{L}_{\infty\omega}$ *formulae that use at most k variables will be denoted by* $\mathcal{L}^k_{\infty\omega}$ *(reminder:* $\mathcal{L}_{\infty\omega}$ *extends* FO *with infinitary conjunctions* $\bigwedge$ *and disjunctions* $\bigvee$*). Finally, we define the finite variable infinitary logic* $\mathcal{L}^\omega_{\infty\omega}$ *by*

$$\mathcal{L}^\omega_{\infty\omega} \;=\; \bigcup_{k\in\mathbb{N}} \mathcal{L}^k_{\infty\omega}.$$

That is, $\mathcal{L}^\omega_{\infty\omega}$ *has formulae of* $\mathcal{L}_{\infty\omega}$ *that only use finitely many variables.*

The quantifier rank $\mathsf{qr}(\cdot)$ *of* $\mathcal{L}^\omega_{\infty\omega}$ *formulae is defined as for* FO *for Boolean connectives and quantifiers; for infinitary connectives, we define*

$$\mathsf{qr}(\bigvee_i \varphi_i) \;=\; \mathsf{qr}(\bigwedge_i \varphi_i) \;=\; \sup_i \mathsf{qr}(\varphi_i).$$

Thus, in general the quantifier rank of an infinitary formula is an *ordinal*. For example, if the φ_n's are FO formulae with $\mathsf{qr}(\varphi_n) = n$, then

$\mathsf{qr}(\bigvee_{n<\omega}\varphi_n) = \omega$, and $\mathsf{qr}(\exists x \bigvee_{n<\omega}\varphi_n) = \omega+1$. When we establish a normal form for $\mathcal{L}^\omega_{\infty\omega}$, we shall see that over finite structures it suffices to consider only formulae of quantifier rank up to ω.

Let us give a few examples of definability in $\mathcal{L}^\omega_{\infty\omega}$. We first consider linear orderings: that is, the vocabulary contains one binary relation <. With the same trick of reusing variables, we define the formulae

$$\begin{aligned} \psi_1(x) &\equiv (x = x) \\ &\ldots\ldots\ldots \\ \psi_{n+1}(x) &\equiv \exists y\, \Big((x > y) \wedge \exists x\, \big(y = x \wedge \psi_n(x)\big)\Big). \end{aligned} \tag{11.4}$$

The formula $\psi_n(a)$ is true in a linear order L iff the set $\{b \mid b \leq a\}$ contains at least n elements. Indeed, $\psi_1(x)$ is true for every x, and $\psi_{n+1}(x)$ says that there is $y < x$ such that there are at least n elements that do not exceed y. Thus, for each n we have a sentence $\Psi_n \equiv \exists x\ \psi_n(x)$ that is true in L iff $|L| \geq n$.

Now let C be an arbitrary subset of $\mathbb{N}$. Consider the sentence

$$\bigvee_{n\in C} (\Psi_n \wedge \neg\Psi_{n+1}).$$

This is a sentence of $\mathcal{L}^2_{\infty\omega}$, as it uses only two variables, x and y, and it is true in L iff $|L| \in C$. Hence, arbitrary cardinalities of linear orderings can be tested in $\mathcal{L}^2_{\infty\omega}$.

Next, consider fixed point computations. Suppose that an FO formula $\varphi(R,\vec{x})$ defines an inductive operator; that is, either φ is monotone in R, or we are considering an inflationary fixed point. We have seen in Chap. 10 that stages of the fixed point computation can be defined by FO formulae $\varphi^n(\vec{x})$; the formulae we used, however, may potentially involve arbitrarily many variables. To be able to express the least fixed point as $\bigvee_n \varphi^n(\vec{x})$, we need to define those formulae $\varphi^n(\vec{x})$ more carefully.

Assume that φ, in addition to $\vec{x} = (x_1,\ldots,x_k)$, uses variables $z_1,\ldots,z_l$. We introduce additional variables $\vec{y} = (y_1,\ldots,y_k)$, and define $\varphi^0(\vec{x})$ as $\neg(x_1 = x_1)$ (i.e., *false*), and then inductively $\varphi^{n+1}(\vec{x})$ as $\varphi(R,\vec{x})$ in which every occurrence of $R(u_1,\ldots,u_k)$, where $u_1,\ldots,u_k$ are variables among $\vec{x}$ and $\vec{z}$, is replaced by

$$\exists\vec{y}\, \Big((\vec{y} = \vec{u}) \;\wedge\; \big(\exists\vec{x}((\vec{x} = \vec{y}) \;\wedge\; \varphi^n(\vec{x}))\big)\Big). \tag{11.5}$$

As usual, $\vec{x} = \vec{y}$ is an abbreviation for $\big((x_1 = y_1)\wedge\ldots\wedge(x_k = y_k)\big)$. Notice that in the resulting formula, variables from $\vec{y}$ cannot appear in any subformula of the form $R(\cdot)$.

The effect of the substitution is that we use φ with R being given the interpretation of the nth stage, so $\bigvee_n \varphi^n(\vec{x})$ does compute the fixed point.

Furthermore, we at most doubled the number of variables in φ. Hence, if $\varphi \in \mathrm{FO}^m$, then both $\mathbf{lfp}_{R,\vec{x}}\varphi$ and $\mathbf{ifp}_{R,\vec{x}}\varphi$ are expressible in $\mathcal{L}^{2m}_{\infty\omega}$.

If we have a complex fixed point formula (e.g., involving nested fixed points), we can then apply the construction inductively, using the same substitution (11.5), since φ^n need not be an FO formula, and can have infinitary connectives. This shows that every LFP or IFP formula is equivalent to a formula of $\mathcal{L}^{\omega}_{\infty\omega}$ (since for every fixed point, we at most double the number of variables). Hence, we have the following.

Theorem 11.2. LFP, IFP, PFP $\subseteq \mathcal{L}^{\omega}_{\infty\omega}$.

Proof. We have proved it already for LFP and IFP; for PFP, the construction is modified slightly: instead of taking the disjunction of all the φ^n's, we define the sentence $good_n$ as $\forall \vec{x}\ (\varphi^n(\vec{x}) \leftrightarrow \varphi^{n+1}(\vec{x}))$ (indicating that the fixed point was reached). Then $[\mathbf{pfp}_{R,\vec{x}}\varphi](\vec{y})$ is expressed by

$$\psi(\vec{y}) \equiv \bigvee_{n\in\mathbb{N}} \Big(good_n \wedge \varphi^n(\vec{x})\Big).$$

Indeed, if there is no n such that $good_n$ holds, then the partial fixed point is the empty set, and $\psi(\vec{y})$ is equivalent to *false*. Otherwise, let n_0 be the smallest natural number n for which $good_n$ holds. Then, for all $m \geq n_0$, we have $\forall \vec{x}\ (\varphi^{n_0}(\vec{x}) \leftrightarrow \varphi^m(\vec{x}))$, and hence $\psi(\vec{y})$ defines the partial fixed point. Therefore, ψ defines $\mathbf{pfp}_{R,\vec{x}}\varphi$, and it at most doubles the number of variables. Using this construction inductively, we see that PFP $\subseteq \mathcal{L}^{\omega}_{\infty\omega}$. ☐

We now revisit the case of orderings. We have shown before that arbitrary cardinalities of linear orderings are definable in $\mathcal{L}^{\omega}_{\infty\omega}$; in other words, every query on finite linear orderings is $\mathcal{L}^{\omega}_{\infty\omega}$-definable. It turns out that this extends to all ordered structures.

Proposition 11.3. *Every query over ordered finite σ-structures is expressible in $\mathcal{L}^{\omega}_{\infty\omega}$. In fact, if m is the maximum arity of a relation symbol in σ, then it suffices to use $\mathcal{L}^{m+1}_{\infty\omega}$.*

Proof. To keep the notation simple, we consider ordered graphs $G = \langle V, E\rangle$, with a linear ordering $<$ on V (i.e., $m = 2$, and in this case we show definability in $\mathcal{L}^{3}_{\infty\omega}$). Recall that we have an $\mathcal{L}^{2}_{\infty\omega}$ formula $\psi_n(x)$, that uses variables x, y, and tests if there are at least n elements in V which do not exceed x in the ordering $<$. Hence, for each n we have an $\mathcal{L}^{2}_{\infty\omega}$ formula $\psi_{=n}(x)$ which holds iff x is the nth element in the ordering $<$. Now, for each G we define a formula χ_G as

$$\forall x \forall z \Big(E(x,z) \leftrightarrow \bigvee_{(i,j)\in E} (\psi_{=i}(x) \wedge \psi_{=j}(z))\Big) \ \wedge\ \exists x\ \psi_p(x) \ \wedge\ \neg\exists x\ \psi_{p+1}(x),$$

viewing the universe V of cardinality p as $\{1, \ldots, p\}$. Here $\psi_{=j}(z)$ is obtained from $\psi_{=j}(x)$ by replacing x by z; that is, this formula uses variables z and y.

Note that $\chi_G \in \mathcal{L}^3_{\infty\omega}$ and $G' \models \chi_G$ iff G' is isomorphic to G (as an ordered graph). Finally, for a class $\mathcal{P}$ of ordered graphs, we let

$$\Phi_{\mathcal{P}} \equiv \bigvee_{G \in \mathcal{P}} \chi_G.$$

Clearly, this formula defines $\mathcal{P}$. □

11.2 Pebble Games

In this section we present Ehrenfeucht-Fraïssé-style games which characterize finite variable logics. There are two elements of these games that we have not seen before. First, these are *pebble* games: the spoiler and the duplicator have a fixed set of pairs of pebbles, and each move consists of placing a pebble on an element of a structure, or removing a pebble and placing it on another element. Second, the game does not have to end in a finite number of rounds (but we can still determine who wins it).

Definition 11.4 (Pebble games). *Let $\mathfrak{A}, \mathfrak{B} \in$ STRUCT$[\sigma]$. A k-pebble game over $\mathfrak{A}$ and $\mathfrak{B}$ is played by the spoiler and the duplicator as follows. The players have a set of pairs of pebbles $\{(p^1_{\mathfrak{A}}, p^1_{\mathfrak{B}}), \ldots, (p^k_{\mathfrak{A}}, p^k_{\mathfrak{B}})\}$. In each move, the following happens:*

- *The spoiler chooses a structure, $\mathfrak{A}$ or $\mathfrak{B}$, and a number $1 \leq i \leq k$.*
  For the description of the other moves, we assume the spoiler has chosen $\mathfrak{A}$. The other case, when the spoiler chooses $\mathfrak{B}$, is completely symmetric.
- *The spoiler places the pebble $p^i_{\mathfrak{A}}$ on some element of $\mathfrak{A}$. If $p^i_{\mathfrak{A}}$ was already placed on $\mathfrak{A}$, this means that the spoiler either leaves it there or removes it and places it on some other element of $\mathfrak{A}$; if $p^i_{\mathfrak{A}}$ was not used, it means that the spoiler picks that pebble and places it on an element of $\mathfrak{A}$.*
- *The duplicator responds by placing $p^i_{\mathfrak{B}}$ on some element of $\mathfrak{B}$.*

We denote the game that continues for n rounds by $\mathrm{PG}^n_k(\mathfrak{A}, \mathfrak{B})$, and the game that continues forever by $\mathrm{PG}^\infty_k(\mathfrak{A}, \mathfrak{B})$.

After each round of the game, the pebbles placed on $\mathfrak{A}$ and $\mathfrak{B}$ define a relation $F \subseteq A \times B$: if $p^i_{\mathfrak{A}}$, for some $i \leq k$, is placed on $a \in A$ and $p^i_{\mathfrak{B}}$ is placed on $b \in B$, then the pair (a, b) is in F.

The duplicator has a winning strategy in $\mathrm{PG}^n_k(\mathfrak{A}, \mathfrak{B})$ if he can ensure that after each round $j \leq n$, the relation F defines a partial isomorphism. That is, F is a graph of a partial isomorphism. In this case we write $\mathfrak{A} \equiv^{\infty\omega}_{k,n} \mathfrak{B}$.

The duplicator has a winning strategy in $\mathrm{PG}^\infty_k(\mathfrak{A}, \mathfrak{B})$ if he can ensure that after every round the relation F defines a partial isomorphism. This is denoted by $\mathfrak{A} \equiv^{\infty\omega}_k \mathfrak{B}$.

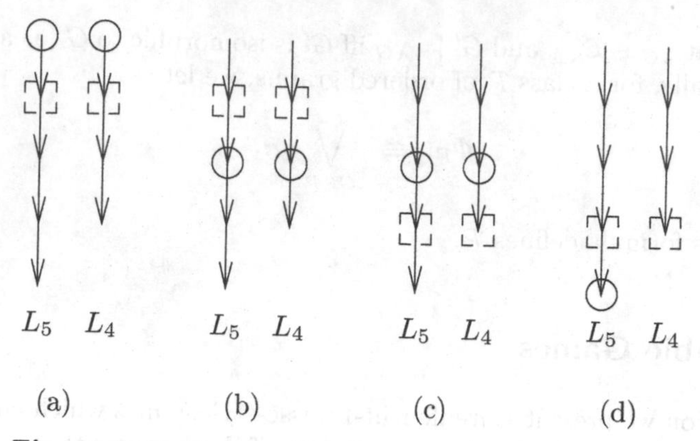

Fig. 11.1. Spoiler winning the pebble game on L_5 and L_4

These games characterize finite variable logics as follows.

Theorem 11.5. *a) Two structures* $\mathfrak{A}, \mathfrak{B} \in \text{STRUCT}[\sigma]$ *agree on all sentences of* $\mathcal{L}^k_{\infty\omega}$ *of quantifier rank up to* n *iff* $\mathfrak{A} \equiv^{\infty\omega}_{k,n} \mathfrak{B}$.

b) Two structures $\mathfrak{A}, \mathfrak{B} \in \text{STRUCT}[\sigma]$ *agree on all sentences of* $\mathcal{L}^k_{\infty\omega}$ *iff* $\mathfrak{A} \equiv^{\infty\omega}_{k} \mathfrak{B}$. □

Before we prove this theorem, we give a few examples of pebble games. First, consider two arbitrary linear orderings L_n, L_m of lengths n and m, $n \neq m$. Here we show that it is the spoiler who wins $\text{PG}^\infty_2(L_n, L_m)$.

The strategy for L_5 and L_4 is shown in Fig. 11.1; the general strategy is exactly the same. We have two pairs of pebbles, and elements pebbled by pebble 1 are shown as circled, and those pebbled by pebble 2 are shown in dashed boxes. The spoiler starts by placing pebble 1 on the top element of L_5; the duplicator is forced to respond by placing the matching pebble on the top element of L_4. Then the spoiler places the second pebble on the second element of L_5, and the duplicator matches it in L_4 (if he does not, he loses in the next round).

This is the configuration shown in Fig. 11.1 (a). Next, the spoiler removes pebble 1 from the top element of L_5 and places it on the third element. The spoiler is forced to mimic the move in L_4, to preserve the order relation. We are now in the position shown in Fig. 11.1 (b). The spoiler then moves the second pebble two levels down; the duplicator matches it. We are now in position (c). At this point the spoiler places pebble 1 on the last element of L_5, and the duplicator has no place for the matching pebble, and thus he loses in the position shown in Fig. 11.1 (d).

Note that we could not have expected any other result here, since we know that all queries over finite linear orderings are expressible in $\mathcal{L}^2_{\infty\omega}$; hence, the duplicator should not be able to win $\text{PG}^\infty_2(L_n, L_m)$ unless $n = m$.

As another example, consider structures of the empty vocabulary: that is, just sets. We claim the following: if $|A|, |B| \geq k$, then the duplicator wins $\mathrm{PG}_k^\infty(A, B)$; in other words, $A \equiv_k^{\infty\omega} B$. Indeed, the strategy for the duplicator is very similar to his strategy in the Ehrenfeucht-Fraïssé game: at all times, he has to maintain the condition that p_A^i and p_A^j are placed on the same element iff p_B^i and p_B^j are placed on the same element. Since both sets have at least k elements, this condition is easily maintained, and the duplicator can win the infinite game. This gives us the following.

Corollary 11.6. *The query* EVEN *is not expressible in* $\mathcal{L}_{\infty\omega}^\omega$.

Proof. Assume, to the contrary, that EVEN is expressible by a sentence Φ of $\mathcal{L}_{\infty\omega}^\omega$. Let k be such that $\Phi \in \mathcal{L}_{\infty\omega}^k$. Choose two sets A and B of cardinalities k and $k+1$, respectively. By the above, $A \equiv_k^{\infty\omega} B$ and hence $A \models \Phi$ iff $B \models \Phi$. This, however, contradicts the assumption that Φ defines EVEN. □

From Corollary 11.6, we derive a result mentioned, but not proved, in Chap. 10.

Corollary 11.7. • LFP $\subsetneq$ (LFP+<)$_{\text{inv}}$.
- IFP $\subsetneq$ (IFP+<)$_{\text{inv}}$.
- PFP $\subsetneq$ (PFP+<)$_{\text{inv}}$.

Proof. Since LFP, IFP, PFP $\subseteq \mathcal{L}_{\infty\omega}^\omega$, none of them defines EVEN; however, over ordered structures these logics capture PTIME and PSPACE, and hence can define EVEN. □

Before proving Theorem 11.5, we make two additional observations. First, consider an infinitary disjunction $\varphi \equiv \bigvee_{i\in I} \varphi_i$, where all φ_i are FO formulae, and assume that $\mathsf{qr}(\varphi) \leq n$. This means that $\mathsf{qr}(\varphi_i) \leq n$ for all $i \in I$. We know that, up to logical equivalence, there are only finitely many different FO formulae of quantifier rank n. Hence, there is a *finite* subset $I_0 \subset I$ such that φ is equivalent to $\bigvee_{i\in I_0} \varphi_i$; that is, to an FO formula. Using this argument inductively on the structure of $\mathcal{L}_{\infty\omega}^\omega$ formulae, we conclude that for every k, every $\mathcal{L}_{\infty\omega}^k$ formula of quantifier rank n is equivalent to an FO^k formula of the same quantifier rank. Hence, if $\mathfrak{A}$ and $\mathfrak{B}$ agree on all FO^k sentences of quantifier rank at most n, then $\mathfrak{A} \equiv_{k,n}^{\infty\omega} \mathfrak{B}$.

Now assume that $\mathfrak{A}$ and $\mathfrak{B}$ agree on all FO^k sentences. That is, for every n, we have $\mathfrak{A} \equiv_{k,n}^{\infty\omega} \mathfrak{B}$. Since $\mathfrak{A}$ and $\mathfrak{B}$ are finite, so is the number of different maps from A^k to B^k, and hence every infinite strategy in $\mathrm{PG}_k^\infty(\mathfrak{A}, \mathfrak{B})$ is completely determined by a finite strategy for sufficiently large n: the one in which all (finitely many) possible configurations of the game appeared. Thus, for sufficiently large n (that depends on $\mathfrak{A}$ and $\mathfrak{B}$), winning $\mathrm{PG}_k^n(\mathfrak{A}, \mathfrak{B})$ implies winning $\mathrm{PG}_k^\infty(\mathfrak{A}, \mathfrak{B})$. We therefore obtain the following.

Proposition 11.8. *For every two structures* $\mathfrak{A}, \mathfrak{B}$, *the following are equivalent:*

1. *$\mathfrak{A}$ and $\mathfrak{B}$ agree on all* FO^k *sentences, and*
2. *$\mathfrak{A}$ and $\mathfrak{B}$ agree on all* $\mathcal{L}^k_{\infty\omega}$ *sentences.* □

The second observation is about formulae with free variables. We write $(\mathfrak{A}, \vec{a}) \equiv^{\infty\omega}_{k,n} (\mathfrak{B}, \vec{b})$ (or $(\mathfrak{A}, \vec{a}) \equiv^{\infty\omega}_{k} (\mathfrak{B}, \vec{b})$), where $|\vec{a}| = |\vec{b}| = m \leq k$, if the duplicator wins the game $\mathrm{PG}^n_k(\mathfrak{A}, \mathfrak{B})$ (or $\mathrm{PG}^\infty_k(\mathfrak{A}, \mathfrak{B})$) from the position where the first m pebbles have been placed on the elements of $\vec{a}$ and $\vec{b}$ respectively. A slight modification of the proof of Theorem 11.5 shows the following.

Corollary 11.9. *Given two structures, $\mathfrak{A}, \mathfrak{B}$, and $\vec{a} \in A^m, \vec{b} \in B^m$, $m \leq k$,*

a) $(\mathfrak{A}, \vec{a}) \equiv^{\infty\omega}_{k,n} (\mathfrak{B}, \vec{b})$ *iff for every* $\varphi(\vec{x}) \in \mathcal{L}^k_{\infty\omega}$ *with* $\mathsf{qr}(\varphi) \leq n$, *it is the case that* $\mathfrak{A} \models \varphi(\vec{a}) \Leftrightarrow \mathfrak{B} \models \varphi(\vec{b})$.

b) $(\mathfrak{A}, \vec{a}) \equiv^{\infty\omega}_{k} (\mathfrak{B}, \vec{b})$ *iff for every* $\varphi(\vec{x}) \in \mathcal{L}^k_{\infty\omega}$, *it is the case that* $\mathfrak{A} \models \varphi(\vec{a}) \Leftrightarrow \mathfrak{B} \models \varphi(\vec{b})$. □

We are now ready to prove Theorem 11.5. As with the Ehrenfeucht-Fraïssé theorem, we shall use a certain back-and-forth property in the proof. We start with a few definitions.

Given a partial map $f : A \to B$, its domain and range will be denoted by $\mathrm{dom}(f)$ and $\mathrm{rng}(f)$; that is, f is defined on $\mathrm{dom}(f) \subseteq A$, and $f(\mathrm{dom}(f)) = \mathrm{rng}(f) \subseteq B$.

We let symbols α and β range over finite and infinite ordinals. Given two structures $\mathfrak{A}$ and $\mathfrak{B}$ and an ordinal β, let $\mathcal{I}_\beta$ be a set of partial isomorphisms between $\mathfrak{A}$ and $\mathfrak{B}$, and let $\mathfrak{I}_\alpha = \{\mathcal{I}_\beta \mid \beta < \alpha\}$. We say that $\mathfrak{I}_\alpha$ has the *k-back-and-forth property* if the following conditions hold:

- Every set $\mathcal{I}_\beta$ is nonempty.
- $\mathcal{I}_{\beta'} \subseteq \mathcal{I}_\beta$ for $\beta < \beta'$.
- Each $\mathcal{I}_\beta$ is downward-closed: if $g \in \mathcal{I}_\beta$ and $f \subseteq g$ (i.e., $\mathrm{dom}(f) \subseteq \mathrm{dom}(g)$, and f and g coincide on $\mathrm{dom}(f)$), then $f \in \mathcal{I}_\beta$.
- If $f \in \mathcal{I}_{\beta+1}$ and $|\mathrm{dom}(f)| < k$, then

forth: for every $a \in A$, there is $g \in \mathcal{I}_\beta$ such that $f \subseteq g$ and $a \in \mathrm{dom}(g)$;
back: for every $b \in B$, there is $g \in \mathcal{I}_\beta$ such that $f \subseteq g$ and $b \in \mathrm{rng}(g)$.

As before, games are nothing but a reformulation of the back-and-forth property. Indeed, for a finite α, having a family $\mathfrak{I}_\alpha$ with the k-back-and-forth property is equivalent to $\mathfrak{A} \equiv^{\infty\omega}_{k,\alpha-1} \mathfrak{B}$: the collection $\mathcal{I}_\beta$ simply consists of configurations from which the duplicator wins with β moves remaining. This also suffices for infinitely long games: as we remarked earlier, for every two finite structures $\mathfrak{A}$ and $\mathfrak{B}$, and for some n, depending on $\mathfrak{A}$ and $\mathfrak{B}$, it is the case that $\mathfrak{A} \equiv^{\infty\omega}_{k,n} \mathfrak{B}$ implies $\mathfrak{A} \equiv^{\infty\omega}_{k} \mathfrak{B}$. Furthermore, if we have a sufficiently long finite chain $\mathfrak{I}_\alpha$, some $\mathcal{I}_\beta$'s will be repeated, as there are only finitely many partial isomorphisms between $\mathfrak{A}$ and $\mathfrak{B}$. Hence, such a chain can then be extended to arbitrary ordinal length.

Therefore, it will be sufficient to establish equivalence between indistinguishability in $\mathcal{L}^k_{\infty\omega}$ and the existence of a family of partial isomorphisms with the k-back-and-forth property. This is done in the following lemma.

Lemma 11.10. *Given two structures $\mathfrak{A}$ and $\mathfrak{B}$, they agree on all sentences of $\mathcal{L}^k_{\infty\omega}$ of quantifier rank $< \alpha$ iff there is a family $\mathfrak{I}_\alpha = \{\mathcal{I}_\beta \mid \beta < \alpha\}$ of partial isomorphisms between $\mathfrak{A}$ and $\mathfrak{B}$ with the k-back-and-forth property.*

In the rest of the section, we prove Lemma 11.10. Suppose $\mathfrak{A}$ and $\mathfrak{B}$ agree on all sentences of $\mathcal{L}^k_{\infty\omega}$ of quantifier rank $< \alpha$. Let $\beta < \alpha$. Define $\mathcal{I}_\beta$ as the set of partial isomorphisms f with $|\operatorname{dom}(f)| \leq k$ such that for every $\varphi \in \mathcal{L}^k_{\infty\omega}$ with $\mathsf{qr}(\varphi) \leq \beta$, and every $\vec{a}$ contained in $\operatorname{dom}(f)$,

$$\mathfrak{A} \models \varphi(\vec{a}) \quad \Leftrightarrow \quad \mathfrak{B} \models \varphi(f(\vec{a})).$$

We show that $\mathfrak{I}_\alpha = \{\mathcal{I}_\beta \mid \beta < \alpha\}$ has the k-back-and-forth property.

Since $\mathfrak{A}$ and $\mathfrak{B}$ agree on all sentences of $\mathcal{L}^k_{\infty\omega}$ of quantifier rank $< \alpha$, each $\mathcal{I}_\beta$ is nonempty as it contains the empty partial isomorphism. The containment $\mathcal{I}_{\beta'} \subseteq \mathcal{I}_\beta$ for $\beta < \beta'$ is immediate from the definition, as is downward-closure. Thus, it remains to prove the back-and-forth property.

Assume, to the contrary, that we found $f \in \mathcal{I}_{\beta+1}$, with $\beta+1 < \alpha$, such that $|\operatorname{dom}(f)| = m < k$, and f violates the *forth* condition. That is, there exists $a \in A$ such that there is no $g \in \mathcal{I}_\beta$ extending f with $a \in \operatorname{dom}(g)$. In this case, by the definition of $\mathcal{I}_\beta$, for every $b \in B$ we can find a formula $\varphi_b(x_0, x_1, \ldots, x_m)$ of quantifier rank at most β such that for some $a_1, \ldots, a_m \in \operatorname{dom}(f)$, we have $\mathfrak{A} \models \varphi_b(a, a_1, \ldots, a_m)$ and $\mathfrak{B} \models \neg\varphi_b(b, f(a_1), \ldots, f(a_m))$.

Now let

$$\varphi(x_1, \ldots, x_m) \;\equiv\; \exists x_0 \bigwedge_{b \in B} \varphi_b(x_0, x_1, \ldots, x_m).$$

Clearly, $\mathfrak{A} \models \varphi(a_1, \ldots, a_m)$, but $\mathfrak{B} \models \neg\varphi(f(a_1), \ldots, f(a_m))$, which contradicts our assumption $f \in \mathcal{I}_{\beta+1}$ (since $\mathsf{qr}(\varphi) \leq \beta + 1$). The case when f violates the *back* condition is handled similarly.

For the other direction, assume that we have a family $\mathfrak{I}_\alpha$ with the k-back-and-forth property. We use (transfinite) induction on β to show that for every $\varphi(x_1, \ldots, x_m) \in \mathcal{L}^k_{\infty\omega}$, $m \leq k$, with $\mathsf{qr}(\varphi) \leq \beta < \alpha$,

$$\begin{array}{l} \text{for every } f \in \mathcal{I}_\beta, \;\; a_1, \ldots, a_m \in \operatorname{dom}(f): \\ \mathfrak{A} \models \varphi(a_1, \ldots, a_m) \;\Leftrightarrow\; \mathfrak{B} \models \varphi(f(a_1), \ldots, f(a_m)). \end{array} \tag{11.6}$$

Clearly, (11.6) suffices, since it implies that $\mathfrak{A}$ and $\mathfrak{B}$ agree on $\mathcal{L}^k_{\infty\omega}$ sentences of quantifier rank $< \alpha$.

The basis case is $\beta = 0$. Then φ is a Boolean combination of atomic formulae (for finite quantifier ranks, as we saw, infinitary connectives are superfluous), and hence (11.6) follows from the assumption that f is a partial isomorphism.

We now use induction on the structure of φ. The case of Boolean combinations is trivial. If $\varphi \equiv \bigvee_i \varphi_i$ and $\mathsf{qr}(\varphi) > \mathsf{qr}(\varphi_i)$ for all i, then β is a limit ordinal and again (11.6) for φ easily follows by applying the hypothesis to all the φ_i's of smaller quantifier rank.

Thus, it remains to consider the case of $\varphi(x_1, \ldots, x_m) \equiv \exists x_0 \; \psi(x_0, \ldots, x_m)$, with $\mathsf{qr}(\varphi) = \beta + 1$ and $\mathsf{qr}(\psi) = \beta$ for some β with $\beta + 1 < \alpha$. We can assume without loss of generality that x_0 is not among $x_1, \ldots, x_m$ (exercise: why?) and hence $m < k$.

Let $f \in \mathcal{I}_{\beta+1}$ and $a_1, \ldots, a_m \in \mathrm{dom}(f)$. Assume that $\mathfrak{A} \models \varphi(a_1, \ldots, a_m)$; that is, for some $a_0 \in A$, $\mathfrak{A} \models \psi(a_0, a_1, \ldots, a_m)$. Since $\mathcal{I}_{\beta+1}$ is downward-closed, we can further assume that $\mathrm{dom}(f) = \{a_1, \ldots, a_m\}$. Since $|\mathrm{dom}(f)| = m < k$, by the k-back-and-forth property we find $g \in \mathcal{I}_\beta$ extending f such that $a_0 \in \mathrm{dom}(g)$. Applying (11.6) inductively to ψ, we derive $\mathfrak{B} \models \psi(g(a_0), g(a_1), \ldots, g(a_m))$. That is, $\mathfrak{B} \models \psi(g(a_0), f(a_1), \ldots, f(a_m))$ since f and g agree on $a_1, \ldots, a_m$. Hence, $\mathfrak{B} \models \varphi(f(a_1), \ldots, f(a_m))$.

The other direction, that $\mathfrak{B} \models \varphi(f(a_1), \ldots, f(a_m))$ implies $\mathfrak{A} \models \varphi(a_1, \ldots, a_m)$, is completely symmetric. This finishes the proof of (11.6), Lemma 11.10, and Theorem 11.5. $\square$

11.3 Definability of Types

For logics like FO and MSO, we have used rank-k *types*, which are collections of all formulae of quantifier rank k that hold in a given structure. An extremely useful feature of types is that they can be defined by formulae of quantifier rank k, and we have used this fact many times.

When we move to finite variable logics, the role of parameter k is played by the number of variables rather than the quantifier rank. We can, therefore, define, FO^k-types, but then it is not immediately clear if every such type is itself definable in FO^k. In this section we prove that this is the case. As with the case of FO or MSO types, this definability result proves very useful, and we derive some interesting corollaries. In particular, we establish a normal form for $\mathcal{L}^k_{\infty\omega}$, and prove that every class of finite structures that is closed under $\equiv^{\infty\omega}_k$ is definable in $\mathcal{L}^k_{\infty\omega}$.

Definition 11.11 (FO^k-types). *Given a structure $\mathfrak{A}$ and a tuple $\vec{a}$, the* FO^k-type *of $(\mathfrak{A}, \vec{a})$ is*

$$\mathrm{tp}_{\mathrm{FO}^k}(\mathfrak{A}, \vec{a}) \;\; = \;\; \{\varphi(\vec{x}) \in \mathrm{FO}^k \;\mid\; \mathfrak{A} \models \varphi(\vec{a})\}.$$

An FO^k*-type is any set of formulae of* FO^k *of the form* $\mathrm{tp}_{\mathrm{FO}^k}(\mathfrak{A}, \vec{a})$.

One could have defined $\mathcal{L}^k_{\infty\omega}$-types as well, as the set of all $\mathcal{L}^k_{\infty\omega}$ formulae that hold in $(\mathfrak{A}, \vec{a})$. This, however, would be unnecessary, since every FO^k-type completely determines the $\mathcal{L}^k_{\infty\omega}$-type: this follows from Proposition 11.8 stating that two structures agree on all $\mathcal{L}^k_{\infty\omega}$ formulae iff they agree on all FO^k formulae.

Note that unlike in the cases of FO and MSO, the number of different FO^k-types need not be finite, since we do not restrict the quantifier rank. In fact we saw in the example of finite linear orderings that there are infinitely many different FO^2-types, since every finite cardinality of a linear ordering can be characterized by an FO^2 sentence.

Each FO^k-type τ is trivially definable in $\mathcal{L}^k_{\infty\omega}$ by $\bigvee_{\varphi\in\tau}\varphi$. More interestingly, we can show that FO^k-types are definable *without* infinitary connectives.

Theorem 11.12. *For every* FO^k*-type* τ*, there is an* FO^k *formula* $\varphi_\tau(\vec{x})$ *such that, for every structure* $\mathfrak{A}$*,*

$$\mathrm{tp}_{\mathrm{FO}^k}(\mathfrak{A}, \vec{a}) = \tau \quad \Leftrightarrow \quad \mathfrak{A} \models \varphi_\tau(\vec{a}).$$

Before we prove Theorem 11.12, let us state a few corollaries. First, restricting our attention to sentences, we obtain the following.

Corollary 11.13. *For every structure* $\mathfrak{A}$*, there is a sentence* $\Psi_{\mathfrak{A}}$ *of* FO^k *such that for any other structure* $\mathfrak{B}$*, we have* $\mathfrak{B} \models \Psi_{\mathfrak{A}}$ *iff* $\mathfrak{A} \equiv^{\infty\omega}_k \mathfrak{B}$*.* □

We know that without restrictions on the number of variables, we can write a sentence that tests if $\mathfrak{B}$ is isomorphic to $\mathfrak{A}$, and this is why the full infinitary logic defines every class of finite structures. Corollary 11.13 shows that, rather than testing isomorphism as in the full infinitary logic, in $\mathcal{L}^k_{\infty\omega}$ one can write a sentence that tests $\equiv^{\infty\omega}_k$-equivalence.

We can also see that closure under $\equiv^{\infty\omega}_k$ is sufficient for definability in $\mathcal{L}^k_{\infty\omega}$.

Corollary 11.14. *If a class* $\mathcal{C}$ *of structures is closed under* $\equiv^{\infty\omega}_k$ *(i.e.,* $\mathfrak{A} \in \mathcal{C}$ *and* $\mathfrak{A} \equiv^{\infty\omega}_k \mathfrak{B}$ *imply* $\mathfrak{B} \in \mathcal{C}$*), then* $\mathcal{C}$ *is definable in* $\mathcal{L}^k_{\infty\omega}$*.*

Proof. Let T be the collection of $\mathcal{L}^k_{\infty\omega}$-types τ such that there is a structure $\mathfrak{A}$ in $\mathcal{C}$ with $\mathrm{tp}_{\mathrm{FO}^k}(\mathfrak{A}) = \tau$. From closure under $\equiv^{\infty\omega}_k$ it follows that $\bigvee_{\tau\in T}\varphi_\tau$ defines $\mathcal{C}$. □

Definability of $\mathcal{L}^k_{\infty\omega}$-types also yields a normal form result, stating that only countable disjunctions of FO^k formulae suffice.

Corollary 11.15. *Every* $\mathcal{L}^k_{\infty\omega}$ *formula is equivalent to a single countable disjunction of* FO^k *formulae.*

Proof. Let $\varphi(\vec{x})$ be an $\mathcal{L}^k_{\infty\omega}$ formula. Consider the set $\mathcal{C}_\varphi = \{(\mathfrak{A}, \vec{a}) \mid \mathfrak{A} \models \varphi(\vec{a})\}$, such that no two elements of $\mathcal{C}_\varphi$ are isomorphic (this ensures that $\mathcal{C}_\varphi$ is countable, since there are only countably many isomorphism types of finite structures). Let $\varphi_{\mathfrak{A},\vec{a}}(\vec{x})$ be the FO^k formula defining $\mathrm{tp}_{\mathrm{FO}^k}(\mathfrak{A}, \vec{a})$. Let

$$\psi(\vec{x}) \equiv \bigvee_{(\mathfrak{A},\vec{a})\in\mathcal{C}_\varphi} \varphi_{(\mathfrak{A},\vec{a})}(\vec{x}).$$

We claim that φ and ψ are equivalent. Suppose $\mathfrak{B} \models \varphi(\vec{b})$. Let $(\mathfrak{B}', \vec{b}')$ be an isomorphic copy of $(\mathfrak{B}, \vec{b})$ present in $\mathcal{C}_\varphi$. Then $\mathfrak{B}' \models \varphi_{(\mathfrak{B}',\vec{b}')}(\vec{b}')$ and thus $\mathfrak{B}' \models \psi(\vec{b}')$ and $\mathfrak{B} \models \psi(\vec{b})$. Conversely, if $\mathfrak{B} \models \psi(\vec{b})$, then for some $\mathfrak{A}$ and $\vec{a}$ with $\mathfrak{A} \models \varphi(\vec{a})$, we have $\mathrm{tp}_{\mathrm{FO}^k}(\mathfrak{A}, \vec{a}) = \mathrm{tp}_{\mathrm{FO}^k}(\mathfrak{B}, \vec{b})$; that is, $(\mathfrak{A}, \vec{a}) \equiv^{\infty\omega}_k (\mathfrak{B}, \vec{b})$. Since φ is an $\mathcal{L}^k_{\infty\omega}$ formula, this implies $\mathfrak{B} \models \varphi(\vec{b})$, showing that φ and ψ are equivalent. ☐

Since the negation of an $\mathcal{L}^k_{\infty\omega}$ formula is an $\mathcal{L}^k_{\infty\omega}$ formula, we obtain a dual result.

Corollary 11.16. *Every $\mathcal{L}^k_{\infty\omega}$ formula is equivalent to a single countable conjunction of* FO^k *formulae.*

We now present the proof of Theorem 11.12. To keep the notation simple, we look at the case when there are no free variables; that is, we deal with $\mathrm{tp}_{\mathrm{FO}^k}(\mathfrak{A})$. Another assumption that we make is that the vocabulary σ is purely relational. Adding free variables and constant symbols poses no problem (Exercise 11.1).

Fix a structure $\mathfrak{A}$, and let $A^{\leq k}$ be the set of all tuples of elements of A of length up to k. For any $\vec{a} = (a_1, \ldots, a_l) \in A^{\leq k}$, where $l \leq k$, we define a formula $\varphi^m_{\vec{a}}(x_1, \ldots, x_l)$. Intuitively, these formulae will have the property that they precisely characterize what one can say about $\vec{a}$ in FO^k, with quantifier rank at most m: that is, $\mathfrak{B} \models \varphi^m_{\vec{a}}(\vec{b})$ iff $(\mathfrak{A}, \vec{a})$ and $(\mathfrak{B}, \vec{b})$ agree on all the FO^k formulae of quantifier rank up to m.

To define these formulae, consider partial functions $h : \{x_1, \ldots, x_k\} \to A$, and first define formulae $\varphi^m_h(\vec{y})$, with free variables $\vec{y}$ being those in $\mathrm{dom}(h)$, as follows:

- $\varphi^0_h(\vec{y})$ is the conjunction of all atomic and negated atomic formulae true in $\mathfrak{A}$ of $h(\vec{y})$.
- To define $\varphi^{m+1}_h(\vec{y})$, consider two cases:
  1. Suppose $|\mathrm{dom}(h)| < k$. Let i be the least index such that $x_i \notin \mathrm{dom}(h)$, and h_a be the extension of h defined on $\mathrm{dom}(h) \cup \{x_i\}$ such that $h_a(x_i) = a$. Then

$$\varphi^{m+1}_h(\vec{y}) \equiv \varphi^m_h(\vec{y}) \wedge \bigwedge_{a\in A} \exists x_i\, \varphi^m_{h_a}(\vec{y}, x_i) \wedge \forall x_i \bigvee_{a\in A} \varphi^m_{h_a}(\vec{y}, x_i).$$

2. Suppose $|\mathrm{dom}(h)| = k$. Let h_i be the restriction of h which is not defined only on x_i. Then

$$\varphi_h^{m+1}(\vec{x}) \equiv \varphi_h^m(\vec{x}) \wedge \bigwedge_{i=1}^{k} \varphi_{h_i}^{m+1}(\vec{x}_i),$$

where $\vec{x}_i$ is $\vec{x}$ with the variable x_i excluded.

Finally, we define $\varphi_{\vec{a}}^m(x_1, \ldots, x_l)$ as $\varphi_h^m(\vec{x})$, where h is given by $h(x_i) = a_i$, for $i = 1, \ldots, l$.

To show that formulae $\varphi_{\vec{a}}^m$ do what they are supposed to do, we show that if they hold, a certain sequence of sets of partial isomorphisms with the k-back-and-forth property must exist.

Lemma 11.17. *Let $\vec{a} = (a_1, \ldots, a_l) \in A^{\leq k}$. Then $\mathfrak{B} \models \varphi_{\vec{a}}^m(\vec{b})$ iff there exists a collection $\mathfrak{I}_m = \{\mathcal{I}_0, \mathcal{I}_1, \ldots, \mathcal{I}_m\}$ of sets of partial isomorphism between $\mathfrak{A}$ and $\mathfrak{B}$ with the k-back-and-forth property such that $\mathcal{I}_m \subseteq \mathcal{I}_{m-1} \subseteq \cdots \subseteq \mathcal{I}_0$, and $g = \{(a_1, b_1), \ldots, (a_l, b_l)\} \in \mathcal{I}_m$.*

Proof of Lemma 11.17. Since $\mathsf{qr}(\varphi_{\vec{a}}^m) = m$ and $\mathfrak{A} \models \varphi_{\vec{a}}^m(\vec{a})$, the existence of $\mathfrak{I}_m$ implies, by Lemma 11.10, that $\mathfrak{B} \models \varphi_{\vec{a}}^m(\vec{b})$.

For the converse, we establish the existence of $\mathfrak{I}_m$ by induction on m.

If $m = 0$, we let $\mathcal{I}_0$ consist of all the restrictions of g. Clearly, $\mathcal{I}_0$ is not empty, and since g is a partial isomorphism (because, by the assumption, $\mathfrak{B} \models \varphi_{\vec{a}}^0(\vec{b})$, and thus $\vec{a}$ and $\vec{b}$ satisfy the same atomic formulae), all elements of $\mathcal{I}_0$ are partial isomorphisms.

For the induction step, to go from m to $m+1$, we distinguish two cases.

Case 1: $l < k$. From $\mathfrak{B} \models \varphi_{\vec{a}}^{m+1}(\vec{b})$ and the definition of $\varphi_{\vec{a}}^{m+1}$ it follows that $\mathfrak{B} \models \varphi_{\vec{a}}^m(\vec{b})$, and thus we have, by the induction hypothesis, a sequence $\mathfrak{I}'_m = \{\mathcal{I}'_0, \ldots, \mathcal{I}'_m\}$ of partial isomorphisms with the k-back-and-forth property such that $g \in \mathcal{I}'_m$.

Looking at the second conjunct of $\varphi_{\vec{a}}^{m+1}$ and applying the induction hypothesis for m, we see that for every $a \in A$ there exists $b \in B$ and a sequence $\mathfrak{I}_m^a = \{\mathcal{I}_0^a, \ldots, \mathcal{I}_m^a\}$ of partial isomorphisms with the k-back-and-forth property such that $g_{a,b} = \{(a_1, b_1), \ldots, (a_l, b_l), (a, b)\} \in \mathcal{I}_m^a$.

We now define:

$$\mathcal{I}_i = \mathcal{I}'_i \cup \bigcup_{a \in A} \mathcal{I}_i^a \quad \text{for } i \leq m$$
$$\mathcal{I}_{m+1} = \{f \mid f \subseteq g\}.$$

It is easy to see that component-wise unions like this preserve the k-back-and-forth property. Furthermore, since $g \in \mathcal{I}'_m$, then $\mathcal{I}_{m+1} \subseteq \mathcal{I}'_m \subseteq \mathcal{I}_m$. Thus, we only have to check the k-back-and-forth property with respect to $\mathcal{I}_{m+1}$ and $\mathcal{I}_m$. But this is guaranteed by the second and the third conjunct of $\varphi_{\vec{a}}^{m+1}$.

Indeed, consider g and $a \in A - \mathrm{dom}(g)$. Since $\mathfrak{B} \models \varphi_{\vec{a}}^{m+1}(\vec{b})$, by the second conjunct we see that $\mathfrak{B} \models \exists x \varphi_{\vec{a}a}^{m}(\vec{b}, x)$ and hence for some $b \in B$, we have $\mathfrak{B} \models \varphi_{\vec{a}a}^{m}(\vec{b}b)$. But then $g \cup \{(a, b)\} \in \mathcal{I}'_m \subseteq \mathcal{I}_m$. The back property is proved similarly. This completes the proof for case 1.

Case 2: $l = k$. By the definition of $\varphi_{\vec{a}}^{m+1}$ for the case of $l = k$, we see that $\mathfrak{B} \models \varphi_{\vec{a}}^{m}(\vec{b})$, and hence by the induction hypothesis, g is a partial isomorphism.

For each $i \leq k$, let g_i be g without the pair (a_i, b_i). Applying the argument for the case $l < k$ to each g_i, we get a sequence of partial isomorphisms $\{\mathcal{I}_0^i, \ldots, \mathcal{I}_{m+1}^i\}$ with the k-back-and-forth property such that $\mathcal{I}_{m+1}^i \subseteq \ldots \subseteq \mathcal{I}_0^i$. Now we define

$$\mathcal{I}_j \quad = \quad \{g\} \ \cup \ \bigcup_{i=1}^{k} \mathcal{I}_j^i, \quad j \leq m+1.$$

One can easily verify all the properties of a sequence of partial isomorphisms with the k-back-and-forth property: in fact, all of the properties are preserved under component-wise union, and since $|\mathrm{dom}(g)| = k$, the k-back-and-forth extension for g is not required. This completes the proof case 2 and Lemma 11.17. □

For each $\vec{a} \in A^{\leq k}$, consider $\varphi_{\vec{a}}^{m}(\mathfrak{A}) = \{\vec{a}_0 \mid \mathfrak{A} \models \varphi_{\vec{a}}^{m}(\vec{a}_0)\}$. By definition, $\varphi_{\vec{a}}^{m+1}$ is of the form $\varphi_{\vec{a}}^{m} \wedge \ldots$, and hence

$$\varphi_{\vec{a}}^{0}(\mathfrak{A}) \supseteq \varphi_{\vec{a}}^{1}(\mathfrak{A}) \supseteq \ldots \supseteq \varphi_{\vec{a}}^{m}(\mathfrak{A}) \supseteq \varphi_{\vec{a}}^{m+1}(\mathfrak{A}) \supseteq \ldots.$$

Since $\mathfrak{A}$ is finite, this sequence eventually stabilizes. Let $m_{\vec{a}}$ be the number such that $\varphi_{\vec{a}}^{m_{\vec{a}}}(\mathfrak{A}) = \varphi_{\vec{a}}^{m}(\mathfrak{A})$ for all $m > m_{\vec{a}}$. Then we define

$$M = \max_{\vec{a} \in A^{\leq k}} m_{\vec{a}}, \quad \text{and}$$

$$\Psi_{\mathfrak{A}} \quad \equiv \quad \varphi_{\epsilon}^{M} \ \wedge \ \bigwedge_{\vec{a} \in A^{\leq k}} \forall x_1 \ldots \forall x_k \Big(\varphi_{\vec{a}}^{M}(\vec{x}) \rightarrow \varphi_{\vec{a}}^{M+1}(\vec{x}) \Big) . \qquad (11.7)$$

Here ϵ stands for the empty sequence. By the definition of M, $\mathfrak{A} \models \Psi_{\mathfrak{A}}$. Furthermore, $\Psi_{\mathfrak{A}} \in \mathrm{FO}^k$.

Thus, to conclude the proof, we show that $\Psi_{\mathfrak{A}}$ defines $\mathrm{tp}_{\mathrm{FO}^k}(\mathfrak{A})$. In other words, we need the following.

Lemma 11.18. *If $\mathfrak{B}$ is a finite structure, then $\mathfrak{B} \models \Psi_{\mathfrak{A}}$ iff $\mathrm{tp}_{\mathrm{FO}^k}(\mathfrak{A}) = \mathrm{tp}_{\mathrm{FO}^k}(\mathfrak{B})$; that is, $\mathfrak{A} \equiv_k^{\infty\omega} \mathfrak{B}$.*

Proof of Lemma 11.18. Since $\Psi_{\mathfrak{A}} \in \mathrm{FO}^k$ and $\mathfrak{A} \models \Psi_{\mathfrak{A}}$, it suffices to show that $\mathfrak{A} \equiv_k^{\infty\omega} \mathfrak{B}$ whenever $\mathfrak{B} \models \Psi_{\mathfrak{A}}$.

Let $\mathfrak{B} \models \Psi_{\mathfrak{A}}$. We define a set G of partial maps between $\mathfrak{A}$ and $\mathfrak{B}$ by

$$\{(a_1, b_1), \ldots, (a_l, b_l)\} \in G \quad \Leftrightarrow \quad \mathfrak{B} \models \varphi_{(a_1, \ldots, a_l)}^{M+1}(b_1, \ldots, b_l).$$

Since $\mathfrak{B} \models \Psi_{\mathfrak{A}}$, the sentence φ_ϵ^{M+1} is true in $\mathfrak{B}$, and thus G is nonempty, as the empty partial map is a member of G.

Applying Lemma 11.17 to each $g = \{(a_1, b_1), \ldots, (a_l, b_l)\} \in G$, we see that there is a sequence $\mathfrak{I}^g = \{\mathcal{I}_0^g, \ldots, \mathcal{I}_{M+1}^g\}$ of partial isomorphisms with the k-back-and-forth property such that $\mathcal{I}_0^g \supseteq \ldots \supseteq \mathcal{I}_{M+1}^g$ and $g \in \mathcal{I}_{M+1}^g$. We now define a family $\mathfrak{I} = \{\mathcal{I}_i \mid i \in \mathbb{N}\}$ by

$$\mathcal{I}_i = \bigcup_{g \in G} \mathcal{I}_i^g \quad \text{for } i \leq M+1$$
$$\mathcal{I}_i = \mathcal{I}_{M+1} \quad \text{for } i > M+1.$$

It remains to show that $\mathfrak{I}$ has the k-back-and-forth property. As we have seen in the proof of Lemma 11.17, the k-back-and-forth property is preserved through component-wise union, and since all $\mathcal{I}_i, i > M+1$, are identical, it suffices to prove that every partial isomorphism in $\mathcal{I}_{M+2}$ can be extended in $\mathcal{I}_{M+1}$.

Fix $f \in \mathcal{I}_{M+2}$ such that $|\text{dom}(f)| < k$. We show the *forth* part; the *back* part is identical. Let $a \in A$. Since $f \in \mathcal{I}_{M+2}$, and the sequence $\{\mathcal{I}_0, \ldots, \mathcal{I}_{M+1}\}$ has the k-back-and-forth property, we can find $f' \in \mathcal{I}_M$ with $f \subseteq f'$ and $a \in \text{dom}(f')$. Let $f' = \{(a_1, b_1), \ldots, (a_l, b_l)\}$. Since f' is a partial isomorphism from $\mathcal{I}_M$, from Lemma 11.17 we conclude that $\mathfrak{B} \models \varphi^M_{(a_1,\ldots,a_l)}(b_1, \ldots, b_l)$. Now from the implication in (11.7), we see that $\mathfrak{B} \models \varphi^{M+1}_{(a_1,\ldots,a_l)}(b_1, \ldots, b_l)$; therefore, $f' \in G$. But then $f' \in \mathcal{I}^{f'}_{M+1}$ and hence $f' \in \mathcal{I}_{M+1}$, which proves the *forth* part. Since the *back* part is symmetric, this concludes the proof of Lemma 11.18 and Theorem 11.12. □

11.4 Ordering of Types

In this section, we show that many interesting properties of types can be expressed in LFP. In particular, consider the following equivalence relation $\approx_{\mathrm{FO}^k}$ on tuples of elements of a structure $\mathfrak{A}$:

$$\vec{a} \approx_{\mathrm{FO}^k} \vec{b} \;\Leftrightarrow\; \mathrm{tp}_{\mathrm{FO}^k}(\mathfrak{A}, \vec{a}) = \mathrm{tp}_{\mathrm{FO}^k}(\mathfrak{A}, \vec{b}).$$

Clearly this relation is definable by an $\mathcal{L}^k_{\infty\omega}$ formula

$$\psi(\vec{x}, \vec{y}) \;\equiv\; \bigvee_\tau \big(\varphi_\tau(\vec{x}) \wedge \varphi_\tau(\vec{y})\big),$$

where τ ranges over all FO^k-types.

It is more interesting, however, that this relation is definable in a weaker logic LFP. Furthermore, it turns out that there is a formula of LFP that defines a certain preorder $\prec_{\mathrm{FO}^k}$ on tuples, such that the equivalence relation induced by this preorder is precisely $\approx_{\mathrm{FO}^k}$. This means that on structures in which all elements have different FO^k-types, we can define a linear order in

LFP, and hence, by the Immerman-Vardi theorem, on such structures LFP captures PTIME.

We start by showing how to define $\approx_{\mathrm{FO}^k}$.

Proposition 11.19. *Fix a vocabulary σ. For every k and $l \leq k$, there is an* LFP *formula $\eta(\vec{x}, \vec{y})$ in $2l$ free variables such that for every $\mathfrak{A} \in \mathrm{STRUCT}[\sigma]$,*

$$\mathfrak{A} \models \eta(\vec{a}, \vec{b}) \quad \Leftrightarrow \quad \vec{a} \approx_{\mathrm{FO}^k} \vec{b}.$$

Proof. The *atomic* FO^k-type of $(\mathfrak{A}, \vec{a})$, with $|\vec{a}| = l \leq k$, is the conjunction of all atomic and negated atomic formulae true of $\vec{a}$ in $\mathfrak{A}$. Since there are finitely many atomic FO^k-formulae, up to logical equivalence, each atomic type is definable by an FO^k formula. Let $\alpha_1(\vec{x}), \ldots, \alpha_s(\vec{x})$ list all such formulae. Then we define

$$\psi_0(\vec{x}, \vec{y}) \equiv \bigvee_{i,j \leq s,\ i \neq j} (\alpha_i(\vec{x}) \wedge \alpha_j(\vec{y})) .$$

This is a formula of quantifier rank 0, and $\mathfrak{A} \models \psi_0(\vec{a}, \vec{b})$ iff the atomic FO^k-types of $\vec{a}$ and $\vec{b}$ are different.

Next, we define a formula ψ in the vocabulary σ expanded with a $2l$-ary relation R:

$$\psi(R, \vec{x}, \vec{y}) \equiv \psi_0(\vec{x}, \vec{y}) \vee \bigvee_{i=1}^{l} \exists x_i \forall y_i R(\vec{x}, \vec{y}) \vee \bigvee_{i=1}^{l} \exists y_i \forall x_i R(\vec{x}, \vec{y}), \qquad (11.8)$$

and let

$$\varphi(\vec{x}, \vec{y}) \equiv [\mathbf{lfp}_{R, \vec{x}, \vec{y}}\ \psi(R, \vec{x}, \vec{y})](\vec{x}, \vec{y}).$$

Consider the fixed point computation for ψ. Initially, we have tuples $(\vec{a}, \vec{b})$ with different atomic types; that is, tuples corresponding to the position in the pebble game in which the spoiler wins. At the next stage, we get all the positions of the pebble game $(\vec{a}, \vec{b})$ such that, in one move, the spoiler can force the winning position. In general, the ith stage consists of positions from which the spoiler can win the pebble game in $i-1$ moves, and hence $\mathfrak{A} \models \varphi(\vec{a}, \vec{b})$ iff from the position $(\vec{a}, \vec{b})$, the spoiler can win the game. In other words, $\mathfrak{A} \models \varphi(\vec{a}, \vec{b})$ iff $(\mathfrak{A}, \vec{a}) \not\equiv_k^{\infty\omega} (\mathfrak{A}, \vec{b})$, or, equivalently, $\mathrm{tp}_{\mathrm{FO}^k}(\mathfrak{A}, \vec{a}) \neq \mathrm{tp}_{\mathrm{FO}^k}(\mathfrak{A}, \vec{b})$. Hence, η can be defined as $\neg\varphi$, which is an LFP formula. □

We now extend this technique to define a preorder $\prec_{\mathrm{FO}^k}$ on tuples, whose associated equivalence relation is precisely $\approx_{\mathrm{FO}^k}$.

Suppose we have a set X partitioned into subsets $X_1, \ldots, X_m$. Consider a binary relation $\prec$ on X given by

$$x \prec y \quad \Leftrightarrow \quad x \in X_i,\ y \in X_j, \text{ and } i < j.$$

We call relations obtained in such a way *strict preorders*. With each strict preorder $\prec$ we associate an equivalence relation whose equivalence classes are precisely $X_1, \ldots, X_m$. It can be defined by the formula $\neg(x \prec y) \wedge \neg(y \prec x)$.

Theorem 11.20. *For every vocabulary σ, and every k, there exists an* LFP *formula $\chi(\vec{x}, \vec{y})$, with $|\vec{x}|=|\vec{y}|=k$, such that on every $\mathfrak{A} \in \mathrm{STRUCT}[\sigma]$, the formula χ defines a strict preorder $\prec_{\mathrm{FO}^k}$ whose equivalence relation is $\approx_{\mathrm{FO}^k}$.*

As we mentioned before, this result becomes useful when one deals with structures $\mathfrak{A}$ such that for every $a, b \in A$, $\mathrm{tp}_{\mathrm{FO}^k}(a) \neq \mathrm{tp}_{\mathrm{FO}^k}(b)$ whenever $a \neq b$. Such structures are called *k-rigid.*

Theorem 11.20 tells us that in a k-rigid structure, there is an LFP-definable strict preorder whose equivalence classes are of size 1: that is, a linear order. Hence, from the Immerman-Vardi theorem we obtain:

Corollary 11.21. *Over k-rigid structures,* LFP *captures* PTIME. □

Now we prove Theorem 11.20. We shall use the following notation. If $\vec{a} = (a_1, \ldots, a_k)$ is a tuple, then $\vec{a}_{i \leftarrow a}$ is the tuple in which a_i was replaced by a, i.e., $(a_1, \ldots, a_{i-1}, a, a_{i+1}, \ldots, a_k)$.

Recall the formula $\psi(\vec{x}, \vec{y})$ (11.8). The fixed point of this formula defined the complement of $\approx_{\mathrm{FO}^k}$, and it follows from the proof of Proposition 11.19 that the jth stage of the fixed point computation for ψ, $\psi^j(\vec{x}, \vec{y})$, defines the set of positions from which the spoiler wins with $j-1$ moves remaining. In other words, $\mathfrak{A} \models \psi^j(\vec{a}, \vec{a})$ iff $(\mathfrak{A}, \vec{a})$ and $(\mathfrak{B}, \vec{b})$ disagree on some FO^k formula of quantifier rank up to $j-1$.

We now use this formula ψ to define a formula $\gamma(S, \vec{x}, \vec{y})$ such that the jth stage of the *inflationary* fixed point computation for γ defines a strict preorder whose equivalence relation is the complement of the relation defined by $\psi^j(\vec{x}, \vec{y})$. In other words, $\gamma^j(\mathfrak{A})$ defines a relation $\prec_j$ on A^k such that the equivalence relation $\sim_j$ associated with this preorder is

$$\vec{a} \sim_j \vec{b} \quad \Leftrightarrow \quad (\mathfrak{A}, \vec{a}) \equiv^{\infty\omega}_{k,j-1} (\mathfrak{A}, \vec{b}).$$

We now explain the idea of the construction. In the beginning, we have to deal with atomic FO^k-types. Since these can be explicitly defined (see the proof of Proposition 11.19), we can choose an arbitrary ordering on them.

Now, suppose we have defined $\prec_j$, the jth stage of the fixed point computation for γ, whose equivalence relation is the set of positions from which the duplicator can play for $j-1$ moves (i.e., the complement of the jth stage of ψ). Let $Y_1, \ldots, Y_s$ be the equivalence classes.

We have to refine $\prec_j$ to come up with a preorder $\prec_{j+1}$. For that, we have to order tuples $(\vec{a}, \vec{b})$ which were equivalent at the jth stage, but become nonequivalent at stage $j+1$. But these are precisely the tuples that get into the fixed point of ψ at stage $j+1$.

Looking at the definition of ψ (11.8), we see that there are two ways for $\psi^{j+1}(\vec{a}, \vec{b})$ to be true (i.e., for $(\vec{a}, \vec{b})$ to get into the fixed point at stage $j+1$):

1. There is $a \in A$ such that $\varphi^j(\vec{a}_{i \leftarrow a}, \vec{b}_{i \leftarrow b})$ holds for every $b \in A$. In other words, the equivalence class of $\vec{a}_{i \leftarrow a}$ contains no tuple of the form $\vec{b}_{i \leftarrow b}$ which is different from $\vec{b}$.

2. Symmetrically, there is $b \in A$ such that the equivalence class of $\vec{b}_{i \leftarrow b}$ contains no tuple of the form $\vec{a}_{i \leftarrow a} \neq \vec{a}$.

Assume that i' is the minimum number $\leq k$ such that either 1 or 2 above, or both, happen. Let Y be the set of all the tuples $\vec{a}_{i' \leftarrow a}$ for case 1 and $\vec{b}_{i' \leftarrow b}$ for case 2. We then consider the smallest, with respect to $\prec_j$, equivalence class Y_p's into which elements of Y may fall. Note that it is impossible that for some a, b, both $\vec{a}_{i' \leftarrow a}$ and $\vec{b}_{i' \leftarrow b}$ are in Y_p. Hence, either

1$'$. for some a, $\vec{a}_{i' \leftarrow a}$ is in Y_p, or
2$'$. for some b, $\vec{b}_{i' \leftarrow b}$ is in Y_p.

In case 1$'$, we let $\vec{a} \prec_{j+1} \vec{b}$, and in case 2$'$, we let $\vec{b} \prec_{j+1} \vec{a}$.

This is the algorithm; it remains to express it in LFP. The formula $\chi(\vec{x}, \vec{y})$ will be defined as $[\mathbf{ifp}_{S,\vec{x},\vec{y}} \gamma(S, \vec{x}, \vec{y})](\vec{x}, \vec{y})$. To express γ, we first deal with the atomic case. Since we have an explicit listing $\alpha_1, \ldots, \alpha_s$ of formulae defining atomic types, we can use

$$\gamma_0(\vec{x}, \vec{y}) \equiv \bigvee_{i<j} (\alpha_i(\vec{x}) \wedge \alpha_j(\vec{y}))$$

to order atomic types.

Next, we define

$$\xi'_i(\vec{x}, \vec{y}) \equiv \forall x_i \exists y_i \, (\neg S(\vec{x}, \vec{y}) \wedge \neg S(\vec{y}, \vec{x})) \wedge \forall y_i \exists x_i \, (\neg S(\vec{x}, \vec{y}) \wedge \neg S(\vec{y}, \vec{x})),$$

$$\xi_i(\vec{x}, \vec{y}) \equiv \big(\bigwedge_{p<i} \xi'_p(\vec{x}, \vec{y}) \big) \wedge \xi'_i(\vec{x}, \vec{y}).$$

The formula $\xi_i(\vec{x}, \vec{b})$ will be used to determine the position i' in the algorithm. To select tuples $\vec{a}_{i \leftarrow a}$ which are inequivalent to all tuples $\vec{b}_{i \leftarrow b}$, we use the formula

$$\delta^1_i(x, \vec{x}, \vec{y}) \equiv \forall y \, \Big(S(\vec{x}_{i \leftarrow x}, \vec{y}_{i \leftarrow y}) \; \vee \; S(\vec{y}_{i \leftarrow y}, \vec{x}_{i \leftarrow x}) \Big),$$

and $\delta^2_i(y, \vec{x}, \vec{y})$ for the symmetric case (in which we reverse the roles of x and y).

Finally, we get the following definition of $\gamma(\vec{x}, \vec{y})$:

$$\gamma_0(\vec{x}, \vec{y}) \; \vee$$
$$\neg S(\vec{y}, \vec{x}) \wedge \bigvee_{i=1}^{l} \left(\begin{array}{l} \xi_i(\vec{x}, \vec{y}) \wedge \\ \exists x \Big(\delta^1_i(x, \vec{x}, \vec{y}) \wedge \forall y \big(\delta^2_i(y, \vec{x}, \vec{y}) \rightarrow S(\vec{x}_{i \leftarrow x}, \vec{y}_{i \leftarrow y}) \big) \Big) \end{array} \right).$$

Notice that γ is not positive in S; however, by the Gurevich-Shelah theorem, $\mathbf{ifp}_{S,\vec{x},\vec{y}} \gamma$ is equivalent to an LFP formula.

We leave it to the reader to complete the proof: that is, to show that γ indeed codes the algorithm described in the beginning of the proof, and to prove by induction that the jth stage of the inflationary fixed point computation for γ defines a preorder whose equivalence relation is $\equiv^{\infty\omega}_{k,j-1}$. □

11.5 Canonical Structures and the Abiteboul-Vianu Theorem

Using definability of a linear ordering on FO^k-types, we show how to convert each structure $\mathfrak{A}$ into another structure $\mathfrak{C}_k(\mathfrak{A})$, which, in essence, captures all the information about $\mathcal{L}^k_{\infty\omega}$-definability over $\mathfrak{A}$. The main application of this construction is the Abiteboul-Vianu theorem, which reduces the problem of separating complexity classes PTIME and PSPACE to separating two logics over *unordered* structures (recall that PTIME and PSPACE are captured by LFP and PFP over structures with a linear ordering).

Fix $k > 0$, and a purely relational vocabulary $\sigma = \{R_1, \ldots, R_l\}$ such that the arity of each R_i is at most k (since we shall be dealing with FO^k formulae, we can impose this additional restriction without loss of generality). We shall use the preorder relation $\prec_{\mathrm{FO}^k}$ defined in the previous section; its equivalence relation is $\vec{a} \approx_{\mathrm{FO}^k} \vec{b}$ given by $\mathrm{tp}_{\mathrm{FO}^k}(\mathfrak{A}, \vec{a}) = \mathrm{tp}_{\mathrm{FO}^k}(\mathfrak{A}, \vec{b})$, for $\vec{a}, \vec{b} \in A^k$. Whenever k and $\mathfrak{A}$ are clear from the context, we shall write $[\vec{a}]$ for the $\approx_{\mathrm{FO}^k}$-equivalence class of $\vec{a}$.

Definition 11.22. *Given a vocabulary $\sigma = \{R_1, \ldots, R_l\}$, where the arities of all the R_i's do not exceed k, and a σ-structure $\mathfrak{A}$, we define a new vocabulary $\mathfrak{c}_k(\sigma)$ and a structure $\mathfrak{C}_k(\mathfrak{A}) \in \mathrm{STRUCT}[\mathfrak{c}_k(\sigma)]$ as follows.*

Let $t = k^k$, and let $\pi_1, \ldots, \pi_t$ enumerate all the functions $\pi : \{1, \ldots, k\} \to \{1, \ldots, k\}$. Then

$$\mathfrak{c}_k(\sigma) \;=\; \{<, U, U_1, \ldots, U_l, S_1, \ldots, S_k, P_1, \ldots, P_t\},$$

where $<$, the S_i's, and the P_j's are binary, and $U, U_1, \ldots, U_l$ are unary.

The universe of $\mathfrak{C}_k(\mathfrak{A})$ is $A^k / \approx_{\mathrm{FO}^k}$, the set of $\approx_{\mathrm{FO}^k}$-equivalence classes of k-tuples from $\mathfrak{A}$. The interpretation of the predicates is as follows (where $\vec{a}$ stands for $(a_1, \ldots, a_k)$):

- *$<$ is interpreted as $\prec_{\mathrm{FO}^k}$.*
- *$U([\vec{a}])$ holds iff $a_1 = a_2$.*
- *$U_i([\vec{a}])$ holds iff $(a_1, \ldots, a_m) \in R_i^{\mathfrak{A}}$, where $m \leq k$ is the arity of R_i.*
- *$S_i([\vec{a}], [\vec{b}])$ holds iff $\vec{a}$ and $\vec{b}$ differ at most in their ith component.*
- *P_π contains pairs $([\vec{a}], [(a_{\pi(1)}, \ldots, a_{\pi(k)})])$ for all $\vec{a} \in A^k$.*

Lemma 11.23. *The structure $\mathfrak{C}_k(\mathfrak{A})$ is well-defined, and $<$ is interpreted as a linear ordering on its universe.*

Proof. Suppose $U([\vec{a}])$ holds and $\vec{b} \in [\vec{a}]$. Then $a_1 = a_2$, and since $\mathrm{tp}_{\mathrm{FO}^k}(\vec{a}) = \mathrm{tp}_{\mathrm{FO}^k}(\vec{b})$, we have $b_1 = b_2$. Since other predicates of $\mathfrak{C}_k(\mathfrak{A})$ are defined in terms of atomic formulae over $\mathfrak{A}$, they are likewise independent of particular representatives of the equivalence classes. Finally, Theorem 11.20 implies that $<$ is a linear ordering on $A^k/\approx_{\mathrm{FO}^k}$. □

The structure $\mathfrak{C}_k(\mathfrak{A})$ can be viewed as a canonical structure in terms of $\mathcal{L}^k_{\infty\omega}$-definability.

Proposition 11.24. *For every* $\mathfrak{A}, \mathfrak{B} \in \mathrm{STRUCT}[\sigma]$,

$$\mathfrak{A} \equiv^{\infty\omega}_k \mathfrak{B} \quad \Leftrightarrow \quad \mathfrak{C}_k(\mathfrak{A}) \cong \mathfrak{C}_k(\mathfrak{B}).$$

Proof sketch. Suppose $\mathfrak{A} \equiv^{\infty\omega}_k \mathfrak{B}$. Since every FO^k-type is definable by an FO^k formula, every type that is realized in $\mathfrak{A}$ is realized in $\mathfrak{B}$. Hence, $|A| = |B|$. Furthermore, since $\prec_{\mathrm{FO}^k}$ is definable by the same formula on all σ-structures, we have an order-preserving map $h : A^k/\approx_{\mathrm{FO}^k} \to B^k/\approx_{\mathrm{FO}^k}$. It is easy to verify that such a map is an isomorphism between $\mathfrak{C}_k(\mathfrak{A})$ and $\mathfrak{C}_k(\mathfrak{B})$.

For the converse, one can use the isomorphism $h : \mathfrak{C}_k(\mathfrak{A}) \to \mathfrak{C}_k(\mathfrak{B})$ together with relations S_i to establish a winning strategy for the duplicator in the k-pebble game. Details are left as an easy exercise for the reader. □

We next show how to translate formulae of LFP and PFP over $\mathfrak{C}_k(\mathfrak{A})$ to formulae over $\mathfrak{A}$, and vice versa. We assume, as throughout most of Chap. 10, that fixed point formulae do not have parameters.

Lemma 11.25. 1. *For every* LFP *or* PFP *formula* $\varphi(\vec{x})$ *over vocabulary* σ *that uses at most k variables, there is an* LFP *(respectively,* PFP*) formula* φ° *over vocabulary* $\mathfrak{c}_k(\sigma)$ *in one free variable such that*

$$\mathfrak{A} \models \varphi(\vec{a}) \quad \Leftrightarrow \quad \mathfrak{C}_k(\mathfrak{A}) \models \varphi^\circ([\vec{a}]). \tag{11.9}$$

2. *For every* LFP *or* PFP *formula* $\varphi(x_1, \ldots, x_m)$ *in the language of* $\mathfrak{c}_k(\sigma)$, *there is an* LFP *(respectively,* PFP*) formula* $\varphi^*(\vec{y})$ *over vocabulary* σ *in* km *free variables such that*

$$\mathfrak{C}_k(\mathfrak{A}) \models \varphi([\vec{a}_1], \ldots, [\vec{a}_m]) \quad \Leftrightarrow \quad \mathfrak{A} \models \varphi^*(\vec{a}_1, \ldots, \vec{a}_m).$$

Before proving Lemma 11.25, we present its main application.

Theorem 11.26 (Abiteboul-Vianu). PTIME = PSPACE *iff* LFP = PFP.

Proof. Suppose PTIME = PSPACE. Let φ be a PFP formula, and let it use k variables. By Lemma 11.25 (1), we have a PFP formula φ° over $\mathfrak{c}_k(\sigma)$. Since φ° is in PFP, it is computable in PSPACE, and thus, by the assumption, in PTIME. Since φ° is defined over ordered structures of the vocabulary $\mathfrak{c}_k(\sigma)$, by the Immerman-Vardi theorem it is definable in LFP over $\mathfrak{c}_k(\sigma)$, by a formula $\psi(x)$. Now applying Lemma 11.25 (2), we get an LFP formula $\psi^*(\vec{x})$ over vocabulary σ which is equivalent to φ. Hence, LFP = PFP.

For the other direction, if LFP = PFP, then LFP+ $<$ = PFP+$<$, and hence PTIME = PSPACE. □

Corollary 11.27. *The following are equivalent:*

- LFP = PFP;
- LFP+< = PFP+<;
- PTIME = PSPACE. □

Notice that this picture differs drastically from what we have seen for logics capturing DLOG, NLOG, and PTIME: while the exact relationships between DETTRCL+< = DLOG, TRCL+< = NLOG, and LFP+< = PTIME are not known, we do know that

$$\text{DetTrCl} \subsetneq \text{TrCl} \subsetneq \text{LFP}.$$

However, for the case of LFP and PFP, we cannot even conclude LFP $\subsetneq$ PFP without resolving the PTIME vs. PSPACE question.

We now prove Lemma 11.25. As the first step, we prove part 1 for the case of φ being an FO^k formula. Note that in general, $\vec{x}$ may have fewer than k variables. However, in this proof we shall treat any such formula as defining a k-ary relation; that is, $\varphi(x_{j_1}, \ldots, x_{j_s})$ defines the relation $\varphi(\mathfrak{A}) = \{(a_1, \ldots, a_k) \mid \mathfrak{A} \models \varphi(a_{j_1}, \ldots, a_{j_s})\}$, and when we write $\mathfrak{A} \models \varphi(\vec{a})$, we actually mean that $\vec{a} \in A^k$ and $\vec{a} \in \varphi(\mathfrak{A})$.

Using this convention, we define φ° by induction on the structure of the formula:

- If φ is $x_i = x_j$, then choose π so that $\pi(1) = i, \pi(2) = j$, and let $\varphi^\circ(x) \equiv \exists y\ (P_\pi(x, y) \wedge U(y))$.
- If φ is an atomic formula of the form $R_i(x_{j_1}, \ldots, x_{j_s})$, choose π so that $\pi(1) = j_1, \ldots, \pi(s) = j_s$, and let $\varphi^\circ(x) \equiv \exists y\ (P_\pi(x, y) \wedge U_i(y))$.
- $(\neg\varphi)^\circ \equiv \neg\varphi^\circ$.
- $(\varphi_1 \vee \varphi_2)^\circ \equiv \varphi_1^\circ \vee \varphi_2^\circ$.
- If φ is $\exists x_i \psi(\vec{x})$, then $\varphi^\circ(x) \equiv\ \exists y\ (S_i(x, y) \wedge \psi^\circ(y))$.

It is routine to verify, by induction on formulae, that the above translation guarantees (11.9). For example, if φ is $x_i = x_j$, then $\mathfrak{A} \models \varphi(\vec{a})$ implies that $a_i = a_j$, and hence $\mathfrak{C}_k(\mathfrak{A}) \models P_\pi([\vec{a}], [\vec{b}])$ for $\pi(i) = 1, \pi(j) = 2$, and $\vec{b} = (a_i, a_j, \ldots)$. Since $\mathfrak{C}_k(\mathfrak{A}) \models U([\vec{b}])$, we conclude that $\mathfrak{C}_k(\mathfrak{A}) \models \varphi^\circ([\vec{a}])$. Conversely, if $\mathfrak{C}_k(\mathfrak{A}) \models P_\pi([\vec{a}], [\vec{b}]) \wedge U([\vec{b}])$ for π as above and some $\vec{b}$, we conclude that there is $\vec{c} \in [\vec{a}]$ with $c_i = c_j$. Since $\text{tp}_{\text{FO}^k}(\vec{a}) = \text{tp}_{\text{FO}^k}(\vec{c})$, it follows that $a_i = a_j$ and $\mathfrak{A} \models \varphi(\vec{a})$. The other basis case is similar.

For the induction step, the only nontrivial case is that of φ being $\exists x_i \psi(\vec{x})$. If $\mathfrak{A} \models \varphi(\vec{a})$, then for some $\vec{a}_i$ that differs from $\vec{a}$ in at most the ith position we have $\mathfrak{A} \models \psi(\vec{a}_i)$, and hence by the induction hypothesis, $\mathfrak{C}_k(\mathfrak{A}) \models S_i([\vec{a}], [\vec{a}_i]) \wedge \psi^\circ([\vec{a}_i])$ and, therefore, $\mathfrak{C}_k(\mathfrak{A}) \models \varphi^\circ([\vec{a}])$. Conversely, assume that for some $\vec{b}$,

$\mathfrak{C}_k(\mathfrak{A}) \models S_i([\vec{a}], [\vec{b}]) \wedge \psi^\circ([\vec{b}])$. Then we can find $\vec{a}_0 \approx_{\mathrm{FO}^k} \vec{a}$ and $\vec{b}_0 \approx_{\mathrm{FO}^k} \vec{b}$ such that $\vec{a}_0$ and $\vec{b}_0$ differ in at most the ith position. Consider the k-pebble game on $(\mathfrak{A}, \vec{a}_0)$ and $(\mathfrak{A}, \vec{a})$. Suppose that in one move the spoiler goes from $(\mathfrak{A}, \vec{a}_0)$ to $(\mathfrak{A}, \vec{b}_0)$. Since the duplicator can play from position $(\vec{a}_0, \vec{a})$, he can respond to this move and find $\vec{b}'$ such that $(\mathfrak{A}, \vec{b}_0) \equiv_k^{\infty\omega} (\mathfrak{A}, \vec{b}')$. Hence, $\vec{b}' \in [\vec{b}]$, and it differs from $\vec{a}$ in at most the ith position. Since $[\vec{b}'] = [\vec{b}]$, by the induction hypothesis we conclude that $\mathfrak{A} \models \psi(\vec{b}')$, which witnesses $\mathfrak{A} \models \varphi(\vec{a})$. This concludes the proof of (11.9) for FO^k formulae.

Furthermore, (11.9) is preserved if we expand the vocabulary by an extra relation symbol R, with a corresponding R' added to $\mathfrak{c}_k(\sigma)$, and interpret R as a relation closed under $\equiv_k^{\infty\omega}$. Since we know that all the stages of **lfp** and **pfp** operators define such relations (see Exercise 11.6), we conclude that (11.9) holds for LFP and PFP formulae.

The proof of part 2 of Lemma 11.25 is by straightforward induction on the formulae, using the fact that $\prec_{\mathrm{FO}^k}$ is definable in LFP (Theorem 11.20). Details are left to the reader as an exercise. □

11.6 Bibliographic Notes

Infinitary logics have been studied extensively in model theory, see, e.g., Barwise and Feferman [18]. The finite variable logic was introduced by Barwise [17], who also defined the notion of a family of partial isomorphisms with the k-back-and-forth property. Pebble games were introduced by Immerman [128] and Poizat [200]. Kolaitis and Vardi [152, 153] studied many aspects of finite variable logics; in particular, they showed that it subsumes fixed point logics, and proved normal forms for $\mathcal{L}_{\infty\omega}^k$.

A systematic study of finite variable logics was undertaken by Dawar, Lindell, and Weinstein [53], and our presentation here is based on that paper. In particular, definability of FO^k-types in FO^k is from [53], as well as the definition of a linear ordering on FO^k-types.

Theorem 11.26 is due to Abiteboul and Vianu [6], but the presentation here is based on the model-theoretic approach of [53] rather than the more computational approach of [6]. The approach of [6] is based on relational complexity. Relational complexity classes are defined using machines that compute directly on structures rather than on their encodings as strings. Abiteboul and Vianu [6] and Abiteboul, Vardi, and Vianu [4] establish a tight connection between fixed point logics and relational complexity classes, and show that questions about containments among standard complexity classes can be translated to questions about containments among relational complexity classes.

Otto's book [191] is a good source for information on finite variable logics over finite models.

Sources for exercises:

Exercises 11.6 and 11.7:	Dawar, Lindell, and Weinstein [53]
Exercises 11.8 and 11.9:	Dawar [49]
Exercise 11.10:	de Rougemont [56]
Exercise 11.11:	Dawar, Lindell, and Weinstein [53]
Exercise 11.12:	Lindell [171]
Exercise 11.13:	Grohe [108]
Exercises 11.14 and 11.15:	Dawar, Lindell, and Weinstein [53]
Exercises 11.16 and 11.17:	Kolaitis and Vardi [154]
Exercise 11.18:	Grohe [110]
Exercise 11.19:	McColm [181]
	Kolaitis and Vardi [153]

11.7 Exercises

Exercise 11.1. Extend the proof of Theorem 11.12 to handle free variables, and constants in the vocabulary.

Exercise 11.2. Fill in the details at the end of the proof of Theorem 11.20.

Exercise 11.3. Complete the proof of Proposition 11.24.

Exercise 11.4. Complete the proof of Lemma 11.25, part 2.

Exercise 11.5. Prove that the FO^k hierarchy is strict: there are properties expressible in FO^{k+1} which are not expressible in FO^k.

Exercise 11.6. The goal of this exercise is to find a tight (as far as the number of variables is concerned) embedding of fixed point logics into $\mathcal{L}^{\omega}_{\infty\omega}$. Let LFP^k, IFP^k, and PFP^k stand for restrictions of LFP, IFP, and PFP to formulae that use at most k distinct variables (we assume that fixed point formulae have no parameters). Prove that $\text{LFP}^k, \text{IFP}^k, \text{PFP}^k \subseteq \mathcal{L}^{k}_{\infty\omega}$.

Hint: Let $\varphi(R, \vec{x})$ be a formula, and let $\varphi^i(\vec{x})$ define the ith stage of a fixed point computation. Show by induction on i that the query defined by φ^i is closed under $\equiv^{\infty\omega}_k$, and use Corollary 11.14.

Exercise 11.7. Prove that if $\mathfrak{A}$ and $\mathfrak{B}$ agree on all FO^k sentences of quantifier rank up to $n^k + k + 1$ and $|A| \leq n$, then $\mathfrak{A} \equiv^{\infty\omega}_k \mathfrak{B}$.

Exercise 11.8. Consider the complete bipartite graph $K_{n,m}$. Show that $K_{k,k} \equiv^{\infty\omega}_k K_{k,k+1}$ for every k. Also show that $K_{n,m}$ is Hamiltonian iff $n = m$. Conclude that Hamiltonicity is not $\mathcal{L}^{\omega}_{\infty\omega}$-definable.

Exercise 11.9. Prove that 3-colorability is not $\mathcal{L}^{\omega}_{\infty\omega}$-definable.

Exercise 11.10. Let I_n be a graph with n isolated vertices and C_m an undirected cycle of length m. For two graphs $G_1 = \langle V_1, E_1 \rangle$ and $G_2 = \langle V_2, E_2 \rangle$ with V_1 and V_2 disjoint, let $G_1 \times G_2$ be the graph whose nodes are $V_1 \cup V_2$, and the edges include E_1, E_2, as well as all the edges (v_1, v_2) for $v_1 \in V_1, v_2 \in V_2$. Prove that for a graph of the form $I_n \times C_m$, it is impossible to test, in $\mathcal{L}^{\omega}_{\infty\omega}$, if $n = m$. Use this result to give another proof (cf. Exercise 11.8) that Hamiltonicity is not $\mathcal{L}^{\omega}_{\infty\omega}$-definable.

Exercise 11.11. A binary tree is balanced if all the leaves are at the same distance from the root. Prove that $\mathcal{L}^4_{\infty\omega}$ defines a Boolean query Q on graphs such that if $Q(G)$ is true, then G is a balanced binary tree.

Exercise 11.12. Prove that there is a PTIME query on balanced binary trees which is not LFP-definable.

Conclude that LFP $\subsetneq \mathcal{L}^\omega_{\infty\omega} \cap$ PTIME.

Exercise 11.13. Prove that the following problems are PTIME-complete for each fixed k.

- Given two σ-structures $\mathfrak{A}$ and $\mathfrak{B}$, is it the case that $\mathfrak{A} \equiv_k^{\infty\omega} \mathfrak{B}$?
- Given a σ-structure $\mathfrak{A}$ and $\vec{a}, \vec{b} \in A^k$, are $\text{tp}_{\text{FO}^k}(\mathfrak{A}, \vec{a})$ and $\text{tp}_{\text{FO}^k}(\mathfrak{A}, \vec{b})$ the same?

Exercise 11.14. Prove that if $\mathfrak{A}$ is a finite rigid structure (i.e., a structure that has no nontrivial automorphisms), then there is a number k such that $\mathfrak{A}$ is k-rigid.

Exercise 11.15. Prove that the structure $\mathfrak{C}_k(\mathfrak{A})$ can be constructed in polynomial time.

Exercise 11.16. Define $\exists\mathcal{L}^k_{\infty\omega}$ as the fragment of $\mathcal{L}^k_{\infty\omega}$ that contains all atomic formulae and is closed under infinitary conjunctions and disjunctions, and existential quantification. Let

$$\exists\mathcal{L}^\omega_{\infty\omega} = \bigcup_k \exists\mathcal{L}^k_{\infty\omega}.$$

Prove that DATALOG $\subseteq \exists\mathcal{L}^\omega_{\infty\omega}$.

Exercise 11.17. Consider the following modification of the k-pebble game. For two structures $\mathfrak{A}$ and $\mathfrak{B}$, the spoiler always plays in $\mathfrak{A}$ and the duplicator always responds in $\mathfrak{B}$. The spoiler wins if at some point, the position $(\vec{a}, \vec{b})$ does not define a partial *homomorphism* (as opposed to a partial isomorphism in the standard game). The duplicator wins (which is denoted by $\mathfrak{A} \lhd_k^{\infty\omega} \mathfrak{B}$) if the spoiler does not win; that is, if after each round the position defines a partial homomorphism.

Prove that the following are equivalent:

- $\mathfrak{A} \lhd_k^{\infty\omega} \mathfrak{B}$.
- If $\Phi \in \exists\mathcal{L}^k_{\infty\omega}$ and $\mathfrak{A} \models \Phi$, then $\mathfrak{B} \models \Phi$.

Exercise 11.18. By an FO^k theory we mean a maximally consistent set of FO^k sentences. Define the k-size of an FO^k theory T as the number of different FO^k-types realized by finite models of T. Prove that there is no recursive bound on the size of the smallest model of an FO^k theory in terms of its k-size. That is, for every k there is a vocabulary σ_k such that is no recursive function f with the property that every FO^k theory T in vocabulary σ_k has a model of size at most $f(n)$, where n is the k-size of T.

Exercise 11.19. Let $\mathcal{C}$ be a class of σ-structures. We call it *bounded* if for every relation symbol $R \notin \sigma$, there exists a number n such that every FO formula $\varphi(R, \vec{x})$ positive in R reaches its least fixed point on any structure in $\mathcal{C}$ in at most n iterations.

Prove that the following are equivalent:

- $\mathcal{C}$ is bounded;
- $\mathcal{L}^\omega_{\infty\omega}$ collapses to FO on $\mathcal{C}$.

Exercise 11.20.* Is the FO^k hierarchy strict over *ordered* structures? That is, are there properties which, over ordered structures, are definable in FO^{k+1} but not in FO^k, for arbitrary k?

12

Zero-One Laws

In this chapter we show that properties expressible in many logics are almost surely true or almost surely false; that is, either they hold for almost all structures, or they fail for almost all structures. This phenomenon is known as the *zero-one law*. We prove it for FO, fixed point logics, and $\mathcal{L}^{\omega}_{\infty\omega}$. We shall also see that the "almost everywhere" behavior of logics is drastically different from their "everywhere" behavior. For example, while satisfiability in the finite is undecidable, it is decidable if a sentence is true in almost all finite models.

12.1 Asymptotic Probabilities and Zero-One Laws

To talk about asymptotic probabilities of properties of finite models, we adopt the convention that the universe of a structure $\mathfrak{A}$ with $|A| = n$ will be $\{0, \ldots, n-1\}$. Let us start by considering the case of undirected graphs. By GR_n we denote the set of all graphs with the universe $\{0, \ldots, n-1\}$. The number of undirected graphs on $\{0, \ldots, n-1\}$ is

$$|\,\mathrm{GR}_n\,| \;=\; 2^{\binom{n}{2}}.$$

Let $\mathcal{P}$ be a property of graphs. We define

$$\mu_n(\mathcal{P}) \;=\; \frac{|\{G \in \mathrm{GR}_n \mid G \text{ has } \mathcal{P}\}|}{|\,\mathrm{GR}_n\,|}.$$

That is, $\mu_n(\mathcal{P})$ is the probability that a randomly chosen graph on the set of nodes $\{0, \ldots, n-1\}$ has $\mathcal{P}$. Randomly here means with respect to the uniform distribution: each graph is equally likely to be chosen.

We then define the *asymptotic probability* of $\mathcal{P}$ as

$$\mu(\mathcal{P}) \;=\; \lim_{n\to\infty} \mu_n(\mathcal{P}), \tag{12.1}$$

if the limit exists. If $\mathcal{P}$ is expressed by a sentence Φ of some logic, then we refer to $\mu_n(\Phi)$ and $\mu(\Phi)$.

In general, we can deal with arbitrary σ-structures. In that case, we can define s_σ^n as the number of different σ-structures with the universe $\{0, \ldots, n-1\}$, and $s_\sigma^n(\mathcal{P})$ as the number of different σ-structures with the universe $\{0, \ldots, n-1\}$ that have the property $\mathcal{P}$, and let

$$\mu_n(\mathcal{P}) \;=\; \frac{s_\sigma^n(\mathcal{P})}{s_\sigma^n}.$$

Then the asymptotic probability $\mu(\mathcal{P})$ is defined again by (12.1).

We now consider a few examples:

- Let $\mathcal{P}$ be the property "there are no isolated nodes". We claim that $\mu(\mathcal{P}) = 1$. For that, we show that $\mu(\bar{\mathcal{P}}) = 0$, where $\bar{\mathcal{P}}$ is: "there is an isolated node". To calculate $\mu_n(\bar{\mathcal{P}})$, note that there are n ways to choose an isolated node, and $2^{\binom{n-1}{2}}$ ways to put edges on the remaining nodes. Hence

$$\mu_n(\bar{\mathcal{P}}) \;\le\; \frac{n \cdot 2^{\binom{n-1}{2}}}{2^{\binom{n}{2}}} \;=\; \frac{n}{2^{n-1}},$$

  and thus $\mu(\bar{\mathcal{P}}) = 0$.
- Let $\mathcal{P}$ be the property of being connected. Again, we show that $\mu(\bar{\mathcal{P}}) = 0$, and thus the asymptotic probability of graph connectivity is 1.

  To calculate $\mu(\bar{\mathcal{P}})$, we have to count the number of graphs with at least two connected components. Assuming the size of one component is k,
  - there are $\binom{n}{k}$ ways to choose a subset $X \subseteq \{0, \ldots, n-1\}$;
  - there are $2^{\binom{k}{2}}$ ways to put edges on X; and
  - there are $2^{\binom{n-k}{2}}$ ways to put edges on the complement of X.

  Hence,

$$\begin{aligned}
\mu_n(\bar{\mathcal{P}}) \;&\le\; \sum_{k=1}^{n-1} \frac{\binom{n}{k} \cdot 2^{\binom{k}{2}} \cdot 2^{\binom{n-k}{2}}}{2^{\binom{n}{2}}} \;=\; \sum_{k=1}^{n-1} \frac{\binom{n}{k}}{2^{k^2+kn}} \\
&=\; \frac{n}{2^{n+1}} + \sum_{k=2}^{n-1} \frac{\binom{n}{k}}{2^{k^2+kn}} \;\le\; \frac{n}{2^{n+1}} + \frac{1}{2^{2n}} \cdot \sum_{k=2}^{n-1} \binom{n}{k} \\
&\le\; \frac{n}{2^{n+1}} + \frac{1}{2^n} \;\to\; 0.
\end{aligned}$$

- Consider the query EVEN. Then

$$\mu_n(\text{EVEN}) = \begin{cases} 1 & \text{if } n \text{ is even,} \\ 0 & \text{if } n \text{ is odd.} \end{cases}$$

  Hence, $\mu(\text{EVEN})$ does not exist.

- The last example is the parity query. If σ has a unary relation U, then $\mathfrak{A}$ satisfies PARITY_U iff $|U^{\mathfrak{A}}| \mod 2 = 0$. Therefore,

$$\mu_n(\text{PARITY}_U) \;=\; \sum_{k\leq n,\ k \text{ even}} \binom{n}{k},$$

and hence $\mu(\text{PARITY}_U) = \frac{1}{2}$.

Thus, for some properties $\mathcal{P}$, the asymptotic probability $\mu(\mathcal{P})$ is 0 or 1, for some, like parity, $\mu(P)$ could be a number between 0 and 1, and for some, like EVEN, it may not even exist.

Definition 12.1 (Zero-one law). *Let $\mathcal{L}$ be a logic. We say that it has the* zero-one law *if for every property $\mathcal{P}$ (i.e., a Boolean query) definable in $\mathcal{L}$, either $\mu(\mathcal{P}) = 0$, or $\mu(\mathcal{P}) = 1$.*

The first property $\mathcal{P}$ for which we proved $\mu(\mathcal{P}) = 1$ was the absence of isolated nodes: this property is FO-definable. Graph connectivity, which also has asymptotic probability 1, is not FO-definable, but it is definable in LFP and hence in $\mathcal{L}^{\omega}_{\infty\omega}$. On the other hand, the EVEN and PARITY_U queries, which violate the zero-one law, are not $\mathcal{L}^{\omega}_{\infty\omega}$-definable, as we saw in Chap. 11. It turns out that $\mu(\mathcal{P})$ is 0 or 1 for *every* property definable in $\mathcal{L}^{\omega}_{\infty\omega}$.

Theorem 12.2. *$\mathcal{L}^{\omega}_{\infty\omega}$ has the zero-one law.*

Corollary 12.3. FO, LFP, IFP, *and* PFP *all have the zero-one law.*

Zero-one laws can be seen as statements that a logic cannot do nontrivial counting. For example, if a logic $\mathcal{L}$ has the zero-one law, then EVEN is not expressible in it, as well as any divisibility properties (e.g., is the size of a certain set congruent to q modulo p?), cardinality comparisons (e.g., is $|X|$ bigger than $|Y|$?), etc.

Note also that while LFP, IFP, PFP, and $\mathcal{L}^{\omega}_{\infty\omega}$ all have the zero-one law, their extensions with ordering no longer have it, since LFP+ $<$ defines EVEN, a PTIME query.

In the presence of a linear order (in fact, even successor), FO fails to have the zero-one law too. To see this, let S be the successor relation, and consider the sentence

$$\forall x \forall y \Big(\forall z\ (\neg S(z,x) \land \neg S(y,z)) \;\rightarrow\; E(x,y)\Big),$$

saying that if x is the initial and y the final element of the successor relation, then there is an edge between them. Since this sentence states the existence of one specific edge, its asymptotic probability is $\frac{1}{2}$.

We shall prove Theorem 12.2 in the next section after we introduce the main tool for the proof: extension axioms.

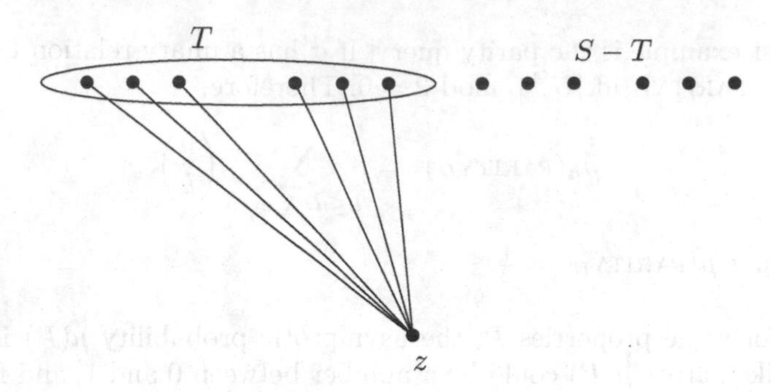

Fig. 12.1. Extension axiom

12.2 Extension Axioms

Extension axioms are statements defined as follows. Let S be a finite set of cardinality n, and let $T \subseteq S$ be of cardinality m. Then the extension axiom $EA_{n,m}$ says that there exists $z \notin S$ such that for all $x \in T$, there is an edge between z and x, and for all $x \in S - T$, there is no edge between z and x. This is illustrated in Fig. 12.1.

Extension axioms can be expressed in FO in the language of graphs. In fact, $EA_{n,m}$ is given by the following sentence:

$$\forall x_1, \ldots, x_n \left(\bigwedge_{i \neq j} x_i \neq x_j \right) \rightarrow \exists z \left(\begin{array}{c} \bigwedge_{i=1}^{n} z \neq x_i \\ \wedge \bigwedge_{i \leq m} E(z, x_i) \\ \wedge \bigwedge_{i > m} \neg E(z, x_j) \end{array} \right). \qquad (12.2)$$

The extension axiom $EA_{n,m}$ is vacuously true in a structure with fewer than n elements, but we shall normally consider it in structures with at least n elements.

We shall be using special cases of extension axioms, when $|S| = 2k$ and $|T|$ is k. Such an extension axiom will be denoted by EA_k. That is, EA_k says if $X \cap Y = \emptyset$, and $|X| = |Y| = k$, then there is z such that there is an edge (x, z) for all $x \in X$ but there is no edge (y, z) for any $y \in Y$.

Proposition 12.4. *$\mu(EA_k) = 1$ for each k.*

Proof. We show instead that $\mu(\neg EA_k) = 0$. Let $n > 2k$. Note that for EA_k to fail, there must be disjoint X and Y of cardinality k such that there is no $z \notin X \cup Y$ with $E(x, z)$ for all $x \in X$ and $\neg E(y, z)$ for all $y \in Y$. We now calculate $\mu_n(\neg EA_k)$, for $n > 2k$.

- There are $\binom{n}{k}$ ways to choose X.

- There are $\binom{n-k}{k}$ ways to choose Y. Therefore, there are at most $\binom{n}{k} \cdot \binom{n-k}{k} \leq n^{2k}$ ways to choose X and Y.
- Since there are no restrictions on edges on $X \cup Y$, there are $2^{\binom{2k}{2}}$ ways to put edges on $X \cup Y$.
- Again, since there are no restrictions on edges outside of $X \cup Y$, there are $2^{\binom{n-2k}{2}}$ ways to put edges outside of $X \cup Y$.
- The only restriction we have is on putting edges between $X \cup Y$ and its complement $\overline{X \cup Y}$: for each of the $n - 2k$ elements $z \in \overline{X \cup Y}$, we can put edges between z and the $2k$ elements of $X \cup Y$ in every possible way except one, where z is connected to every member of X and not connected to any member of Y. Hence, for each z there are $2^{2k} - 1$ ways of putting edges between z and $X \cup Y$, and therefore the number of ways to put edges between $X \cup Y$ and $\overline{X \cup Y}$ is $(2^{2k} - 1)^{n-2k}$.

Thus,

$$\mu_n(\neg EA_k) \leq \frac{n^{2k} \cdot 2^{\binom{2k}{2}} \cdot 2^{\binom{n-2k}{2}} \cdot (2^{2k} - 1)^{n-2k}}{2^{\binom{n}{2}}}. \tag{12.3}$$

A simple calculation shows that

$$\frac{2^{\binom{2k}{2}} \cdot 2^{\binom{n-2k}{2}}}{2^{\binom{n}{2}}} \leq \frac{1}{2^{2k(n-2k)}}. \tag{12.4}$$

Combining (12.3) and (12.4) we obtain

$$\mu_n(\neg EA_k) \leq n^{2k} \cdot \left(1 - \frac{1}{2^{2k}}\right)^{n-2k} \to 0,$$

proving that $\mu(\neg EA_k) = 0$ and $\mu(EA_k) = 1$. □

Corollary 12.5. $\mu(EA_{n,m}) = 1$, *for any n and $m \leq n$.*

Proof. For graphs of size $> 2n$, EA_n implies $EA_{n,m}$ for any $m \leq n$. □

Corollary 12.6. *Each EA_k has arbitrarily large finite models.* □

Notice that it is not immediately obvious from the statement of EA_k that there are finite graphs with at least $2k$ elements satisfying it. However, Proposition 12.4 tells us that we can find such graphs; in fact, almost all graphs satisfy EA_k.

We now move to the proof of the zero-one law for $\mathcal{L}^{\omega}_{\infty\omega}$. First, we need a lemma.

Lemma 12.7. *Let G_1, G_2 be finite graphs such that $G_1, G_2 \models EA_{n,m}$ for all $m \leq n \leq k$. Then $G_1 \equiv_k^{\infty\omega} G_2$.*

Proof. The extension axioms provide the strategy. Suppose we have a position in the game where $(a_1, \ldots, a_k)$ have been played in G_1 and $(b_1, \ldots, b_k)$ in G_2. Let the spoiler move the ith pebble from a_i to some element a. Let $I \subseteq \{1, \ldots, k\} - \{i\}$ be all the indices such that there is an edge from a to a_j, for all $j \in I$. Then by the extension axioms we can find $b \in G_2$ such that there is an edge from b to every b_j, for $j \in I$, and there are no edges from b to any b_l, for $l \notin I$. Hence, the duplicator can play b as the response to a. This shows that the pebble game can continue indefinitely, and thus $G_1 \equiv_k^{\infty\omega} G_2$. □

And finally, we prove the zero-one law. Let Φ be from $\mathcal{L}_{\infty\omega}^k$. Suppose there is a model G of EA_k, of size at least $2k$, that is also a model of Φ. Suppose G' is a graph that satisfies EA_k and has at least $2k$ elements. Then, by Lemma 12.7, we have $G' \equiv_k^{\infty\omega} G$ and hence $G' \models \Phi$. Therefore, $\mu(\varphi) \geq \mu(EA_k) = 1$. Conversely, assume that no model of EA_k of size $\geq 2k$ is a model of Φ. Then $\mu(\Phi) \leq \mu(\neg EA_k) = 0$. □

We now revisit the example of graph connectivity, for which the asymptotic probability was shown to be 1. If we look at EA_2, then for graphs with at least four nodes it implies that, for any $x \neq y$, there exists z such that $E(x, z)$ and $E(y, z)$ hold. Hence, every graph with at least four nodes satisfying EA_2 is connected, and thus $\mu(\text{connectivity}) = 1$.

As another example of using extension axioms for computing asymptotic probabilities, consider EA_2 and an edge (x, y). As before, we can find a node z such that $E(x, z)$ and $E(y, z)$ hold, and hence a graph satisfying EA_2 has a cycle (x, y, z). This means that $\mu(\text{acyclicity}) = 0$.

Finally, we explain how to state the extension axioms for an arbitrary vocabulary σ that contains only relation symbols. Given variables $x_1, \ldots, x_n$, let $A_\sigma(x_1, \ldots, x_n)$ be the collection of all atomic σ-formulae of the form $R(x_{i_1}, \ldots, x_{i_m})$, where R ranges over relations from σ, and m is the arity of R. Let $F \subseteq A_\sigma(x_1, \ldots, x_n)$. With F, we associate a formula $\chi_F(x_1, \ldots, x_n)$ (called a *complete description*) given by

$$\bigwedge_{\varphi \in F} \varphi \wedge \bigwedge_{\psi \in A_\sigma(x_1, \ldots, x_n) - F} \neg\psi.$$

That is, a complete description states precisely which atomic formulae in $x_1, \ldots, x_n$ are true, and which are not.

Let F now be a subset of $A_\sigma(x_1, \ldots, x_n)$, and G a subset of $A_\sigma(x_1, \ldots, x_n, x_{n+1})$ such that G extends F; that is, $F \subseteq G$. Then the extension axiom $EA_{F,G}$ is the sentence

$$\forall x_1 \ldots x_n \left(\begin{array}{l} \left((\bigwedge_{i \neq j} x_i \neq x_j) \ \wedge \ \chi_F(x_1, \ldots, x_n) \right) \rightarrow \\ \qquad \qquad \exists x_{n+1} \left(\bigwedge_{i \leq n} (x_{n+1} \neq x_i) \ \wedge \ \chi_G(x_1, \ldots, x_n) \right) \end{array} \right)$$

saying that every complete description in n variables can be extended to every consistent complete description in $n+1$ variables. A similar argument shows that $\mu(EA_{F,G}) = 1$. Therefore, the zero-one law holds for arbitrary finite structures, not only graphs.

12.3 The Random Graph

In this section we deal with a certain infinite structure. This structure, called the *random graph*, has an interesting FO theory: it consists of precisely all the sentences Φ for which $\mu(\Phi) = 1$. By analyzing the random graph, we prove that it is decidable, for an FO sentence Φ, whether $\mu(\Phi) = 1$.

First, recall the BIT predicate: $\mathrm{BIT}(i, j)$ is true iff the jth bit of the binary expansion of i is 1.

Definition 12.8. *The* random graph *is defined as the infinite (undirected) graph* $\mathcal{RG} = \langle \mathbb{N}, E \rangle$ *where there is an edge between* i *and* j*, for* $j < i$*, iff* $\mathrm{BIT}(i, j)$ *is true.*

Why is this graph called *random*? After all, the construction is completely deterministic. It turns out there is a probabilistic construction that results in this graph. Suppose someone wants to randomly build a countable graph whose nodes are natural numbers. When reaching a new node n, this person would look at all nodes $k < n$, and for each of them will toss a coin to decide if there is an edge between k and n. What kind of graph does one get as the result? It turns out that with probability 1, the constructed graph is isomorphic to $\mathcal{RG}$.

However, for our purposes, we do not need the probabilistic construction. What is important to us is that the random graph satisfies all the extension axioms. Indeed, to see that $\mathcal{RG} \models EA_{n,m}$, let $S \subset \mathbb{N}$ be of size n and $X \subseteq S$ be of size m. Let l be a number which, when given in binary, has ones in positions from X, and zeros in positions from $S - X$. Furthermore, assume that l has a one in some position whose number is higher than the maximal number in S. Then l witnesses $EA_{n,m}$ for S and T. To give a concrete example, if $S = \{0, 1, 2, 3, 4\}$ and $X = \{0, 2, 3\}$, then the number l is 45, or 101101 in binary.

Next, we define a theory

$$\boldsymbol{EA} \quad = \quad \{EA_k \mid k \in \mathbb{N}\}. \qquad (12.5)$$

Recall that a theory T (a set of sentences over vocabulary σ) is *complete* if for each sentence Φ, either $T \models \Phi$ or $T \models \neg\Phi$; it is ω*-categorical* if, up to isomorphism, it has only one countable model, and *decidable*, if it is decidable whether $T \models \Phi$.

Theorem 12.9. ***EA*** *is complete, ω-categorical, and decidable.*

Proof. For ω-categoricity, we claim that up to isomorphism, $\mathcal{RG}$ is the only countable model of $\boldsymbol{EA}$. Suppose that $\mathcal{G}$ is another model of $\boldsymbol{EA}$ (and thus it satisfies all the extension axioms $EA_{n,m}$). We claim that $\mathcal{RG} \equiv_\omega \mathcal{G}$; that is, the duplicator can play countably many moves of the Ehrenfeucht-Fraïssé game on $\mathcal{RG}$ and $\mathcal{G}$. Indeed, suppose after round r we have a position $((a_1, \ldots, a_r), (b_1, \ldots, b_r))$ defining a partial isomorphism, and suppose the spoiler plays a_{r+1} in $\mathcal{RG}$. Let $I = \{i \leq r \mid \mathcal{RG} \models E(a_{r+1}, a_i)\}$. Since $\mathcal{G} \models \boldsymbol{EA}$, by the appropriate extension axiom we can find b_{r+1} such that $\mathcal{G} \models E(b_{r+1}, b_i)$ iff $i \in I$. Thus, the resulting position $((a_1, \ldots, a_r, a_{r+1}), (b_1, \ldots, b_r, b_{r+1}))$ still defines a partial isomorphism.

If we have two countable structures such that $\mathfrak{A} \equiv_\omega \mathfrak{B}$, then $\mathfrak{A} \cong \mathfrak{B}$. Indeed, if $A = \{a_i \mid i \in \mathbb{N}\}$ and $B = \{b_i \mid i \in \mathbb{N}\}$, let the spoiler play, in each even round, the smallest unused element of A, and in each odd round the smallest unused element of B. Then the union of the sequence of partial isomorphisms generated by this play is an isomorphism between $\mathfrak{A}$ and $\mathfrak{B}$.

Thus, we have shown that $\mathcal{G} \models \boldsymbol{EA}$ implies $\mathcal{G} \cong \mathcal{RG}$ and hence $\boldsymbol{EA}$ is ω-categorical.

The next step is to show completeness of $\boldsymbol{EA}$. Suppose that we have a sentence Φ such that neither $\boldsymbol{EA} \models \Phi$ nor $\boldsymbol{EA} \models \neg\Phi$. Thus, both theories $\boldsymbol{EA} \cup \{\Phi\}$ and $\boldsymbol{EA} \cup \{\neg\Phi\}$ are consistent. By the Löwenheim-Skolem theorem, we get two countable models $\mathcal{G}', \mathcal{G}''$ of $\boldsymbol{EA}$ such that $\mathcal{G}' \models \Phi$ and $\mathcal{G}'' \models \neg\Phi$. However, by ω-categoricity, this means that $\mathcal{G}' \cong \mathcal{G}'' \cong \mathcal{RG}$. This contradiction proves that $\boldsymbol{EA}$ is complete.

Finally, a classical result in model theory says that a recursively axiomatizable complete theory is decidable. Since (12.5) provides a recursive axiomatization, we conclude that $\boldsymbol{EA}$ is decidable. □

Corollary 12.10. *If Φ is an* FO *sentence, then $\mathcal{RG} \models \Phi$ iff $\mu(\Phi) = 1$.*

Proof. Let $\mathcal{RG} \models \Phi$. Since $\boldsymbol{EA}$ is complete, $\boldsymbol{EA} \models \Phi$, and hence, by compactness, for some $k > 0$, $\{EA_i \mid i \leq k\} \models \Phi$. Thus, $EA_k \models \Phi$ and hence $\mu(\Phi) \geq \mu(EA_k) = 1$. Conversely, if $\mathcal{RG} \models \neg\Phi$, then $\mu(\neg\Phi) = 1$ and $\mu(\Phi) = 0$. Hence, for any Φ with $\mu(\Phi) = 1$, we have $\mathcal{RG} \models \varphi$. □

Combining Corollary 12.10 and decidability of $\boldsymbol{EA}$, we obtain the following.

Corollary 12.11. *For an* FO *sentence Φ it is decidable whether $\mu(\Phi) = 1$.*

Thus, Trakhtenbrot's theorem tells us that it is undecidable whether a sentence is true in all finite models, but now we see that it is *decidable* whether a sentence is true in almost all finite models.

12.4 Zero-One Law and Second-Order Logic

We have proved the zero-one law for the finite variable logic $\mathcal{L}^{\omega}_{\infty\omega}$ and its fragments such as FO and fixed point logics. It is natural to ask what other logics have it. Since the zero-one law can be seen as a statement saying that a logic cannot count, counting logics cannot have it. Another possibility is second-order logic and its fragments. Even such a simple fragment as ∃SO, the existential second-order logic, does not have the zero-one law: since ∃SO equals NP, the query EVEN is in ∃SO. But we shall see that some nontrivial restrictions of ∃SO have the zero-one law.

One way to obtain such restrictions is to look at quantifier prefixes of the first-order part. Recall that an ∃SO sentence can be written as

$$\exists X_1 \ldots \exists X_n Q_1 x_1 \ldots Q_m x_m \; \varphi(X_1, \ldots, X_n, x_1, \ldots, x_m), \tag{12.6}$$

where each Q_i is $\forall$ or $\exists$, and φ is quantifier-free. If r is a regular expression over the alphabet $\{\exists, \forall\}$, by $\exists\text{SO}(r)$ we denote the set of all sentences (12.6) such that the string $Q_1 \ldots Q_m$ is in the language denoted by r. For example, $\exists\text{SO}(\exists^*\forall^*)$ is a fragment of ∃SO that consists of sentences (12.6) for which the first-order part has all existential quantifiers in front of the universal quantifiers.

Theorem 12.12. $\exists\text{SO}(\exists^*\forall^*)$ *has the zero-one law.*

Proof. To keep the notation simple, we shall prove this for undirected graphs, but the result is true for arbitrary vocabularies that contain only relation symbols. The result will follow from two lemmas.

Lemma 12.13. *Let $S_1, \ldots, S_m$ be relation symbols, and φ an* FO *sentence of vocabulary $\{S_1, \ldots, S_m, E\}$ such that*

$$\mathcal{RG} \models \forall S_1 \ldots \forall S_m \; \varphi(S_1, \ldots, S_m).$$

Then there is an FO *sentence Φ of vocabulary $\{E\}$ such that $\mu(\Phi) = 1$ and $\Phi \to \forall \vec{S} \; \varphi$ is a valid sentence.*

Lemma 12.14. *Let $S_1, \ldots, S_m$ be relation symbols, and $\varphi(\vec{x}, \vec{y})$ a quantifier-free* FO *formula of vocabulary $\{S_1, \ldots, S_m, E\}$ such that*

$$\mathcal{RG} \models \exists S_1 \ldots \exists S_m \; \exists \vec{x} \; \forall \vec{y} \; \varphi(\vec{S}, \vec{x}, \vec{y}).$$

Then there is an FO *sentence Ψ of vocabulary $\{E\}$ such that $\mu(\Phi) = 1$ and $\Phi \to \exists \vec{S} \; \exists \vec{x} \; \forall \vec{y} \; \varphi$ is a finitely valid sentence.*

First, these lemmas imply the theorem. Indeed, assume that we are given an $\exists\text{SO}(\exists^*\forall^*)$ sentence $\Theta \equiv \exists \vec{S} \; \exists \vec{x} \; \forall \vec{y} \; \varphi$. Let $\mathcal{RG} \models \Theta$. Then, by Lemma 12.14, there is a sentence Φ with $\mu(\Phi) = 1$ such that Θ is true in every finite model of Φ, and hence $\mu(\Theta) = 1$. Conversely, assume $\mathcal{RG} \models \neg\Theta$. Since $\neg\Theta$ is an $\forall$SO sentence, by Lemma 12.13 we find a sentence Φ with $\mu(\Phi) = 1$ such that $\neg\Theta$ is true in every model of Φ, and thus $\mu(\neg\Theta) = 1$ and $\mu(\Theta) = 0$. Hence, $\mu(\Theta)$ is either 0 or 1. It remains to prove the lemmas.

Proof of Lemma 12.13. Assume that $\mathcal{RG} \models \forall\vec{S}\varphi(\vec{S})$, but for every FO sentence Φ with $\mu(\Phi) = 1$, it is the case that $(\Phi \to \forall\vec{S}\ \varphi)$ is not a valid sentence (i.e., $\Phi \wedge \exists\vec{S}\neg\varphi(\vec{S})$ has a model).

Consider the theory $T = \boldsymbol{EA} \cup \{\neg\varphi\}$ of vocabulary $\{S_1, \ldots, S_m, E\}$. Since every finite conjunction of extension axioms has asymptotic probability 1, by compactness we conclude that T is consistent, and by the Löwenheim-Skolem theorem, it has a countable model $\mathfrak{A}$. Since $\boldsymbol{EA}$ is ω-categorical, the $\{E\}$-reduct of $\mathfrak{A}$ is isomorphic to $\mathcal{RG}$. But then $\mathcal{RG} \models \exists\vec{S}\neg\varphi(\vec{S})$, a contradiction. This proves Lemma 12.13.

Proof of Lemma 12.14. Let $|\vec{S}| = m$ and $|\vec{x}| = n$. Let $A_1, \ldots, A_m$ witness the second-order quantifiers, and let $a_1, \ldots, a_n$ be the elements of $\mathcal{RG}$ witnessing FO existential quantifiers. Let $\mathcal{RG}_0$ be the finite subgraph of $\mathcal{RG}$ with the universe $\{a_1, \ldots, a_n\}$. We can find finitely many extension axioms $\{EA_{k,l}\}$ such that their conjunction implies the existence of a subgraph isomorphic to $\mathcal{RG}_0$. Let Φ be the conjunction of all such extension axioms. Let $\mathfrak{A}$ be a finite model of Φ. By the extension axioms, there is a subgraph $\mathcal{RG}_{\mathfrak{A}}$ of $\mathcal{RG}$ that is isomorphic to $\mathfrak{A}$ and contains $\mathcal{RG}_0$. Now we claim that $\mathcal{RG}_{\mathfrak{A}} \models \exists\vec{S}\exists\vec{x}\forall\vec{y}\ \varphi$. To witness the second-order quantifiers, we take the restrictions of the A_i's to $\mathcal{RG}_{\mathfrak{A}}$; as witnesses of FO existential quantifiers we take $a_1, \ldots, a_n$. Since universal sentences are preserved under substructures, we conclude that $\mathcal{RG}_{\mathfrak{A}} \models \forall\vec{y}\ \varphi(\vec{A}, \vec{a}, \vec{y})$, and thus $\mathcal{RG}_{\mathfrak{A}} \models \exists\vec{S}\exists\vec{x}\forall\vec{y}\ \varphi$. Therefore, $\mathfrak{A} \models \exists\vec{S}\exists\vec{x}\forall\vec{y}\ \varphi$, which proves the lemma. ☐

There are more results concerning zero-one laws for fragments of SO, but they are significantly more complicated, and we present them without proofs. One other prefix class which admits the zero-one law is $\exists^*\forall\exists^*$; that is, exactly one universal quantifier is present.

Theorem 12.15. $\exists$SO$(\exists^*\forall\exists^*)$ *has the zero-one law.* ☐

Going to two universal quantifiers, however, creates problems.

Theorem 12.16. $\exists$SO$(\forall\forall\exists)$ *does not have the zero-one law, even if the* FO *part does not use equality.* ☐

For some prefix classes, the failure of the zero-one law is fairly easy to show. Consider, for example, the sentence

$$\exists S\ \forall x \exists y \forall z \begin{pmatrix} S(x,y) \wedge \neg S(x,x) \\ \wedge\ S(x,z) \to y = z \\ \wedge\ S(x,z) \leftrightarrow S(z,x) \end{pmatrix}.$$

This in an $\exists$SO($\forall\exists\forall$) sentence saying there is a permutation S in which every element has order 2; that is, this sentence expresses EVEN and thus $\exists$SO($\forall\exists\forall$) fails the zero-one law. A similar sentence can be written in $\exists$SO($\forall\forall\forall\exists$). The result can further be strengthened to show that both $\exists$SO($\forall\exists\forall$) and $\exists$SO($\forall\forall\forall\exists$) fail to have the zero-one law even if the FO order part does not mention equality.

12.5 Almost Everywhere Equivalence of Logics

In this short section, we shall prove a somewhat surprising result that on almost all structures, there is no difference between FO, LFP, PFP, and $\mathcal{L}^{\omega}_{\infty\omega}$.

Definition 12.17. *Given a logic $\mathcal{L}$, its fragment $\mathcal{L}'$, and a vocabulary σ, we say that $\mathcal{L}$ and $\mathcal{L}'$ are* almost everywhere equivalent *over σ, if there is a class $\mathcal{C}$ of finite σ-structures such that $\mu(\mathcal{C}) = 1$ and for every $\mathcal{L}$ formula φ, there is an $\mathcal{L}'$ formula ψ such that φ and ψ coincide on structures from $\mathcal{C}$.*

Theorem 12.18. *$\mathcal{L}^{\omega}_{\infty\omega}$ and* FO *are almost everywhere equivalent over σ, for any purely relational vocabulary σ.*

Proof sketch. For simplicity, we deal with undirected graphs. Let $\mathcal{C}_k$ be the class of finite graphs satisfying EA_k. We claim that on $\mathcal{C}_k$, every $\mathcal{L}^{k}_{\infty\omega}$ formula is equivalent to an FO^k formula. Indeed, for a tuple $\vec{a} = (a_1, \ldots, a_k)$ in a structure $\mathfrak{A} \in \mathcal{C}_k$, its FO^k type $\text{tp}_{\text{FO}^k}(\mathfrak{A}, \vec{a})$ is completely determined by the atomic type of $\vec{a}$; that is, by the atomic formulae $E(a_i, a_j)$ that hold for $\vec{a}$. To see this, notice that if $\vec{a}$ and $\vec{b}$ have the same atomic type, then $(\vec{a}, \vec{b})$ is a partial isomorphism, and by EA_k from the position $(\vec{a}, \vec{b})$ the duplicator can play indefinitely in the k-pebble game; hence, $(\mathfrak{A}, \vec{a}) \equiv^{\infty\omega}_{k} (\mathfrak{A}, \vec{b})$.

Therefore, there are only finitely many FO^k types, and each $\mathcal{L}^{k}_{\infty\omega}$ formula is a disjunction of those, and thus equivalent to an FO^k formula. (In fact, we proved a stronger statement that on $\mathcal{C}_k$, every $\mathcal{L}^{k}_{\infty\omega}$ formula is equivalent to a quantifier-free FO^k formula.)

We now consider the classes $\mathcal{C}_1 \subseteq \mathcal{C}_2 \subseteq \ldots$, and observe that since each $\mu(\mathcal{C}_k)$ is 1, then for any sequence $\epsilon_1 > \epsilon_2 > \ldots > 0$ such that $\lim_{n\to\infty} \epsilon_n = 0$, we can find an increasing sequence of numbers $n_1 < n_2 < \ldots < n_k < \ldots$ such that

$$\mu_n(\mathcal{C}_k \cap \text{GR}_n) \ > \ 1 - \epsilon_k, \quad \text{for} \quad n > n_k.$$

We then define

$$\mathcal{C} \ = \ \{\mathfrak{A} \in \text{STRUCT}[\{E\}] \mid \text{if } |A| \geq n_k, \text{ then } \mathfrak{A} \in \mathcal{C}_k\}.$$

One can easily check that $\mu(\mathcal{C}) = 1$. We claim that every $\mathcal{L}^{\omega}_{\infty\omega}$ formula is equivalent to an FO formula on $\mathcal{C}$. Indeed, let φ be an $\mathcal{L}^{k}_{\infty\omega}$ formula. We know that on $\mathcal{C}_k$, it is equivalent to an FO^k formula φ'. Thus, to find a formula ψ

to which φ is equivalent on $\mathcal{C}$, one explicitly enumerates all the structures of cardinality up to n_k and evaluates φ on them. Then, one writes an FO formula ψ_k saying that if $\mathfrak{A}$ is one of the structures with $|A| < n_k$, then $\psi_k(\mathfrak{A}) = \varphi(\mathfrak{A})$, and for all the structures with $|A| \geq n_k$, ψ_k agrees with φ'. Since the number of structures of cardinality up to n_k is fixed, this can be done in FO. □

This result has complexity-theoretic implications. While we know that LFP and PFP queries have respectively PTIME and PSPACE data complexity, Theorem 12.18 shows that their complexity can be reduced to AC^0 on almost all structures.

12.6 Bibliographic Notes

That FO has the zero-one law was proved first by Glebskii et al. [92] in 1969, and independently by Fagin (announced in 1972, but the journal version [73] appeared in 1976). Fagin used extension axioms introduced by Gaifman [87]. Blass, Gurevich, and Kozen [22] and – independently – Talanov and Knyazev [227] proved that LFP has the zero-one law, and the result for $\mathcal{L}^\omega_{\infty\omega}$ is due to Kolaitis and Vardi [152].

The random graph was discovered by Erdös and Rényi [67] (the probabilistic construction); the deterministic construction used here is due to Rado [203]. In fact, $\mathcal{RG}$ is sometimes referred to as the Rado graph. This is also a standard construction in model theory (the Fraïssé limit of finite graphs, see [125]). The results about the theory of the random graph are from Gaifman [87]. Fagin [74] offers some additional insights into the history of extension axioms. The fact that the infinite Ehrenfeucht-Fraïssé game implies isomorphism of countable structures is from Karp [143].

The study of the zero-one law for fragments of ∃SO was initiated by Kolaitis and Vardi [150], where they proved Theorem 12.12. Theorem 12.15 is from Kolaitis and Vardi [151], and Theorem 12.16 is from Le Bars [163]. A good survey on zero-one laws and SO is Kolaitis and Vardi [155] (in particular, it explains how to prove that the zero-one law fails for ∃SO(∀∃∀) and ∃SO(∀∀∀∃) without equality).

Theorem 12.18 is from Hella, Kolaitis, and Luosto [122]. For related results in the context of database query evaluation, see Abiteboul, Compton, and Vianu [1].

Sources for exercises:

Exercises 12.3 and 12.4:	Fagin [73]
Exercise 12.5:	Lynch [173]
Exercise 12.6:	Kaufmann and Shelah [144] and Le Bars [164]
Exercise 12.7:	Grandjean [104]
Exercise 12.8:	Hodges [125]
Exercise 12.9 (b):	Cameron [31]

Exercise 12.11: Le Bars [163]
Exercise 12.12: Kolaitis and Vardi [150]
Kolaitis and Vardi [155]
Blass, Gurevich, and Kozen [22]

12.7 Exercises

Exercise 12.1. Calculate $\mu(\mathcal{P})$ for the following properties $\mathcal{P}$:

- rigidity;
- 2-colorability;
- being a tree;
- Hamiltonicity;
- having diameter 2.

Exercise 12.2. Prove the zero-one law for arbitrary vocabularies, using extension axioms $EA_{F,G}$.

Exercise 12.3. Instead of $\mu_n(\mathcal{P})$, consider $\nu_n(\mathcal{P})$ as the ratio of the number of different isomorphism types of graphs on $\{0, \ldots, n-1\}$ that have $\mathcal{P}$ and the number of all different isomorphism types of graphs on $\{0, \ldots, n-1\}$. Let $\nu(\mathcal{P})$ be defined as the limit of $\nu_n(\mathcal{P})$. Prove that if $\mathcal{P}$ is an FO-definable property, then $\nu(\mathcal{P}) = \mu(\mathcal{P})$, and thus is either 0 or 1.

Exercise 12.4. If constant or function symbols are allowed in the vocabulary, the zero-one law may not be true. Specifically, prove that:

- if c is a constant symbol and U a unary predicate symbol, then $U(c)$ has asymptotic probability $\frac{1}{2}$;
- if f is a unary function symbol, then $\forall x \, \neg(x = f(x))$ has asymptotic probability $\frac{1}{e}$.

Exercise 12.5. Instead of the usual successor relation, consider a *circular* successor: a relation of the form $\{(a_1, a_2), (a_2, a_3), \ldots, (a_{n-1}, a_n), (a_n, a_1)\}$. Prove that in the presence of a circular successor, FO continues to have the zero-one law.

Exercise 12.6. Prove that MSO does not have the zero-one law.

Hint: choose a vocabulary σ to consist of several binary relations, and prove that there is an FO formula $\varphi(x, y)$ of vocabulary $\sigma \cup \{U\}$, where U is unary, such that the MSO sentence $\exists U \, \varphi'$ almost surely holds, where φ' states that the set of pairs for (x, y) for which $\varphi(x, y)$ holds is a linear ordering.

Then the failure of the zero-one law follows since we know that MSO+ $<$ can define EVEN.

Prove a stronger version of this failure, for the vocabulary of one binary relation.

Exercise 12.7. Prove that for vocabularies with bounded arities, the problem of deciding whether $\mu(\Phi) = 1$, where Φ is FO, is PSPACE-complete.

Exercise 12.8. Prove that the random graph admits quantifier elimination: that is, every formula $\varphi(\vec{x})$ is equivalent to a quantifier-free formula $\varphi'(\vec{x})$.

Exercise 12.9. (a) Consider the following undirected graph $\mathcal{G}$: its universe is $\mathbb{N}_+ = \{n \in \mathbb{N} \mid n > 0\}$ and there is an edge between n and m, for $n > m$, iff n is divisible by p_m, the mth prime. Prove that $\mathcal{G}$ is isomorphic to the random graph $\mathcal{RG}$.

Hint: the proof does not require any number theory, and is a simple application of extension axioms.

(b) Consider another countable graph $\mathcal{G}'$ whose universe is the set of primes congruent to 1 modulo 4. Put an edge between p and q if p is a quadratic residue modulo q. Prove that $\mathcal{G}'$ is isomorphic to the random graph $\mathcal{RG}$.

Exercise 12.10. Let Φ be an arbitrary $\exists$SO sentence. Prove that it is undecidable whether $\mu(\Phi) = 1$.

Exercise 12.11. Prove that the restriction of $\exists$SO, where the first-order part is a formula of FO^2, does not have the zero-one law.

Exercise 12.12. Prove that for vocabularies with bounded arities, the problem of deciding whether $\mu(\Phi) = 1$ is

- NEXPTIME-complete, if Φ is an $\exists\text{SO}(\exists^*\forall^*)$ sentence, or an $\exists\text{SO}(\exists^*\forall\exists^*)$ sentence;
- EXPTIME-complete, if Φ is an LFP sentence.

Exercise 12.13.* Does $\exists\text{SO}(\forall\forall\exists)$ have the zero-one law over graphs?

13

Embedded Finite Models

In finite model theory, we deal with logics over finite structures. In embedded finite model theory, we deal with logics over finite structures embedded into infinite ones. For example, one assumes that nodes of graphs are numbers, and writes sentences like

$$\exists x \exists y \ (E(x, y) \wedge (x \cdot y = x \cdot x + 1))$$

saying that there is an edge (x, y) in a graph with $xy = x^2 + 1$. The infinite structure in this case could be $\langle \mathbb{R}, +, \cdot \rangle$, or $\langle \mathbb{N}, +, \cdot \rangle$, or $\langle \mathbb{Q}, +, \cdot \rangle$.

What kinds of queries can one write in this setting? We shall see in this chapter that the answer depends heavily on the properties of the infinite structure into which the finite structures are embedded: for example, queries such as EVEN and graph connectivity turn out to be expressible on structures embedded into $\langle \mathbb{N}, +, \cdot \rangle$, or $\langle \mathbb{Q}, +, \cdot \rangle$, but not $\langle \mathbb{R}, +, \cdot \rangle$.

The main motivation for embedded finite models comes from database theory. Relational calculus – that is, FO – is the basic relational query language. However, databases store *interpreted* elements such as numbers or strings, and queries in all practical languages use domain-specific operations, like arithmetic operations for numbers, or concatenation and prefix comparison for strings, etc. Embedded finite model theory studies precisely these kinds of languages over finite models, where the underlying domain is potentially infinite, and operations over that domain can be used in formulae.

13.1 Embedded Finite Models: the Setting

Assume that we have two vocabularies, Ω and σ, where σ is finite and relational. Let $\mathfrak{M}$ be an infinite Ω-structure $\langle U, \Omega \rangle$, where U is an infinite set. For example, if Ω contains two binary functions $+$ and $\cdot$, then $\langle \mathbb{R}, +, \cdot \rangle$ and $\langle \mathbb{N}, +, \cdot \rangle$ are two possible infinite Ω-structures, with $+$ and $\cdot$ interpreted, in both cases, as addition and multiplication respectively.

Definition 13.1. *Let* $\mathfrak{M} = \langle U, \Omega\rangle$ *be an infinite* Ω*-structure, and let* $\sigma = \{R_1, \ldots, R_m\}$. *Suppose the arity of each* R_i *is* $p_i > 0$. *Then an* embedded finite model *(i.e., a* σ*-structure embedded into* $\mathfrak{M}$*) is a structure*

$$\mathfrak{A} \;=\; \langle A, R_1^{\mathfrak{A}}, \ldots, R_l^{\mathfrak{A}}\rangle,$$

where each $R_i^{\mathfrak{A}}$ *is a* finite *subset of* U^{p_i}, *and* A *is the set of all the elements of* U *that occur in the relations* $R_1^{\mathfrak{A}}, \ldots, R_l^{\mathfrak{A}}$. *The set* A *is called the* active domain *of* $\mathfrak{A}$, *and is denoted by* $adom(\mathfrak{A})$. □

So far this is not that much different from the usual finite model, except that the universe comes from a given infinite set U. What makes the setting different, however, is the presence of the underlying structure $\mathfrak{M}$, which makes it possible to use rich logics for defining queries on embedded finite models. That is, instead of just FO over $\mathfrak{A}$, we shall use FO over

$$(\mathfrak{M}, \mathfrak{A}) \;=\; (U, \Omega, R_1^{\mathfrak{A}}, \ldots, R_l^{\mathfrak{A}}),$$

making use of operations available on $\mathfrak{M}$.

Before we define this logic, denoted by $\mathrm{FO}(\mathfrak{M}, \sigma)$, we shall address the issue of quantification. The universe of $(\mathfrak{M}, \mathfrak{A})$ is U, so saying $\exists x \varphi(x)$ means that there is an element of U that witnesses φ. But while we are dealing with finite structures $\mathfrak{A}$ embedded into $\mathfrak{M}$, quantification over the entire set U is not always very convenient.

Consider, for example, the simple property of reflexivity. In the usual finite model theory context, to state that a binary relation E is reflexive, we would say $\forall x\, E(x, x)$. However, if the interpretation of $\forall x$ is "for all $x \in U$", this sentence would be false in all embedded finite models! What we really want to say here is: "for all x in the active domain, $E(x, x)$ holds".

The definition of $\mathrm{FO}(\mathfrak{M}, \sigma)$ thus provides additional syntax to quantify over elements of the active domain.

Definition 13.2. *Given* $\mathfrak{M} = \langle U, \Omega\rangle$ *and a relational vocabulary* σ, first-order logic (FO) over $\mathfrak{M}$ and σ, *denoted by* $\mathrm{FO}(\mathfrak{M}, \sigma)$, *is defined as follows:*

- *Any atomic FO formula in the language of* $\mathfrak{M}$ *is an atomic* $\mathrm{FO}(\mathfrak{M}, \sigma)$ *formula. For any* p*-ary symbol* R *from* σ *and terms* $t_1, \ldots, t_p$ *in the language of* $\mathfrak{M}$, $R(t_1, \ldots, t_p)$ *is an atomic* $\mathrm{FO}(\mathfrak{M}, \sigma)$ *formula.*
- *Formulae of* $\mathrm{FO}(\mathfrak{M}, \sigma)$ *are closed under the Boolean connectives* $\vee$, $\wedge$, *and* $\neg$.
- *If* φ *is an* $\mathrm{FO}(\mathfrak{M}, \sigma)$ *formula, then the following are* $\mathrm{FO}(\mathfrak{M}, \sigma)$ *formulae:*
  - $\exists x\ \varphi$,
  - $\forall x\ \varphi$,
  - $\exists x \in adom\ \varphi$, *and*
  - $\forall x \in adom\ \varphi$.

The class of first-order formulae in the language of $\mathfrak{M}$ will be denoted by $\mathrm{FO}(\mathfrak{M})$ (i.e., the formulae built up from atomic $\mathfrak{M}$-formulae by Boolean connectives and quantification $\exists, \forall$). The class of formulae not using the symbols from Ω will be denoted by $\mathrm{FO}(\sigma)$ (in this case all four quantifiers are allowed).

The notions of free and bound variables are the usual ones. To define the semantics, we need to define the relation $(\mathfrak{M}, \mathfrak{A}) \models \varphi(\vec{a})$, for a formula $\varphi(\vec{x})$ and a tuple $\vec{a}$ over U of values of free variables. All the cases are standard, except quantification. If we have a formula $\varphi(x, \vec{y})$, and a tuple of elements $\vec{b}$ (values for $\vec{y}$), then

$$(\mathfrak{M}, \mathfrak{A}) \models \exists x\ \varphi(x, \vec{b}) \quad \text{iff} \quad (\mathfrak{M}, \mathfrak{A}) \models \varphi(a, \vec{b}) \text{ for some } a \in U.$$

On the other hand,

$$(\mathfrak{M}, \mathfrak{A}) \models \exists x \in adom\ \varphi(x, \vec{b}) \quad \text{iff} \quad (\mathfrak{M}, \mathfrak{A}) \models \varphi(a, \vec{b}) \text{ for some } a \in adom(\mathfrak{A}).$$

The definitions for the universal quantification are:

$$(\mathfrak{M}, \mathfrak{A}) \models \forall x\ \varphi(x, \vec{b}) \text{ iff } (\mathfrak{M}, \mathfrak{A}) \models \varphi(a, \vec{b}) \text{ for all } a \in U$$
$$(\mathfrak{M}, \mathfrak{A}) \models \forall x \in adom\ \varphi(x, \vec{b}) \text{ iff } (\mathfrak{M}, \mathfrak{A}) \models \varphi(a, \vec{b}) \text{ for all } a \in adom(\mathfrak{A}).$$

Since $\mathfrak{M}$ is most of the time clear from the context, we shall often write $\mathfrak{A} \models \varphi(\vec{a})$ instead of the more formal $(\mathfrak{M}, \mathfrak{A}) \models \varphi(\vec{a})$.

The quantifiers $\exists x \in adom\ \varphi$ and $\forall x \in adom\ \varphi$ are called *active-domain* quantifiers. We shall sometimes refer to the usual quantifies $\exists$ and $\forall$ as *unrestricted* quantifiers.

From the point of view of expressive power, active-domain quantifiers are a mere convenience: since $adom(\mathfrak{A})$ is definable with unrestricted quantification, so are these quantifiers. But we use them separately in order to define an important sublogic of $\mathrm{FO}(\mathfrak{M}, \sigma)$.

Definition 13.3. *By* $\mathrm{FO}_{\mathrm{act}}(\mathfrak{M}, \sigma)$ *we denote the fragment of* $\mathrm{FO}(\mathfrak{M}, \sigma)$ *that only uses quantifiers* $\exists x \in adom$ *and* $\forall x \in adom$*. Formulae in this fragment are called the* active-domain formulae.

Before moving on to the expressive power of $\mathrm{FO}(\mathfrak{M}, \sigma)$, we briefly discuss evaluation of such formulae. Since quantification is no longer restricted to a finite set, it is not clear a priori that formulae of $\mathrm{FO}(\mathfrak{M}, \sigma)$ can be evaluated – and, indeed, in some cases there is no algorithm for evaluating them. However, there is one special case when evaluation of formulae is "easy" (that is, easy to explain, not necessarily easy to evaluate).

Suppose we have a sentence Φ of $\mathrm{FO}(\mathfrak{M}, \sigma)$, and an embedded finite model $\mathfrak{A}$. We further assume that every element $c \in adom(\mathfrak{A})$ is definable over $\mathfrak{M}$: that is, there is an $\mathrm{FO}(\mathfrak{M})$ formula $\alpha_c(x)$ such that $\mathfrak{M} \models \alpha_c(x)$ iff $x = c$.

In such a case, we replace every occurrence of an atomic formula $R(t_1(\vec{x}), \ldots, t_m(\vec{x}))$, where $R \in \sigma$ and the t_i's are terms, by

$$\bigvee_{(c_1,\ldots,c_m)\in R^{\mathfrak{A}}} \alpha_{c_1}(t_1(\vec{x})) \wedge \ldots \wedge \alpha_{c_m}(t_m(\vec{x})).$$

That is, we say that the tuple of values of the $t_i(\vec{x})$'s is one of the tuples in $R^{\mathfrak{A}}$. Thus, if $\Phi^{\mathfrak{A}}$ is the sentence obtained from Φ by such a replacement, then

$$(\mathfrak{M}, \mathfrak{A}) \models \Phi \quad \Leftrightarrow \quad \mathfrak{M} \models \Phi^{\mathfrak{A}}. \tag{13.1}$$

Notice that $\Phi^{\mathfrak{A}}$ is an $\mathrm{FO}(\mathfrak{M})$ sentence, since all the σ-relations disappeared. Now using (13.1) we can propose the following evaluation algorithm: given Φ, construct $\Phi^{\mathfrak{A}}$, and check if $\mathfrak{M} \models \Phi^{\mathfrak{A}}$. The last is possible if the theory of $\mathfrak{M}$ is decidable.

13.2 Analyzing Embedded Finite Models

When we briefly looked at the standard model-theoretic techniques in Chap. 3, we noticed that they are generally inapplicable in the setting of finite model theory. For embedded finite models, we mix the finite and the infinite: we study logics over pairs $(\mathfrak{M}, \mathfrak{A})$, where $\mathfrak{M}$ is infinite and $\mathfrak{A}$ is finite. So the question arises: can we use techniques of either finite or infinite model theory?

It turns out that we cannot use finite or infinite model-theoretic techniques directly; as we are about to show, in general, they fail over embedded finite models. Then we outline a new kind of tools that is used with embedded finite models: by using infinite model-theoretic techniques, we reduce questions about embedded finite models to questions about finite models, for which the preceding 12 chapters give us plenty of answers. In general, we shall see that the behavior of $\mathrm{FO}(\mathfrak{M}, \sigma)$ depends heavily on model-theoretic properties of the underlying structure $\mathfrak{M}$.

We now discuss standard (finite) model-theoretic tools and their applicability to the study of embedded finite models.

First, notice that compactness fails over embedded finite models for the same reason as for finite models. One can write sentences λ_n, $n \geq 0$, stating that $adom(\mathfrak{A})$ contains at least n elements. Then $T = \{\lambda_n \mid n \geq 0\}$ is finitely consistent: every finite set of sentences has a finite model. However, T itself does not have a finite model.

One tool that definitely applies in the embedded setting is Ehrenfeucht-Fraïssé games. However, playing a game is very hard. Assume, for example, that $\mathfrak{M}$ is the real field $\langle \mathbb{R}, +, \cdot \rangle$. Suppose σ is empty, and we want to show that the query EVEN, testing if $|\,adom(\mathfrak{A})\,|$ is even, is not expressible (which, as we shall see later, is a true statement). As in the proof given in Chap. 3, suppose EVEN is expressible by a sentence Φ of quantifier rank k. Before, we picked two structures, $\mathfrak{A}_1$ of cardinality k and $\mathfrak{A}_2$ of cardinality $k+1$, and showed that $\mathfrak{A}_1 \equiv_k \mathfrak{A}_2$. Our problem now is that showing $\mathfrak{A}_1 \equiv_k \mathfrak{A}_2$ no longer suffices, as we have to prove

$$(\mathfrak{M}, \mathfrak{A}_1) \equiv_k (\mathfrak{M}, \mathfrak{A}_2) \tag{13.2}$$

instead. For example, in the old strategy for winning the game on $\mathfrak{A}_1$ and $\mathfrak{A}_2$, if the spoiler plays any point a_1 in $\mathfrak{A}_1$ in the first move, the duplicator can respond by any point $\mathfrak{A}_2$. But now we have to account for additional atomic formulae such as $p(x) = 0$, where p is a polynomial. So if we know that $p(a_1) = 0$ for some given p, the strategy must also ensure that $p(a_2) = 0$. It is not at all clear how one can play a game like that, to satisfy (13.2).

The next obvious approach is to try finite model-theoretic techniques that avoid Ehrenfeucht-Fraïssé games, such as locality and zero-one laws. This approach, however, cannot be used for all structures $\mathfrak{M}$, as the following example shows.

Let $\mathfrak{N}$ be the well-known structure $\langle \mathbb{N}, +, \cdot \rangle$; that is, natural numbers with the usual arithmetic operations. A σ-structure over $\mathfrak{N}$ is a σ-structure whose active domain is a finite subset of $\mathbb{N}$, and hence it can be encoded by some reasonable encoding (e.g., a slight modification of the encoding of Chap. 6, where in addition all numbers in the active domain are encoded in binary). A *Boolean query* on σ-structures embedded into $\mathfrak{N}$ is a function Q from such structures into $\{true, false\}$. It is *computable* if there is a computable function $f_Q : \{0,1\}^* \to \{0,1\}$ such that $f_Q(s) = 1$ iff s is an encoding of a structure $\mathfrak{A}$ such that $Q(\mathfrak{A}) = true$.

Proposition 13.4. *Every computable Boolean query on σ-structures embedded into $\mathfrak{N}$ can be expressed in* $\mathrm{FO}(\mathfrak{N}, \sigma)$.

Proof. Without loss of generality, we assume that σ contains a single binary relation E. We use the following well-known fact about $\mathfrak{N}$: every computable predicate $P \subseteq \mathbb{N}^m$ is definable by an $\mathrm{FO}(\mathfrak{N})$ formula, which we shall denote by $\psi_P(x_1, \ldots, x_m)$. The idea of the proof then is to code finite σ-structures with numbers. For a query Q, the sentence defining it will be

$$\Phi_Q \;=\; \exists x \, \big(\chi(x) \wedge \psi_{P_Q}(x)\big), \tag{13.3}$$

where $\chi(x)$ says that the input structure $\mathfrak{A}$ is coded by the number x, and the predicate P_Q is the computable predicate such that $P_Q(n)$ holds iff n is the code of a structure $\mathfrak{A}$ with $Q(\mathfrak{A}) = true$.

Thus, we have to show how to code structures. Let p_n denote the nth prime, with the numeration starting at $p_0 = 2$. Suppose we have a structure $\mathfrak{A}$ with $adom(\mathfrak{A}) = \{n_1, \ldots, n_k\}$. We first code the active domain by

$$\mathrm{code}_0(\mathfrak{A}) \;=\; \prod_{i=1}^{k} p_{n_i}.$$

There is a formula $\chi_0(x)$ of $\mathrm{FO}(\mathfrak{N}, \sigma)$ such that $\mathfrak{A} \models \chi_0(n)$ iff $\mathrm{code}_0(\mathfrak{A}) = n$. Such a formula states the following condition:

- for each $l \in adom(\mathfrak{A})$, n is divisible by p_l but not divisible by p_l^2, and
- if n is divisible by a prime number p, then p is of the form p_l for some $l \in adom(\mathfrak{A})$.

Since the binary relation $\{(n, p_n) \mid n \geq 0\}$ is computable and thus definable in FO($\mathfrak{N}$), χ_0 can be expressed as an FO($\mathfrak{N}, \sigma$) formula.

We next code the edge relation E. Let pair : $\mathbb{N} \times \mathbb{N} \to \mathbb{N}$ be the standard pairing function. We then code $E^{\mathfrak{A}}$ by

$$\text{code}_1(\mathfrak{A}) \quad = \prod_{(n_i,n_j)\in E^{\mathfrak{A}}} p_{\text{pair}(n_i,n_j)}.$$

As in the case of coding the active domain, there exists a formula $\chi_1(x)$ such that $\mathfrak{A} \models \chi_1(n)$ iff $\text{code}_1(\mathfrak{A}) = n$ – the proof is the same as for χ_0. Finally, we code the whole structure by

$$\text{code}(\mathfrak{A}) \quad = \quad \text{pair}\big(\text{code}_0(\mathfrak{A}), \text{code}_1(\mathfrak{A})\big).$$

Clearly, $\mathfrak{A} \neq \mathfrak{B}$ implies $\text{code}(\mathfrak{A}) \neq \text{code}(\mathfrak{B})$, so we did define a coding function. Moreover, since χ_0 and χ_1 are FO($\mathfrak{N}, \sigma$) formulae, the formula $\chi(x)$ can be defined as $\exists y \exists z\ \chi_0(y) \wedge \chi_1(z) \wedge \psi_P(y, z, x)$, where P is the graph of the pairing function. This completes the coding scheme, and thus shows that (13.3) defines Q on structures embedded into $\mathfrak{N}$. ☐

Therefore, in FO($\mathfrak{N}, \sigma$) we can express queries that violate locality notions (e.g., connectivity) and queries that do not obey the zero-one law (e.g., parity).

Hence, we need a totally different set of techniques for proving bounds on the expressive power of FO($\mathfrak{M}, \sigma$). If we want to prove results about FO($\mathfrak{M}, \sigma$), perhaps we can relate this logic to something we know how to deal with: the pure finite model theory setting. In our new terminology, this would be $\text{FO}_{\text{act}}(\mathfrak{U}_\emptyset, \sigma)$, where $\mathfrak{U}_\emptyset = \langle U, \emptyset \rangle$ is a structure of the empty vocabulary. That is, there are no functions or predicates from $\mathfrak{M}$ used in formulae, and all quantification is restricted to the finite universe $adom(\mathfrak{A})$. (Notice that the setting of $\text{FO}_{\text{act}}(\mathfrak{U}_\emptyset, \sigma)$ is in fact a bit more restrictive than the usual finite model theory setting: in the latter, we quantify over a finite universe that may be larger than the active domain.)

For technical reasons that will become clear a bit later, we shall deal not with $\mathfrak{U}_\emptyset$ but rather with $\mathfrak{U}_< = \langle U, < \rangle$, where $<$ is a linear order on U. Then $\text{FO}_{\text{act}}(\mathfrak{U}_<, \sigma)$ corresponds to what we called FO+$<$ in the finite model theory setting. We know a number of results about this logic: in particular, it cannot express the query EVEN (Theorem 3.6) nor can it express graph connectivity (Theorem 5.8).

We now present the first of our two new tools. First, we need the following. Suppose Ω' expands Ω by adding some (perhaps infinitely many) predicate symbols. We call a structure $\mathfrak{M}' = \langle U, \Omega' \rangle$ a *definitional expansion* of $\mathfrak{M} = \langle U, \Omega \rangle$ if for every predicate $P \in \Omega' - \Omega$, there exists a formula $\psi_P(\vec{x})$ in the language of $\mathfrak{M}$ such that $P^{\mathfrak{M}'} = \{\vec{a} \mid \mathfrak{M} \models \psi_P(\vec{a})\}$.

Definition 13.5. *We say that* $\mathfrak{M}$ *admits the* restricted quantifier collapse, *or* RQC, *if there exists a definitional expansion* $\mathfrak{M}'$ *of* $\mathfrak{M}$ *such that*

$$\mathrm{FO}(\mathfrak{M}, \sigma) \quad = \quad \mathrm{FO}_{\mathrm{act}}(\mathfrak{M}', \sigma)$$

for every σ.

The notion of RQC can be formulated without using a definitional expansion as follows. For every $\mathrm{FO}(\mathfrak{M}, \sigma)$ formula $\varphi(\vec{x})$, there is an equivalent formula $\varphi'(\vec{x})$ such that no σ-relation appears within the scope of an unrestricted quantifier $\exists$ or $\forall$ (i.e., σ-relations only appear within the scope of restricted quantifiers $\exists x \in adom$ and $\forall x \in adom$).

There is one special form of the restricted quantifier collapse, which arises for structures $\mathfrak{M}$ that have the collapse and also have quantifier elimination (that is, every $\mathrm{FO}(\mathfrak{M})$ formula is equivalent to a quantifier-free one). In this case, if $\mathrm{FO}_{\mathrm{act}}(\mathfrak{M}', \sigma)$ refers to a definable predicate $P \in \Omega' - \Omega$, we know that P is definable by a quantifier-free formula over $\mathfrak{M}$. Hence, using the definition of P, we obtain an equivalent $\mathrm{FO}(\mathfrak{M}, \sigma)$ formula. Thus, we have:

Proposition 13.6. *If* $\mathfrak{M}$ *admits the restricted quantifier collapse (RQC) and has quantifier elimination, then*

$$\mathrm{FO}(\mathfrak{M}, \sigma) \quad = \quad \mathrm{FO}_{\mathrm{act}}(\mathfrak{M}, \sigma). \tag{13.4}$$

The condition in (13.4) is usually called the *natural-active collapse*, since the standard unrestricted interpretation of quantifiers is sometimes called the "natural interpretation".

Using RQC, or the natural-active collapse, eliminates quantification outside of the active domain. To reduce the expressiveness of $\mathrm{FO}(\mathfrak{M}, \sigma)$ to that of $\mathrm{FO}_{\mathrm{act}}(\mathfrak{U}_<, \sigma)$, we would also like to eliminate all references to $\mathfrak{M}$ functions and predicates, except possibly order. This, however, in general is impossible: how could one express a query like $\exists x \in adom \exists y \in adom\ E(x, y) \wedge x \cdot y = x + 1$?

To deal with this problem, we use the notion of *genericity* which comes from the classical relational database setting. Informally, it states the following: when one evaluates formulae on embedded finite models, exact values of elements in the active domain do not matter. For example, the answer to the query "Does a graph have diameter 2?" is the same for the graph $\{(1,2), (1,3), (1,4)\}$ and for the graph $\{(5,10), (5,15), (5,20)\}$, which is obtained by the mapping $1 \mapsto 5, 2 \mapsto 10, 3 \mapsto 15, 4 \mapsto 20$.

In general, generic queries commute with permutations of the universe. Queries expressible in $\mathrm{FO}(\mathfrak{M}, \sigma)$ need not be generic: for example, the query given by $\exists x \in adom \exists y \in adom\ E(x, y) \wedge x \cdot y = x + 1$ is true on $E = \{(1, 2)\}$ but false on $E = \{(1, 3)\}$. However, as all queries definable in standard logics over finite structures are generic, to reduce questions about $\mathrm{FO}(\mathfrak{M}, \sigma)$ to those in ordinary finite model theory, it suffices to restrict one's attention to generic queries.

We now define genericity for queries (which map a finite σ-structure $\mathfrak{A}$ to a finite subset of A^m, $m \geq 0$). Given a function $\pi : U \to U$, we extend it to finite σ-structures $\mathfrak{A}$ by replacing each occurrence of $a \in adom(\mathfrak{A})$ with $\pi(a)$.

Definition 13.7.
- *A query Q is* generic *if for every partial injective function $\pi : U \to U$ which is defined on $adom(\mathfrak{A})$, it is the case that $Q(\mathfrak{A}) = Q(\pi(\mathfrak{A}))$.*
- *The class of generic queries definable in* $\mathrm{FO}(\mathfrak{M}, \sigma)$ *or* $\mathrm{FO}_{\mathrm{act}}(\mathfrak{M}, \sigma)$ *is denoted by* $\mathrm{FO}^{\mathrm{gen}}(\mathfrak{M}, \sigma)$ *or* $\mathrm{FO}^{\mathrm{gen}}_{\mathrm{act}}(\mathfrak{M}, \sigma)$*, respectively.*

While it is undecidable in general if an $\mathrm{FO}(\mathfrak{M}, \sigma)$ query is generic, most queries whose inexpressibility we want to prove are generic.

Definition 13.8. *We say that $\mathfrak{M}$ admits the* active-generic collapse, *if*

$$\mathrm{FO}^{\mathrm{gen}}_{\mathrm{act}}(\mathfrak{M}, \sigma) \subseteq \mathrm{FO}_{\mathrm{act}}(\mathfrak{U}_<, \sigma).$$

Now using the different notions of collapse together, we come up with the following methodology of proving bounds on $\mathrm{FO}(\mathfrak{M}, \sigma)$.

Proposition 13.9. *Let $\mathfrak{M}$ admit both the restricted-quantifier collapse (RQC) and the active-generic collapse. Then every generic query expressible in* $\mathrm{FO}(\mathfrak{M}, \sigma)$ *is also expressible in* $\mathrm{FO}_{\mathrm{act}}(\mathfrak{U}_<, \sigma)$. ☐

For example, it would follow from Theorem 3.6 that for $\mathfrak{M}$ as in the proposition above, EVEN is not expressible in $\mathrm{FO}(\mathfrak{M}, \sigma)$. Furthermore, for such $\mathfrak{M}$, every query in $\mathrm{FO}^{\mathrm{gen}}(\mathfrak{M}, \sigma)$ is Gaifman-local, by Proposition 13.9 and Theorem 5.8.

Thus, our next goal is to see for what structures collapse results can be established. We start with the active-generic collapse, and prove, in the next section, that it holds for *all* structures.

The situation with RQC is not nearly as simple. We shall see that it fails for $\langle \mathbb{N}, +, \cdot \rangle$ and $\langle \mathbb{Q}, +, \cdot \rangle$, but we shall prove it for the ordered real field $\langle \mathbb{R}, +, \cdot, <, 0, 1 \rangle$. This structure motivated much of the initial work on embedded finite models due to its database applications; this will be explained in Sect. 13.6. More examples of RQC (or its failure) are given in the exercises. We shall also revisit the random graph of the previous chapter and relate queries over it to those definable in MSO.

13.3 Active-Generic Collapse

Our goal is to prove the following result.

Theorem 13.10. *Every infinite structure $\mathfrak{M}$ admits the active-generic collapse.* ☐

We shall assume that $\mathfrak{M}$ is ordered: that is, one of its predicates is $<$ interpreted as a linear order on its universe U. If this were not the case, we could have expanded $\mathfrak{M}$ to $\mathfrak{M}_<$ by adding a linear order. Since $\mathrm{FO}(\mathfrak{M},\sigma) \subseteq \mathrm{FO}(\mathfrak{M}_<,\sigma)$, the active-generic collapse for $\mathfrak{M}_<$ would imply the collapse for $\mathfrak{M}$:

$$\mathrm{FO}^{\mathrm{gen}}_{\mathrm{act}}(\mathfrak{M},\sigma) \subseteq \mathrm{FO}^{\mathrm{gen}}_{\mathrm{act}}(\mathfrak{M}_<,\sigma) \subseteq \mathrm{FO}_{\mathrm{act}}(\mathfrak{U}_<,\sigma).$$

The idea behind the proof of Theorem 13.10 is as follows: we show that for each formula, its behavior on some infinite set is described by a first-order formula which only uses $<$ and no other symbol from the vocabulary of $\mathfrak{M}$. This is called the Ramsey property. We then show how genericity and the Ramsey property imply the collapse.

Definition 13.11. *Let $\mathfrak{M} = \langle U, \Omega\rangle$ be an ordered structure. We say that an $\mathrm{FO}_{\mathrm{act}}(\mathfrak{M},\sigma)$ formula $\varphi(\vec{x})$ has the* Ramsey property *if the following is true:*

> *Let X be an infinite subset of U. Then there exists an infinite set $Y \subseteq X$ and an $\mathrm{FO}_{\mathrm{act}}(\mathfrak{U}_<,\sigma)$ formula $\psi(\vec{x})$ such that for every σ-structure $\mathfrak{A}$ with $adom(\mathfrak{A}) \subset Y$, and for every $\vec{a}$ over Y, it is the case that $\mathfrak{A} \models \varphi(\vec{a}) \leftrightarrow \psi(\vec{a})$.*

We now prove the Ramsey property for an arbitrary ordered $\mathfrak{M}$. The following simple lemma will often be used as a first step in proofs of collapse results. Before stating it, note that for an $\mathrm{FO}(\mathfrak{M},\sigma)$ formula $(x = y)$ can be viewed as both an atomic $\mathrm{FO}(\sigma)$ formula and an atomic $\mathrm{FO}(\mathfrak{M})$ formula. We choose to view it as an atomic $\mathrm{FO}(\mathfrak{M})$ formula; that is, atomic $\mathrm{FO}(\sigma)$ formulae are only those of the form $R(\cdots)$ for $R \in \sigma$.

Lemma 13.12. *Let $\varphi(\vec{x})$ be an $\mathrm{FO}(\mathfrak{M},\sigma)$ formula. Then there exists an equivalent formula $\psi(\vec{x})$ such that every atomic subformula of ψ is either an $\mathrm{FO}(\sigma)$ formula, or an $\mathrm{FO}(\mathfrak{M})$ formula. Furthermore, it can be assumed that none of the free variables $\vec{x}$ occurs in an $\mathrm{FO}(\sigma)$-atomic subformula of $\psi(\vec{x})$. If φ is an $\mathrm{FO}_{\mathrm{act}}(\mathfrak{M},\sigma)$ formula, then ψ is also an $\mathrm{FO}_{\mathrm{act}}(\mathfrak{M},\sigma)$ formula.*

Proof. Introduce m fresh variables $z_1,\ldots,z_m$, where m is the maximal arity of a relation in σ, and replace any atomic formula of the form $R(t_1(\vec{y}),\ldots,t_l(\vec{y}))$, where $l \leq m$ and the t_i's are $\mathfrak{M}$-terms, by $\exists z_1 \in adom \ldots \exists z_l \in adom \bigwedge_i (z_i = t_i(\vec{y})) \wedge R(z_1,\ldots,z_l)$. Similarly use existential quantifiers to eliminate the free $\vec{x}$-variables from $\mathrm{FO}(\sigma)$-atomic formulae. □

The key in the inductive proof of the Ramsey property is the case of $\mathrm{FO}(\mathfrak{M})$ subformulae. For this, we first recall the infinite version of Ramsey's theorem, in the form most convenient for our purposes.

Theorem 13.13 (Ramsey). *Given an infinite ordered set X, and any partition of the set of all ordered m-tuples $\langle x_1,\ldots,x_m\rangle$, $x_1 < \ldots < x_m$, of elements of X into l classes $A_1,\ldots,A_l$, there exists an infinite subset $Y \subseteq X$ such that all ordered m-tuples of elements of Y belong to the same class A_i.* □

The following is a standard model-theoretic result that we prove here for the sake of completeness.

Lemma 13.14. *Let $\varphi(\vec{x})$ be an* $\mathrm{FO}(\mathfrak{M})$ *formula. Then φ has the Ramsey property.*

Proof. Consider a (finite) enumeration of all the ways in which the variables $\vec{x}$ may appear in the order of $\mathcal{U}$. For example, if $\vec{x} = (x_1, \ldots, x_4)$, one possibility is $x_1 = x_3, x_2 = x_4$, and $x_1 < x_2$. Let P be such an arrangement, and $\zeta(P)$ a first-order formula that defines it ($x_1 = x_3 \wedge x_2 = x_4 \wedge x_1 < x_2$ in the above example). Note that there are finitely many such arrangements P; let $\mathcal{P}$ be the set of all of those. Each P induces an equivalence relation on $\vec{x}$: for example, $\{(x_1, x_3), (x_2, x_4)\}$ for P above. Let $\vec{x}^P$ be a subtuple of $\vec{x}$ containing a representative for each class (e.g., (x_1, x_4)) and let $\varphi^P(\vec{x}^P)$ be obtained from φ by replacing all variables from an equivalence class by the chosen representative. Then $\varphi(x)$ is equivalent to

$$\bigvee_{P \in \mathcal{P}} \zeta(P) \wedge \varphi^P(\vec{x}^P).$$

We now show the following. Let $\mathcal{P}' \subseteq \mathcal{P}$ and $P_0 \in \mathcal{P}'$. Let $X \subseteq U$ be an infinite set. Assume that $\psi(\vec{x})$ is given by

$$\bigvee_{P \in \mathcal{P}'} \zeta(P) \wedge \varphi^P(\vec{x}^P).$$

Then there exists an infinite set $Y \subseteq X$ and a quantifier-free formula $\gamma_{P_0}(\vec{x})$ of the vocabulary $\{<\}$ such that ψ is equivalent to

$$\gamma_{P_0}(\vec{x}) \vee \bigvee_{P \in \mathcal{P}' - \{P_0\}} \zeta(P) \wedge \varphi^P(\vec{x}^P)$$

for tuples $\vec{x}$ of elements of Y.

To see this, suppose that P_0 has m equivalence classes. Consider a partition of tuples of X^m ordered according to P_0 into two classes: A_1 of those tuples for which $\varphi^{P_0}(\vec{x}^{P_0})$ is true, and A_2 of those for which $\varphi^{P_0}(\vec{x}^{P_0})$ is false. By Ramsey's theorem, for some infinite set $Y \subseteq X$ either all ordered tuples over Y^m are in A_1, or all are in A_2. In the first case, ψ is equivalent to $\zeta(P_0) \vee \bigvee_{P \in \mathcal{P}' - \{P_0\}} \zeta(P) \wedge \varphi^P(\vec{x}^P)$, and in the second case ψ is equivalent to $\bigvee_{P \in \mathcal{P}' - \{P_0\}} \zeta(P) \wedge \varphi^P(\vec{x}^P)$, proving the claim.

The lemma now follows by applying this claim inductively to every partition $P \in \mathcal{P}$, passing to smaller infinite sets, while getting rid of all the formulae containing symbols other than $=$ and $<$. At the end we have an infinite set over which φ is equivalent to a quantifier-free formula in the vocabulary $\{<\}$. □

The next lemma lifts the Ramsey property from $\mathrm{FO}(\mathfrak{M})$ formulae to arbitrary $\mathrm{FO}_{\mathrm{act}}(\mathfrak{M}, \sigma)$ formulae.

Lemma 13.15. *Every* $\mathrm{FO}_{\mathrm{act}}(\mathfrak{M}, \sigma)$ *formula has the Ramsey property.*

Proof. By Lemma 13.12, we assume that every atomic subformula is an $\mathrm{FO}_{\mathrm{act}}(\sigma)$ formula or an $\mathrm{FO}(\mathfrak{M})$ formula. The base cases for the induction are the case of $\mathrm{FO}_{\mathrm{act}}(\sigma)$ formulae, where there is no need to change the formula or find a subset, and the case of $\mathrm{FO}(\mathfrak{M})$ atomic formulae, which is given by Lemma 13.14.

Let $\varphi(\vec{x}) \equiv \varphi_1(\vec{x}) \wedge \varphi_2(\vec{x})$, where $X \subseteq U$ is infinite. First, find ψ_1, $Y_1 \subseteq X$, such that for every $\mathfrak{A}$ and $\vec{a}$ over Y_1, it is the case that $\mathfrak{A} \models \varphi_1(\vec{a}) \leftrightarrow \psi_1(\vec{a})$. Next, by using the hypothesis for φ_2 and Y_1, find an infinite set $Y_2 \subseteq Y_1$ such that for every $\mathfrak{A}$ and $\vec{a}$ over Y_2, it is the case that $\mathfrak{A} \models \varphi_2(\vec{a}) \leftrightarrow \psi_2(\vec{a})$. Then take $\psi \equiv \psi_1 \wedge \psi_2$ and $Y = Y_2$.

The case of $\varphi = \neg\varphi'$ is trivial.

For the existential case, let $\varphi(\vec{x}) \equiv \exists y \in adom\ \varphi_1(y, \vec{x})$. By the hypothesis, find $Y \subseteq X$ and $\psi_1(y, \vec{x})$ such that for every $\mathfrak{A}$ and $\vec{a}$ over Y and every $b \in Y$ we have $\mathfrak{A} \models \varphi_1(b, \vec{a}) \leftrightarrow \psi_1(b, \vec{a})$. Let $\psi(\vec{x}) \equiv \exists y \in adom\ \psi_1(y, \vec{x})$. Then, for every $\mathfrak{A}$ and $\vec{a}$ over Y, $\mathfrak{A} \models \psi(\vec{a})$ iff $\mathfrak{A} \models \psi_1(b, \vec{a})$ for some $b \in adom(\mathfrak{A})$ iff $\mathfrak{A} \models \varphi_1(b, \vec{a})$ for some $b \in adom(\mathfrak{A})$ iff $\mathfrak{A} \models \varphi_1(\vec{a})$, thus finishing the proof. □

To finish the proof of Theorem 13.10, we have to show the following.

Lemma 13.16. *Assume that every* $\mathrm{FO}_{\mathrm{act}}(\mathfrak{M}, \sigma)$ *formula has the Ramsey property. Then* $\mathfrak{M}$ *admits the active-generic collapse.*

Proof. Let Q be a generic query definable in $\mathrm{FO}_{\mathrm{act}}(\mathfrak{M}, \sigma)$. By the Ramsey property, we find an infinite $X \subseteq U$ and an $\mathrm{FO}_{\mathrm{act}}(\mathfrak{U}_<, \sigma)$-definable Q' that coincides with Q on X. We claim they coincide everywhere. Let $\mathfrak{A}$ be a σ-structure. Since X is infinite, there exists a partial monotone injective function π from $adom(\mathfrak{A})$ into X such that for every pair of elements $a < a'$ of $adom(\mathfrak{A})$, there exist $x_1, x_2, x_3 \in X - \pi(adom(\mathfrak{A}))$ with the property that $x_1 < \pi(a) < x_2 < \pi(a') < x_3$.

By the genericity of Q, we have $\pi(Q(\mathfrak{A})) = Q(\pi(\mathfrak{A}))$. Thus, $Q(\pi(\mathfrak{A}))$ coincides with the restriction of $Q'(\pi(\mathfrak{A}))$ to X. We now notice that Q' does not extend its active domain. Indeed, if $adom(Q'(\pi(\mathfrak{A})))$ contained an element $b \notin \pi(adom(\mathfrak{A}))$, we could have replaced this element by $b' \in X - \pi(adom(\mathfrak{A}))$ such that for every $a \in \pi(adom(\mathfrak{A}))$, $a < b$ iff $a < b'$. Since Q' is $\mathrm{FO}_{\mathrm{act}}(\mathfrak{U}_<, \sigma)$-definable, this would imply that $b' \in adom(Q'(\pi(\mathfrak{A})))$, which contradicts the fact that over X, the queries Q and Q' coincide.

Hence, $\pi(Q(\mathfrak{A})) = Q(\pi(\mathfrak{A})) = Q'(\pi(\mathfrak{A}))$. Again, since Q' is $\mathrm{FO}_{\mathrm{act}}(\mathfrak{U}_<, \sigma)$-definable, it commutes with any monotone injective map, and thus $Q'(\pi(\mathfrak{A})) = \pi(Q'(\mathfrak{A}))$. We have shown that $\pi(Q(\mathfrak{A})) = \pi(Q'(\mathfrak{A}))$, from which $Q(\mathfrak{A}) = Q'(\mathfrak{A})$ follows. □

This completes the proof of Theorem 13.10.

Thus, no matter what functions and predicates there are in $\mathfrak{M}$, FO cannot express more generic active-domain semantics queries over it than just $\mathrm{FO}_{\mathrm{act}}(\mathfrak{U}_<, \sigma)$. In particular, we have the following.

Corollary 13.17. *Let $\mathfrak{M}$ be an arbitrary structure. Then queries such as* EVEN, PARITY, *majority, connectivity, transitive closure, and acyclicity are not definable in* $\mathrm{FO}_{\mathrm{act}}(\mathfrak{M}, \sigma)$. □

13.4 Restricted Quantifier Collapse

One part of our program for establishing bounds on $\mathrm{FO}(\mathfrak{M}, \sigma)$ has been very successful: we prove the active-generic collapse for arbitrary structures. Can we hope to achieve the same success with the restricted-quantifier collapse (RQC)? The answer is clearly negative.

Corollary 13.18. *The restricted-quantifier collapse fails over* $\mathfrak{N} = \langle \mathbb{N}, +, \cdot \rangle$.

Proof. By Corollary 13.17, parity is not definable in $\mathrm{FO}_{\mathrm{act}}(\mathfrak{N}, \sigma)$, but by Proposition 13.4, it is expressible in $\mathrm{FO}(\mathfrak{N}, \sigma)$. □

Furthermore, RQC fails over $\langle \mathbb{Q}, +, \cdot \rangle$, since it is possible to define natural numbers within this structure, and then emulate the proof of Proposition 13.4 to show that every computable query is expressible.

However, the situation becomes very different when we move to the real numbers. We shall consider the real ordered field: that is, the structure

$$\mathbf{R} \;=\; \langle \mathbb{R}, +, \cdot, <, 0, 1 \rangle.$$

This is the structure that motivated much of the initial development in embedded finite models, due to its close connections with questions about the expressiveness of languages for geographical databases.

Consider the following $\mathrm{FO}(\mathbf{R}, \{E\})$ sentence, where E is a binary relation symbol:

$$\exists u \exists v \forall x \in adom \forall y \in adom \left(E(x, y) \rightarrow y = u \cdot x + v \right), \tag{13.5}$$

saying that all elements of $E \subset \mathbb{R}^2$ lie on a line. Notice that it is essential that the first two quantifiers range over the entire set $\mathbb{R}$. For example, if E is interpreted as $\{(2,2), (3,3), (4,4)\}$, then the sentence (13.5) is true, and the witnesses for the existential quantifiers are $u = 1$ and $v = 0$. But neither 0 nor 1 is in the active domain of E.

Nevertheless, (13.5) can be expressed by an $\mathrm{FO}_{\mathrm{act}}(\mathbf{R}, \{E\})$ sentence. To see this, notice that E lies on a line iff every three points in E are collinear. This can be expressed as

$$\begin{array}{l} \forall x_1 \in adom \forall y_1 \in adom \forall x_2 \in adom \forall y_2 \in adom \forall x_3 \in adom \forall y_3 \in adom \\ \quad \Big(\big(E(x_1, y_1) \wedge E(x_2, y_2) \wedge E(x_3, y_3) \big) \;\rightarrow\; \mathrm{collinear}(\vec{x}, \vec{y}) \Big) \end{array} \tag{13.6}$$

where collinear$(\vec{x}, \vec{y})$ is a formula, over $\mathbf{R}$, stating that (x_1, y_1), (x_2, y_2), and (x_3, y_3) are collinear. It is easy to check that collinear$(\vec{x}, \vec{y})$ can be written as a quantifier-free formula (in fact, due to the quantifier elimination for the real field, every formula over $\mathbf{R}$ is equivalent to a quantifier-free formula, but the condition for collinearity can easily be expressed directly). Hence, (13.6) is an $\mathrm{FO}_{\mathrm{act}}(\mathbf{R}, \{E\})$ formula, equivalent to (13.5).

This example is an instance of a much more general result, stating that the real field $\mathbf{R}$ admits RQC. In fact, we show the natural-active collapse for $\mathbf{R}$ (since $\mathbf{R}$ has quantifier elimination). Moreover, the proof is constructive.

Theorem 13.19. *The real field* $\mathbf{R} = \langle \mathbb{R}, +, \cdot, <, 0, 1 \rangle$ *admits the restricted quantifier collapse. That is, for every* $\mathrm{FO}(\mathbf{R}, \sigma)$ *formula* $\varphi(\vec{x})$*, there is an equivalent* $\mathrm{FO}_{\mathrm{act}}(\mathbf{R}, \sigma)$ *formula* $\varphi_{\mathrm{act}}(\vec{x})$*. Moreover, there is an algorithm that constructs* φ_{act} *from* φ*.*

Proof. The proof of this result is by induction on the structure of the formula. We shall always assume, by Lemma 13.12, that all atomic $\mathrm{FO}(\sigma)$ formulae are of the form $S(\vec{y})$, where $\vec{y}$ contains only variables. Thus, the base cases of the induction are as follows:

- $\varphi(\vec{x})$ is $S(\vec{x})$. In this case $\varphi_{\mathrm{act}} \equiv \varphi$.
- $\varphi(\vec{x})$ is an atomic $\mathrm{FO}(\mathbf{R})$ formula. Again, $\varphi_{\mathrm{act}} \equiv \varphi$ in this case.

The cases of Boolean operations are simple:

- If $\varphi \equiv \psi \vee \chi$, then $\varphi_{\mathrm{act}} \equiv \psi_{\mathrm{act}} \vee \chi_{\mathrm{act}}$;
- if $\varphi \equiv \neg\psi$, then $\psi_{\mathrm{act}} \equiv \neg\psi_{\mathrm{act}}$.

We now move to the case of an unrestricted existential quantifier. We shall first treat the case of σ-structures $\mathfrak{A}$ with $adom(\mathfrak{A}) \neq \emptyset$; at the end of the proof, we shall explain how to deal with empty structures.

Suppose $\varphi(\vec{x}) \equiv \exists z\, \beta(\vec{x}, z)$. By the induction hypothesis, β can be assumed to be of the form

$$\beta(\vec{x}, z) \;\equiv\; \mathbf{Q} y_1 \in adom \ldots \mathbf{Q} y_m \in adom \;\mathrm{BC}\big(\alpha_i(\vec{x}, \vec{y}, z)\big),$$

where each $\mathbf{Q}$ is either $\exists$ or $\forall$, and:

1. $\mathrm{BC}\big(\alpha_i(\vec{x}, \vec{y}, z)\big)$ is a Boolean combination of atomic formulae $\alpha_1, \ldots, \alpha_s$;
2. each $\mathrm{FO}(\sigma)$ atomic formula is of the form $S(\vec{u})$, where $\vec{u} \subseteq \vec{y}$;
3. all atomic $\mathrm{FO}(\mathbf{R})$ formulae are of the form $p(\vec{x}, \vec{y}, z) = 0$ or $p(\vec{x}, \vec{y}, z) > 0$, where p is a polynomial; and

4. $n, m > 0$, and at least one of the FO(**R**) atomic formulae involves a multivariate polynomial $p(\vec{x}, \vec{y}, z) = y_i - z$ for some y_i.

   The reason for this is that, under the assumption $adom(\mathfrak{A}) \neq \emptyset$, we can always replace β by

$$\beta \wedge \Big(\exists y \in adom\ (y - y = 0) \wedge ((y - z = 0) \vee \neg(y - z = 0))\Big).$$

   Putting the resulting formula in the prenex normal form fulfills the conditions listed in this item.

We now assume that $\alpha_i(\vec{x}, \vec{y}, z)$, $1 \leq i \leq n$, are FO(**R**) atomic formulae $p_i(\vec{x}, \vec{y}, z)\left\{\begin{matrix} = \\ > \end{matrix}\right\}0$, and α_i, $n < i \leq s$, are FO(σ) atomic formulae. We let d_i be the degree, in z, of p_i. For each $\vec{a}, \vec{b}$, by $p_i^{\vec{a},\vec{b}}(z)$ we denote the univariate polynomial $p_i(\vec{a}, \vec{b}, z)$. Note that the degree of $p_i^{\vec{a},\vec{b}}$ is at most d_i. We let $d = \max_i d_i$. Whenever we refer to the jth root of a univariate polynomial p, we mean its jth real root in the usual ordering, if such a root exists, and 0 otherwise. Note that there exists an FO(**R**) formula $\text{root}_p^j(x)$ which holds iff x is the jth root of p.

We now prove the following.

Lemma 13.20. *Let $\varphi(\vec{x})$ be as above, where the assumptions 1–4 hold. Let $\mathfrak{A}$ be such that $adom(\mathfrak{A}) \neq \emptyset$. Fix a tuple of real numbers $\vec{a}$. Then $(\mathbf{R}, \mathfrak{A}) \models \varphi(\vec{a})$ iff there exist $i, k \leq n$, and $j, l \leq d$ and two tuples $\vec{b}, \vec{c}$ over $adom(\mathfrak{A})$ of length $|\vec{y}|$, such that*

$$(\mathbf{R}, \mathfrak{A}) \models \beta\left(\vec{a}, \frac{r_{ij}^{\vec{a},\vec{b}} + r_{kl}^{\vec{a},\vec{c}}}{2}\right) \ \vee\ \beta(\vec{a}, r_{ij}^{\vec{a},\vec{b}} + 1) \ \vee\ \beta(\vec{a}, r_{ij}^{\vec{a},\vec{b}} - 1),$$

where $r_{ij}^{\vec{a},\vec{b}}$ is the jth root of $p_i^{\vec{a},\vec{b}}$ and $r_{kl}^{\vec{a},\vec{c}}$ is the kth root of $p_j^{\vec{a},\vec{c}}$.

Proof of Lemma 13.20. One direction is trivial: if there is a witness of a given form, then there is a witness. For the other direction, assume that $(\mathbf{R}, \mathfrak{A}) \models \varphi(\vec{a})$. We then must show that there exists $a_0 \in \mathbb{R}$ of the form $\frac{r_{ij}^{\vec{a},\vec{b}} + r_{kl}^{\vec{a},\vec{c}}}{2}$ or $r_{ij}^{\vec{a},\vec{b}} \pm 1$ such that $(\mathbf{R}, \mathfrak{A}) \models \beta(\vec{a}, a_0)$.

Let $\vec{b}_1, \ldots, \vec{b}_M$ be the enumeration of all the tuples of length $|\vec{y}|$ consisting of elements of $adom(\mathfrak{A})$. Consider all univariate polynomials $p_i^{\vec{a},\vec{b}_j}(z)$, and let r_{ijk} be the kth root of $p_i^{\vec{a},\vec{b}_j}(z)$, for $k \leq d$. Let S be the family of all elements of the form $r_{ijk}, i \leq n, j \leq M, k \leq d$. It follows from our assumptions that $S \neq \emptyset$ and $adom(\mathfrak{A}) \subseteq S$, since one of the polynomials is $y_i - z$. We let $r_{\min}$ and $r_{\max}$ be the minimum and the maximum elements of S, respectively.

Suppose $(\mathbf{R}, \mathfrak{A}) \models \beta(\vec{a}, a_0)$. If $a_0 \in S$, then there is a polynomial p_i, a tuple $\vec{b}$, and $j \leq d$ such that $a_0 = r_{ij}^{\vec{a},\vec{b}}$. By selecting $\vec{c} = \vec{b}, k = i, l = j$, we see that a_0 is of the required form.

Assume $a_0 \notin S$. There are three possible cases:

1. $a_0 < r_{\min}$, or
2. $a_0 > r_{\max}$, or
3. there exist $r_1, r_2 \in S$ such that $r_1 < a_0 < r_2$, and there is no other $r \in S$ with $r_1 < r < r_2$.

We claim that for every p_i and every $\vec{b}_j$:

$$\begin{aligned} \operatorname{sign}\big(p_i^{\vec{a},\vec{b}_j}(a_0)\big) &= \operatorname{sign}\big(p_i^{\vec{a},\vec{b}_j}(r_{\min}-1)\big) \text{ in case 1} \\ \operatorname{sign}\big(p_i^{\vec{a},\vec{b}_j}(a_0)\big) &= \operatorname{sign}\big(p_i^{\vec{a},\vec{b}_j}(r_{\max}+1)\big) \text{ in case 2} \\ \operatorname{sign}\big(p_i^{\vec{a},\vec{b}_j}(a_0)\big) &= \operatorname{sign}\left(p_i^{\vec{a},\vec{b}_j}\big(\frac{r_1+r_2}{2}\big)\right) \text{ in case 3.} \end{aligned} \tag{13.7}$$

Indeed, in the third case, suppose $\operatorname{sign}\big(p_i^{\vec{a},\vec{b}_j}(a_0)\big) \neq \operatorname{sign}\big(p_i^{\vec{a},\vec{b}_j}(\frac{r_1+r_2}{2})\big)$. Then the interval $[a_0, \frac{r_1+r_2}{2}]$ contains a real root of $p_i^{\vec{a},\vec{b}_j}(z)$, which then must be in S. We conclude that there is an element of S between r_1 and r_2, a contradiction. The other two cases are similar.

Let a_1 be $(r_{\min}-1)$ for case 1, $(r_{\max}+1)$ for case 2, and $\frac{r_1+r_2}{2}$ for case 3. Then for every tuple $\vec{b}_j, j \leq M$, and every atomic formula α_i, we have

$$\alpha_i(\vec{a}, \vec{b}_j, a_0) \leftrightarrow \alpha_i(\vec{a}, \vec{b}_j, a_1). \tag{13.8}$$

This follows from (13.7) and the fact that FO(σ) atomic formulae may not contain variable z.

We can now use (13.8) to conclude that $\beta(\vec{a}, a_0) \leftrightarrow \beta(\vec{a}, a_1)$. Clearly, the equivalence (13.8) propagates through Boolean combinations of formulae. Furthermore, notice that if for a finite set A and $m > 0$, $\alpha(\vec{a}, b, \vec{b}, a_0) \leftrightarrow \alpha(\vec{a}, b, \vec{b}, a_1)$ for every $b \in A$ and every $\vec{b} \in A^m$, then

$$(\exists x \in A\ \alpha(\vec{a}, x, \vec{b}, a_0)) \leftrightarrow (\exists x \in A\ \alpha(\vec{a}, x, \vec{b}, a_1))$$

for every $\vec{b} \in A^m$. This shows that (13.8) propagates through active-domain quantification, and hence $\beta(\vec{a}, a_0) \leftrightarrow \beta(\vec{a}, a_1)$.

Thus, if $(\mathbf{R}, \mathfrak{A}) \models \beta(\vec{a}, a_0)$, then $(\mathbf{R}, \mathfrak{A}) \models \beta(\vec{a}, a_1)$. Since a_1 is of the right form (either $r-1$, or $r+1$ for $r \in S$, or $\frac{r+r'}{2}$ for $r, r' \in S$), this concludes the proof of the lemma.

To conclude the proof of the theorem, we note that Lemma 13.20 can be translated into an FO definition as follows. For each FO($\mathbf{R}$) atomic formula $\alpha(\vec{x}, \vec{y}, z)$, and for any two tuples $\vec{u}, \vec{v}$ of the same length as $\vec{y}$, we define the following formulae:

- $\alpha^{1/2}_{ikjl}(\vec{x},\vec{y},\vec{u},\vec{v})$, for $i,k \leq n, j,l \leq d$, says that $\alpha(\vec{x},\vec{y},z)$ holds when z is equal to $\frac{r^{\vec{x},\vec{u}}_{ij}+r^{\vec{x},\vec{v}}_{kl}}{2}$. That is,

$$\exists z \exists z_1 \exists z_2 \left(\text{root}^j[p^{\vec{x},\vec{u}}_i](z_1) \wedge \text{root}^l[p^{\vec{x},\vec{v}}_k](z_2) \wedge (2z = z_1 + z_2) \wedge \alpha(\vec{x},\vec{y},z)\right).$$

- $\alpha^{+}_{ij}(\vec{x},\vec{y},\vec{u})$ for $i \leq n, j \leq d$, says that $\alpha(\vec{x},\vec{y},z)$ holds for $z = r^{\vec{x},\vec{u}}_{ij} + 1$; that is,

$$\exists z \exists z_1 \left(\text{root}^j[p^{\vec{x},\vec{u}}_i](z_1) \wedge (z = z_1 + 1) \wedge \alpha(\vec{x},\vec{y},z)\right).$$

- $\alpha^{-}_{ij}(\vec{x},\vec{y},\vec{u})$ for $i \leq n, j \leq d$, says that $\alpha(\vec{x},\vec{y},z)$ holds for $z = r^{\vec{x},\vec{u}}_{ij} - 1$; the FO definition is similar to the one given above, except that we use a conjunct $z = z_1 - 1$.

Note that by quantifier elimination for $\mathbf{R}$, we may assume that all formulae $\alpha^{1/2}_{ikjl}(\vec{x},\vec{y},\vec{u},\vec{v})$, $\alpha^{+}_{ij}(\vec{x},\vec{y},\vec{u})$, and $\alpha^{-}_{ij}(\vec{x},\vec{y},\vec{u})$ are quantifier-free.

For $i,k \leq n$, and $j,l \leq d$, let $\gamma^{1/2}_{ikjl}(\vec{x},\vec{y},\vec{u},\vec{v})$ be the Boolean combination $\text{BC}(\alpha_s)$ where each atomic $\text{FO}(\mathbf{R})$ formula α is replaced by $\alpha^{1/2}_{ikjl}(\vec{x},\vec{y},\vec{u},\vec{v})$. Let $\beta^{1/2}_{ijkl}(\vec{x},\vec{u},\vec{v})$ be

$$\mathbf{Q}y_1 \in adom \ldots \mathbf{Q}y_m \in adom \ \gamma^{1/2}_{ikjl}(\vec{x},\vec{y},\vec{u},\vec{v}).$$

Likewise, we define $\gamma^{+}_{ij}(\vec{x},\vec{y},\vec{u})$ to be the Boolean combination $\text{BC}(\alpha_s)$ where each atomic $\text{FO}(\mathbf{R})$ formula α is replaced by $\alpha^{+}_{ij}(\vec{x},\vec{y},\vec{u})$, and let $\beta^{+}_{ij}(\vec{x},\vec{u})$ be $\gamma^{+}_{ij}(\vec{x},\vec{y},\vec{u})$ preceded by the quantifier prefix of β. Finally, we define $\beta^{-}_{ij}(\vec{x},\vec{u})$ as $\beta^{+}_{ij}(\vec{x},\vec{u})$, except by using formulae $\alpha^{-}_{ij}(\vec{x},\vec{y},\vec{u})$.

Now Lemma 13.20 says that $\exists z\ \beta(\vec{x},z)$ is equivalent to

$$\exists \vec{u} \in adom \exists \vec{v} \in adom \bigvee_{i,k \leq n} \bigvee_{j,l \leq d} \left(\beta^{1/2}_{ijkl}(\vec{x},\vec{u},\vec{v}) \ \vee \ \beta^{+}_{ij}(\vec{x},\vec{u}) \ \vee \ \beta^{-}_{ij}(\vec{x},\vec{u})\right),$$

which is an $\text{FO}_{\text{act}}(\mathbf{R},\sigma)$ formula.

This completes the proof of the translation for the case of structures $\mathfrak{A}$ with $adom(\mathfrak{A}) \neq \emptyset$. To deal with empty structures $\mathfrak{A}$, consider a formula $\varphi(\vec{x})$, and let $\varphi'(\vec{x})$ be an $\text{FO}(\mathbf{R})$ formula obtained from $\varphi(\vec{x})$ by replacing each atomic $\text{FO}(\sigma)$ subformula by *false*. Note that if $adom(\mathfrak{A}) = \emptyset$, then $(\mathbf{R},\mathfrak{A}) \models \varphi(\vec{a})$ iff $\mathbf{R} \models \varphi'(\vec{a})$. By quantifier elimination, we may assume that φ' is quantifier-free. Hence, φ is equivalent to

$$\left(\neg \exists y \in adom(y = y) \wedge \varphi'(\vec{x})\right) \ \vee \ \left(\exists y \in adom(y = y) \wedge \varphi_{\text{act}}(\vec{x})\right), \tag{13.9}$$

where φ_{act} is constructed by the algorithm for the case of nonempty structures. Clearly, (13.9) will work for both empty and nonempty structures. Since (13.9) is an $\text{FO}_{\text{act}}(\mathbf{R},\sigma)$ formula, this completes the proof. □

Corollary 13.21. *Every generic query in* $\mathrm{FO}(\mathbf{R}, \sigma)$ *is expressible in* $\mathrm{FO}_{\mathrm{act}}(\langle \mathbb{R}, < \rangle, \sigma)$*. In particular, every such query is local, and* EVEN *is not expressible in* $\mathrm{FO}(\mathbf{R}, \sigma)$. □

What other structures have RQC? There are many known examples, some of them presented as exercises at the end of the chapter. It follows immediately from Theorem 13.19 that $\langle \mathbb{R}, +, < \rangle$ has RQC. Another example is given by $\langle \mathbb{R}, +, \cdot, e^x \rangle$, the expansion of the real field with the function $x \mapsto e^x$. The field of complex numbers is known to have RQC, as well as several structures on finite strings. See Exercises 13.10 – 13.14.

13.5 The Random Graph and Collapse to MSO

The real field is a structure with a decidable theory. So is the structure $\mathfrak{Z} = \langle \mathbb{Z}, +, < \rangle$, which also admits RQC (see Exercise 13.10). In fact both admit quantifier elimination: for $\mathfrak{Z}$, one has to add all the definable relations $(x - y) \bmod k = 0$, as well as constant 1.

Could it be true that one can guarantee RQC for every structure $\mathfrak{M}$ with decidable theory? We give a negative answer here, which establishes a different kind of collapse: of $\mathrm{FO}(\mathfrak{M}, \sigma)$ to MSO under the active-domain semantics.

The structure is the random graph $\mathcal{RG} = \langle U, E \rangle$, introduced in Chap. 12. This is any undirected graph on a countably infinite set U that satisfies every sentence that is true in almost all finite undirected graphs. Recall that the set of all such sentences forms a complete theory with infinite models, and that this theory is decidable and ω-categorical.

The random graph satisfies the extension axioms $EA_{n,m}$ (12.2), for each $n \geq m \geq 0$. These say that for every finite n-element subset S of U, and an m-element subset T of S, there exists $z \notin S$ such that $(z, x) \in E$ for all $x \in T$, and $(z, x) \notin E$ for all $x \in S - T$.

Recall that MSO (see Chap. 7), is a restriction of second-order logic in which second-order variables range over sets. We define $\mathrm{MSO}_{\mathrm{act}}(\mathfrak{M}, \sigma)$ as MSO over the vocabulary that consists of both Ω and σ, every first-order quantifier is an active-domain quantifier (i.e., $\exists x \in adom$ or $\forall x \in adom$), and every MSO quantifier is restricted to the active domain. We write such MSO quantifiers as $\exists X \subseteq adom$ or $\forall X \subseteq adom$. The semantics is as follows: $(\mathfrak{M}, \mathfrak{A}) \models \exists X \subseteq adom\ \varphi(X, \cdot)$ if for some set $C \subseteq adom(\mathfrak{A})$, it is the case that $(\mathfrak{M}, \mathfrak{A}) \models \varphi(C, \cdot)$.

Theorem 13.22. *For every* σ,

$$\mathrm{FO}(\mathcal{RG}, \sigma) = \mathrm{MSO}_{\mathrm{act}}(\mathcal{RG}, \sigma).$$

Proof. The idea is to use the extension axioms to model MSO queries. Consider an $\mathrm{MSO}_{\mathrm{act}}$ formula $\varphi(\vec{x})$

$$\mathbf{Q}X_1 \subseteq adom \ldots \mathbf{Q}X_m \subseteq adom\ \mathbf{Q}y_1 \in adom \ldots \mathbf{Q}y_n \in adom\ \alpha(\vec{X}, \vec{x}, \vec{y}),$$

where the X_i's are second-order variables, the y_j's are first-order variables, and α is a Boolean combination of σ- and $\mathcal{RG}$-formulae in variables $\vec{x}, \vec{y}$, and formulae $X_i(x_j)$ and $X_i(y_j)$. Construct a new FO($\mathcal{RG}, \sigma$) formula $\varphi'(\vec{x})$ by replacing each $\mathbf{Q}X_i \subseteq adom$ with $\mathbf{Q}z_i \notin adom \cup \vec{x}$ (which is FO-definable), and changing every atomic subformula $X_i(u)$ to $E(z_i, u)$. In other words, a subset X_i of the active domain is identified by an element z_i from which there are edges to all elements of X_i, and no edges to the elements of the active domain which do not belong to X_i. It is then easy to see, from the extension axioms, that φ' is equivalent to φ. Hence, $\text{MSO}_{\text{act}}(\mathcal{RG}, \sigma) \subseteq \text{FO}(\mathcal{RG}, \sigma)$.

For the other direction, proceed by induction on the FO($\mathcal{RG}, \sigma$) formulae. The only nontrivial case is that of unrestricted existential quantification. Suppose we have an $\text{MSO}_{\text{act}}(\mathcal{RG}, \sigma)$ formula

$$\varphi(\vec{x}, z) \equiv \mathbf{Q}\vec{X} \subseteq adom\ \mathbf{Q}\vec{y} \in adom\ \alpha(\vec{X}, \vec{x}, \vec{y}, z),$$

where $\vec{x} = (x_1, \ldots, x_n)$, and α again is a Boolean combination of atomic σ- and $\mathcal{RG}$-formulae, as well as formulae $X_i(u)$, where u is one of the first-order variables $z, \vec{x}, \vec{y}$. We want to find an MSO_{act} formula equivalent to $\exists z\ \varphi$.

Such a formula is a disjunction of the form

$$\exists z \in adom\ \varphi \ \vee\ \bigvee_i \varphi(\vec{x}, x_i) \ \vee\ \exists z \notin adom\ \varphi.$$

Both $\exists z \in adom\ \varphi$ and $\varphi(\vec{x}, x_i)$ are $\text{MSO}_{\text{act}}(\mathcal{RG}, \sigma)$ formulae. To eliminate z from $\exists z \notin adom\ \varphi$, all we have to know about z is its connections to $\vec{x}$ and to the active domain in the random graph; the former is taken care of by a disjunction listing all subsets of $\{1, \ldots, n\}$, and the latter by a second-order quantifier over the active domain. For $I \subseteq \{1, \ldots, n\}$, let $\chi_I(\vec{x})$ be a quantifier-free formula saying that no x_i, x_j with $i \in I, j \notin I$, could be equal. We introduce a new second-order variable Z and define an MSO_{act} formula $\psi(\vec{x})$ as

$$\exists Z \subseteq adom \bigvee_{I \subseteq \{1,\ldots,n\}} \Big(\chi_I(\vec{x}) \wedge \mathbf{Q}\vec{X} \subseteq adom\ \mathbf{Q}\vec{y} \in adom\ \alpha_I^Z(\vec{X}, Z, \vec{x}, \vec{y})\Big),$$

where $\alpha_I^Z(\vec{X}, Z, \vec{x}, \vec{y})$ is obtained from α by:

1. replacing each $E(z, x_i)$ by *true* for $i \in I$ and *false* for $i \notin I$,
2. replacing each $E(z, y_j)$ by $Z(y_j)$, and
3. replacing each $X_i(z)$ by *false*.

The extension axioms then ensure that ψ is equivalent to $\exists z \notin adom\ \varphi$. □

The active-generic collapse, as it turns out, can be extended to MSO.

Proposition 13.23. *Every generic query in* $\text{MSO}_{\text{act}}(\mathcal{RG}, \sigma)$ *is expressible in* MSO *over* σ*-structures.*

Proof. First, we notice that there exists an infinite subset Z of $\mathcal{RG}$ such that for every pair $a, b \in Z$, there is no edge between a and b (such a subset is easy to construct using one of the concrete representations of the random graph). Next, we show by induction on the formulae that for every $\text{MSO}_{\text{act}}(\mathcal{RG}, \sigma)$ formula $\varphi(\vec{X}, \vec{x})$ and every infinite set $Z' \subseteq Z$, there is an infinite set $Z'' \subseteq Z$ and an MSO formula $\varphi'(\vec{X}, \vec{x})$ of vocabulary σ such that for every σ-structure $\mathfrak{A}$, and an interpretation of $\vec{x}, \vec{X}$ as $\vec{c}, \vec{C}$ over $adom(\mathfrak{A})$,

$$(\mathcal{RG}, \mathfrak{A}) \models \varphi(\vec{C}, \vec{c}) \leftrightarrow \varphi'(\vec{C}, \vec{c}).$$

Indeed, atomic formulae $E(x, y)$ can be replaced by *false*. The rest of the proof is exactly the same as the proof of Lemma 13.15: the active-domain MSO quantifiers are handled exactly as the active-domain FO quantifiers.

Next, the same proof as in Lemma 13.16 shows that if φ defines a generic query, then it is equivalent to φ' over all σ-structures. This proves the proposition. □

Corollary 13.24. *The class of generic queries expressible in* $\text{FO}(\mathcal{RG}, \sigma)$ *is precisely the class of queries definable in* MSO *over* σ*-structures.* □

Thus, $\mathcal{RG}$ provides an example of a structure with quantifier elimination and decidable first-order theory (see Exercise 12.8) that does not admit RQC, but at the same time, one can establish meaningful bounds on the expressiveness of queries. For example, each generic query in $\text{FO}(\mathcal{RG}, \sigma)$ can be evaluated in PH, and string languages definable in $\text{FO}(\mathcal{RG}, \sigma)$ are precisely the regular languages.

13.6 An Application: Constraint Databases

The framework of constraint databases can be described formally as the logic $\text{FO}(\mathfrak{M}, \sigma)$, where each m-relation S in σ is interpreted not as a finite set, but as a *definable* subset of U^m. That is, there is a formula $\alpha_S(x_1, \ldots, x_m)$ of $\text{FO}(\mathfrak{M})$ such that S is the set $\{\vec{a} \mid \mathfrak{M} \models \alpha_S(\vec{a})\}$.

The main application of constraint databases is in querying spatial information. The key idea of constraint databases is that regions are represented by FO formulae over some underlying structure: typically either the real field $\mathbf{R}$, or $\mathbf{R}_{\text{lin}} = \langle \mathbb{R}, +, -, 0, 1, < \rangle$. That is, they are described by polynomial or linear constraints over the reals.

To illustrate how linear constraints can be used to describe a specific spatial database, consider the following example, representing an approximate map of Belgium (a real map will have many more constraints, but the basic ideas are the same). Fig. 13.1 shows the map itself, while Fig. 13.2 shows how regions and cities are described by constraints.

One can then use $\text{FO}(\mathbf{R}, \sigma)$ or $\text{FO}(\mathbf{R}_{\text{lin}}, \sigma)$ to query those databases as if they were usual relational databases that store infinitely many points. For

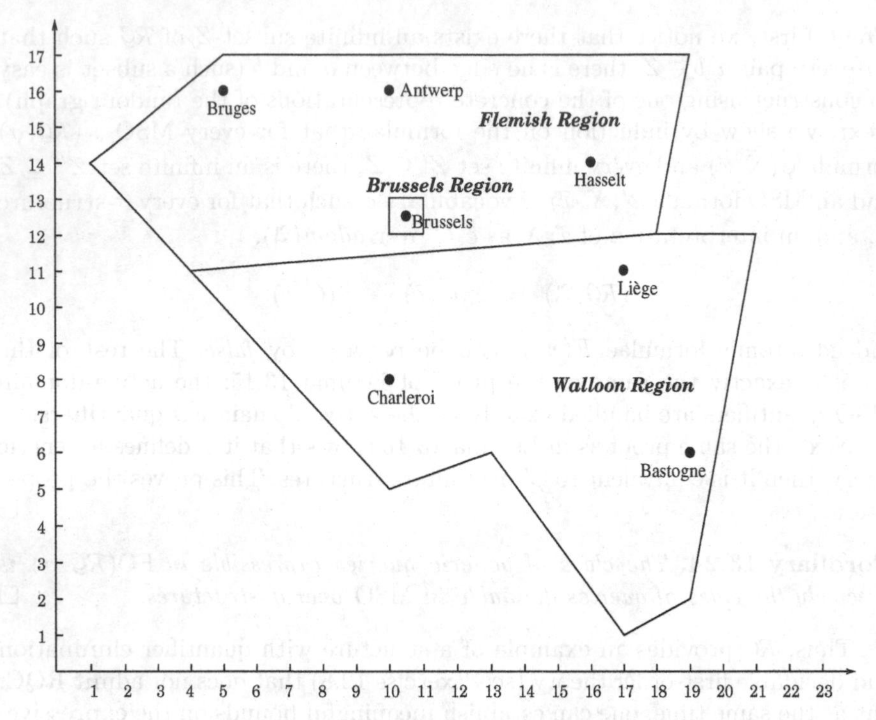

Fig. 13.1. Spatial information map of Belgium

example, to find all points in the Walloon region that are east of Hasselt one would write

$$\varphi(x, y) \;=\; \text{Walloon}(x, y) \wedge \exists u, v \left(\text{Hasselt}(u, v) \wedge x > u\right). \tag{13.10}$$

To find all the points in the Walloon region that are on the direct line from Hasselt to Liège, one writes a formula $\varphi(x, y)$ as the conjunction of Walloon(x, y) and

$$\exists u, v, s, t, \lambda \begin{pmatrix} \text{Hasselt}(u, v) \wedge \text{Liége}(s, t) \\ \wedge\, 0 \leq \lambda \;\wedge\; \lambda \leq 1 \\ \wedge\, x = \lambda u + (1 - \lambda) s \\ \wedge\, y = \lambda v + (1 - \lambda) t \end{pmatrix}. \tag{13.11}$$

In these examples, (13.10) is an $\mathrm{FO}(\langle \mathbb{R}, < \rangle, \sigma)$ query, while (13.11) needs to be expressed in the more expressive language $\mathrm{FO}(\mathbf{R}, \sigma)$.

We now give one simple application of embedded finite models to constraint databases. A basic property of regions is their topological connectivity. Most regions represented in geographical databases are connected (and the few examples of unconnected ones to be rather well known, as they usually lead to nasty political problems). But can we test this property in FO-based query

Cities

Name	*Geometry*
Antwerp	$(x = 10) \wedge$ $(y = 16)$
Bastogne	$(x = 19) \wedge$ $(y = 6)$
Bruges	$(x = 5) \wedge$ $(y = 16)$
Brussels	$(x = 10.5) \wedge$ $(y = 12.5)$
Charleroi	$(x = 10) \wedge$ $(y = 8)$
Hasselt	$(x = 16) \wedge$ $(y = 14)$
Liège	$(x = 17) \wedge$ $(y = 11)$

Regions

Name	*Geometry*
Brussels	$(y \leq 13) \wedge (x \leq 11) \wedge$ $(y \geq 12) \wedge (x \geq 10)$
Flanders	$(y \leq 17) \wedge (5x - y \leq 78) \wedge$ $(x - 14y \leq -150) \wedge$ $(x + y \geq 45) \wedge$ $(3x - 4y \geq -53) \wedge$ $(\neg((y \leq 13) \wedge (x \leq 11) \wedge$ $\wedge (y \geq 12) \wedge (x \geq 10)))$
Walloon	$((x - 14y \geq -150) \wedge (y \leq 12) \wedge$ $(19x + 7y \leq 375) \wedge$ $(x - 2y \leq 15) \wedge (x \geq 13) \wedge$ $(5x + 4y \geq 89)) \vee$ $((3y - x \geq 5) \wedge (x + y \geq 45) \wedge$ $(x - 14y \geq -150) \wedge (x \geq 13))$

Fig. 13.2. A spatial database of Belgium

languages? We now give a simple proof of the negative answer, by reduction to collapse results.

Theorem 13.25. *Topological connectivity is not expressible in* FO$(\mathbf{R}, \sigma)$.

Proof. Assume, to the contrary, that topological connectivity of sets in $\mathbb{R}^3$ is definable (one can show that connectivity of sets on the plane is undefinable as well; the proof involves a slightly more complicated reduction and is the subject of Exercise 13.5). We show that graph connectivity is then definable.

Suppose we have a finite undirected graph G with $adom(G) \subset \mathbb{R}$. For each edge (a, b) in G, we define the segment $s(a, b)$ in $\mathbb{R}^3$ between $(a, 1, 0)$ and $(0, 0, b)$. Each point in $s(a, b)$ is of the form $(\lambda a, \lambda, (1 - \lambda)b)$ for some $0 \leq \lambda \leq 1$. Note that this implies that $s(a, b) \cap s(c, d) \neq \emptyset$ can only happen if $a = c$ or $b = d$, since $(\lambda a, \lambda, (1 - \lambda)b) = (\mu c, \mu, (1 - \mu)d)$ implies $\lambda = \mu$ and thus for $\lambda \neq 0, 1$ we have $a = c$ and $b = d$, for $\lambda = 0$ we get $b = d$, and for $\lambda = 1$ we get $a = c$.

Now we encode each edge (a, b) by the set $e(a, b) = s(a, b) \cup s(b, a) \cup s(a, a) \cup s(b, b)$ (see Fig. 13.3). Note that $e(a, b)$ is a connected set, and that $e(a, b) \cap e(c, d) \neq \emptyset$ iff the edges (a, b) and (c, d) have a common node.

We then define a new set X_G in $\mathbb{R}^3$ as

$$X_G = \bigcup_{(a,b) \in G} e(a, b).$$

It follows that X_G is topologically connected iff G is connected as a graph. Since the transformation $G \to X_G$ is definable in FO$(\mathbf{R}, \sigma)$, the assumption

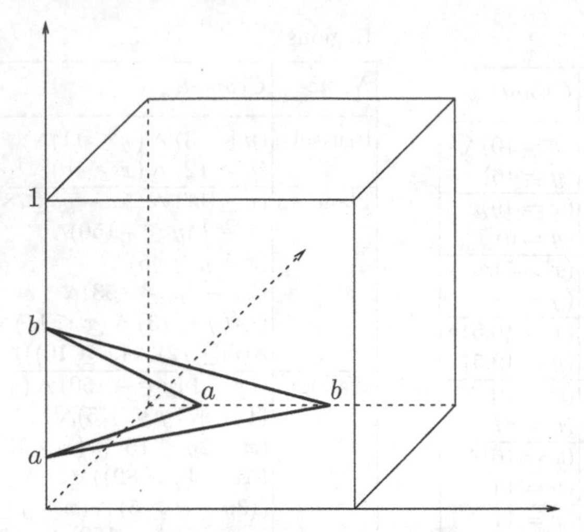

Fig. 13.3. Embedding an edge (a, b) into $\mathbb{R}^3$

that topological connectivity is definable implies that so is graph connectivity. However, we know from Corollary 13.21 that graph connectivity cannot be expressed. This contradiction proves the theorem. $\square$

13.7 Bibliographic Notes

The framework of embedded finite models originated in database theory, in connection with attempts to understand query languages that use interpreted operations, as well as query languages for constraint databases. Constraint databases were introduced by Kanellakis, Kuper, and Revesz [142] (see also the surveys by Kuper, Libkin, and Paredaens [158], Libkin [168], and Van den Bussche [242]).

Soon after [142] was published, it became clear that many questions about languages for constraint databases reduce to questions about embedded finite models. For example, Grumbach and Su [115] present many reductions to the finite case.

Collapse results as a technique for proving bounds on $\mathrm{FO}(\mathfrak{M}, \sigma)$ were introduced by Paredaens, Van den Bussche, and Van Gucht [197], where the restrcited-quantifier collapse for $\mathbf{R}_{\text{lin}}$ was proved. The collapse for the real field was shown by Benedikt and Libkin [19] (in fact the proof in [19] applies to a larger class of *o-minimal* structures; see [243]). The active-generic collapse was shown by Otto and Van den Bussche [193]; the proof given here follows [19]. For the basics of Ramsey theory, see Graham, Rothschild, and Spencer [103]. The collapse to MSO over the random graph is from [168], although one direction was proved earlier by [193].

Inexpressibility of connectivity by reduction to the finite case was first shown in [115]; for a different approach that characterizes topological properties expressible in FO($\mathbf{R}$, $\{S\}$), where S is binary, see Kuijpers, Paredaens, and Van den Bussche [157]. For a study of these problems over complex numbers, we refer to Chapuis and Koiran [36]. See also Exercise 13.6.

Although we said in the beginning of the chapter that no collapse results were proved with the help of Ehrenfeucht-Fraïssé games, results by Fournier [83] show how to use games to establish bounds on the quantifier rank for expresssing certain properties over embedded finite models. An example is presented in Exercise 13.8.

In this chapter we used a number of well-known results in classical model theory, such as decidability and quantifier elimination for the real field $\mathbf{R}$ (see Tarski [229]) and undecidability of the FO theory of $\langle \mathbb{Q}, +, \cdot \rangle$ (see Robinson [206]).

Sources for exercises:

Exercise 13.4:	Benedikt and Libkin [19]
Exercise 13.6:	Chapuis and Koiran [36]
Exercise 13.7:	Grumbach and Su [115]
Exercise 13.8:	Fournier [83]
Exercise 13.9:	Hull and Su [127]
Exercise 13.10:	Flum and Ziegler [82] (see also [168] for a self-contained proof)
Exercise 13.11:	Benedikt and Libkin [19]
Exercise 13.12:	Flum and Ziegler [82]
Exercise 13.13:	Barrington et al. [15]
Exercises 13.14–13.16:	Benedikt et al. [21]

13.8 Exercises

Exercise 13.1. Give an example of a noncomputable query expressible in FO($\mathfrak{N}, \sigma$).

Exercise 13.2. Prove that it is undecidable if a query expressible in FO($\mathfrak{M}, \sigma$) is generic (even if the theory of $\mathfrak{M}$ is decidable).

Exercise 13.3. Suppose that S is a binary relation symbol, and R is a ternary one, and both are interpreted as sets definable over the real field $\mathbf{R} = \langle \mathbb{R}, +, \cdot, 0, 1, < \rangle$. Show how to express the following in FO($\mathbf{R}$, $\{S, R\}$):

- S is a graph of a function $f : \mathbb{R} \to \mathbb{R}$;
- S is a graph of a continuous function $f : \mathbb{R} \to \mathbb{R}$;
- S is a graph of a differentiable function $f : \mathbb{R} \to \mathbb{R}$;
- R is a trajectory of an object: that is, a triple $(x, y, t) \in R$ gives a position (x, y) at time t;
- a formula $\varphi(x, y, v)$ which holds iff v is the speed of the object at time t (assuming that R defines a trajectory).

Exercise 13.4. Prove a generalization of the Ramsey property (i.e., each active-semantics sentence expressing a generic query can be written using just the order relation) for SO, ∃SO, FO(**Cnt**), and a fixed point logic of your choice. Also prove that $\mathcal{L}^{\omega}_{\infty\omega}$ does not have such a generalized Ramsey property.

Exercise 13.5. Use a reduction different from the one in the proof of Theorem 13.25 to show that topological connectivity of subsets of $\mathbb{R}^2$ is not definable in FO($\mathbf{R}, \{S\}$), where S is binary.

Exercise 13.6. Prove that topological connectivity of subsets of $\mathbb{C}^2$ which are definable in $\langle \mathbb{C}, +, -, \cdot, 0, 1\rangle$ cannot be expressed in FO($\langle \mathbb{C}, +, -, \cdot, 0, 1\rangle, \{S\}$), where S is binary.

Exercise 13.7. Prove that if S and S' are interpreted as subsets of $\mathbb{R}^2$ definable in $\mathbf{R}$, then none of the following is expressible in FO($\mathbf{R}, \{S, S'\}$):

- S contains at least one hole (assuming S is a closed set).
- S has a Eulerian traversal. That is, if S is a union of line segments, then it has a traversal going through each line segment exactly once.
- S and S' are homeomorphic.

Use reductions to the finite case for all three problems.

Exercise 13.8. Show that in FO($\mathbf{R}, \sigma$) one can express EVEN for sets of cardinality up to n using a sentence of quantifier rank $O(\sqrt{\log n})$.

Exercise 13.9. Prove the natural-active collapse for $\mathfrak{U}_\emptyset = \langle U, \emptyset\rangle$.

Exercise 13.10. Prove the restricted quantifier collapse for $\langle \mathbb{Z}, +, <\rangle$.

Exercise 13.11. An ordered structure $\mathfrak{M} = \langle U, \Omega, <\rangle$ is called *o-minimal* if every definable subset of U is a finite union of points and open intervals $(a, b), (-\infty, a), (a, \infty)$.

Prove the restricted quantifier collapse for an arbitrary o-minimal structure.

Hint: you will need the following uniform bounds result of Pillay and Steinhorn [198]. If $\varphi(x, \vec{y})$ is an FO($\mathfrak{M}$) formula, then there exists a constant k such that for every $\vec{b}$, the set $\{a \mid \mathfrak{M} \models \varphi(a, \vec{b})\}$ is a union of fewer than k points and open intervals.

One can use this result to infer that $\langle \mathbb{R}, +, \cdot, e^x\rangle$ admits the restricted quantifier collapse, since Wilkie [248] proved that it is o-minimal.

Exercise 13.12. We say that a structure $\mathfrak{M}$ has the *finite cover property* if there is a formula $\varphi(x, \vec{y})$ such that for every $n > 0$, one can find tuples $\vec{a}_1, \ldots, \vec{a}_n$ such that $\exists x \bigwedge_{j \neq i} \varphi(x, \vec{a}_j)$ holds for each $i \leq n$, but $\exists x \bigwedge_{j \leq n} \varphi(x, \vec{a}_j)$ does not hold.

- Prove that if $\mathfrak{M}$ does *not* have the finite cover property, then it admits the restricted quantifier collapse.
- Conclude that $\langle \mathbb{C}, +, \cdot\rangle$ and $\langle \mathbb{N}, \mathit{succ}\rangle$ admit the restricted quantifier collapse.

Exercise 13.13. We say that a language $L \subseteq \Sigma^*$ has a *neutral letter* if there exists $a \in \Sigma$ such that for every two strings $s, s' \in \Sigma^*$, we have $s \cdot s' \in L$ iff $s \cdot a \cdot s' \in L$.

Now let Ω be a set of arithmetic predicates. We say that a language L is FO(Ω)-definable if there is an FO sentence Φ_L of vocabulary $\sigma_\Sigma \cup \Omega$ such that $M_s^\Omega \models \Phi_L$ iff $s \in L$. Here M_s^Ω is the structure M_s expanded with the interpretation of Ω-predicates on its universe.

The following statement is known as the *Crane Beach conjecture* for Ω: if L is FO(Ω)-definable and has a neutral letter, then it is star-free.

- Use Exercise 13.10 to prove that the Crane Beach conjecture is true when $\Omega = \{+\}$ (the graph of the addition operation).
- Prove that the Crane Beach conjecture is false when $\Omega = \{+, \times\}$ (*hint:* use Theorem 6.12).

Exercise 13.14. Consider the structure $\langle \Sigma^*, \prec, (f_a)_{a \in \Sigma} \rangle$, where $\prec$ is the prefix relation, and $f_a : \Sigma^* \to \Sigma^*$ is defined by $f_a(x) = x \cdot a$. Prove that this structure has the restricted quantifier collapse. Prove that it still has the restricted quantifier collapse when augmented with the following:

- The predicate P_L, for each regular language L, that is true of s iff s is in L.
- The functions $g_a : \Sigma^* \to \Sigma^*$ defined by $g_a(x) = a \cdot x$.

Exercise 13.15. Suppose S is an infinite set, and $\mathcal{C} \subseteq 2^S$ is a family of subsets of S. Let $F \subset S$ be finite; we say that $\mathcal{C}$ *shatters* F if the collection $\{F \cap C \mid C \in \mathcal{C}\}$ is $\wp(F)$, the powerset of F. The *Vapnik-Chervonenkis (VC) dimension* of $\mathcal{C}$ is the maximal cardinality of a finite set shattered by $\mathcal{C}$. If arbitrarily large finite sets are shattered by $\mathcal{C}$, we let the VC dimension be ∞.

If $\mathfrak{M}$ is a structure and $\varphi(\vec{x}, \vec{y})$ is an FO($\mathfrak{M}$) formula, with $|\vec{x}| = n$, $|\vec{y}| = m$, then for each $\vec{a} \in U^n$, we define $\varphi(\vec{a}, \mathfrak{M}) = \{\vec{b} \in U^m \mid \mathfrak{M} \models \varphi(\vec{a}, \vec{b})\}$, and let $F_\varphi(\mathfrak{M})$ be $\{\varphi(\vec{a}, \mathfrak{M}) \mid \vec{a} \in \mathcal{U}^n\}$. Families of sets arising in such a way are called *definable families.* We say that $\mathfrak{M}$ has finite VC dimension if every definable family in $\mathfrak{M}$ has finite VC dimension.

Prove that if $\mathfrak{M}$ admits the restricted quantifier collapse, then it has finite VC dimension.

Exercise 13.16. Consider an expansion $\mathfrak{M}$ of $\langle \Sigma^*, \prec, (f_a)_{a \in \Sigma} \rangle$ with the predicate $\mathrm{el}(x, y)$ which is true iff $|x| = |y|$. We have seen this structure in Chap. 7 (Exercise 7.20); it defines precisely the regular relations.

Prove that FO($\mathfrak{M}, \sigma$) cannot express EVEN.

Exercise 13.17.* For the structure $\mathfrak{M}$ of Exercise 13.16, is $\mathrm{FO}^{\mathrm{gen}}_{\mathrm{act}}(\mathfrak{M}, \sigma)$ contained in $\mathrm{FO}_{\mathrm{act}}(\mathfrak{U}_<, \sigma)$?

14

Other Applications of Finite Model Theory

In this final chapter, we briefly outline three different application areas of finite model theory. In mathematical logic, finite models are used as a tool for proving decidability results for satisfiability of FO sentences. In the area of temporal logics and verification, one analyzes the behavior of certain logics on some special finite structures (Kripke structures). And finally, it was recently discovered that many constraint satisfaction problems can be reduced to the existence of a homomorphism between two finite structures.

14.1 Finite Model Property and Decision Problems

The classical decision problem in mathematical logic is the satisfiability problem for FO sentences: that is,

Given a first-order sentence Φ, does it have a model?

We know that in general, satisfiability is undecidable. However, a complete classification of decidable fragments in terms of quantifier-prefix classes exists. For the rest of the section, we assume that the vocabulary is purely relational.

We have already seen classes of formulae defined by their quantifier prefixes in Sect. 12.4. For a regular expression r over the alphabet $\{\exists, \forall\}$, we denote by $\mathrm{FO}(r)$ the set of all prenex sentences

$$Q_1 x_1 \ldots Q_n x_n \; \varphi(x_1, \ldots, x_n),$$

where the string $Q_1 \ldots Q_n$ is in the language denoted by r. Here, each Q_i is either $\exists$ or $\forall$, and φ is quantifier-free.

It is known that there are precisely two maximal prefix classes for which the satisfiability problem is decidable: these are $\mathrm{FO}(\exists^*\forall^*)$ (known as the Bernays-Schönfinkel class), and $\mathrm{FO}(\exists^*\forall\exists^*)$ (known as the Ackermann class).

The proof technique in both cases relies on the following property.

Definition 14.1. *We say that a class $\mathcal{K}$ of sentences has the* finite model property *if for every sentence Φ in $\mathcal{K}$, either Φ is unsatisfiable, or it has a finite model.*

In other words, in a class $\mathcal{K}$ that has the finite model property, every satisfiable sentence has a finite model.

It turns out that both $\text{FO}(\exists^*\forall^*)$ and $\text{FO}(\exists^*\forall\exists^*)$ have the finite model property, and, furthermore, there is an upper bound on the size of a finite model of Φ in terms of $\| \Phi \|$, the size of Φ. We prove this for the Bernays-Schönfinkel class.

Proposition 14.2. *If Φ is a satisfiable sentence of* $\text{FO}(\exists^*\forall^*)$*, then it has a model whose size is at most linear in* $\| \Phi \|$.

Proof. Let Φ be

$$\exists x_1 \ldots \exists x_n \forall y_1 \ldots \forall y_m \; \varphi(\vec{x}, \vec{y}),$$

where φ is quantifier-free. Let $\psi(\vec{x})$ be $\forall \vec{y} \; \varphi(\vec{x}, \vec{y})$.

Since Φ is satisfiable, it has a model $\mathfrak{A}$. Let $a_1, \ldots, a_n$ witness the existential quantifiers: that is, $\mathfrak{A} \models \psi(\vec{a})$. Let $\mathfrak{A}'$ be the finite substructure of $\mathfrak{A}$ whose universe is $\{a_1, \ldots, a_n\}$. Since ψ is a universal formula, it is preserved under taking substructures. Hence, $\mathfrak{A}' \models \psi(\vec{a})$, and therefore, $\mathfrak{A}' \models \Phi$. Thus, we have shown that Φ has a model whose universe has at most n elements. □

This immediately gives us the decision procedure for the class $\text{FO}(\exists^*\forall^*)$: given a sentence Φ with n existential quantifiers, look at all nonisomorphic structures whose universes are of size up to n, and check if any of them is a model of Φ. This algorithm also suggests a complexity bound: one can guess a structure $\mathfrak{A}$ with $|A| \leq n$, and check if $\mathfrak{A} \models \Phi$. Notice that in terms of $\| \Phi \|$, the size of such a structure could be *exponential*. For each relation symbol R of arity m, there could be up to n^m different tuples in $R^{\mathfrak{A}}$. Since there is no a priori bound on the arity of R, it may well depend on $\| \Phi \|$, which gives us an exponential upper bound on $\| \mathfrak{A} \|$. Hence, the algorithm runs in nondeterministic exponential time.

It turns out that one cannot improve this bound.

Theorem 14.3. *The satisfiability problem for* $\text{FO}(\exists^*\forall^*)$ *is* NEXPTIME-*complete.* □

If we have a vocabulary of bounded arity (i.e., there is a constant k such that every relation symbol has arity at most k), then the size of a structure on n elements is at most polynomial in n. Thus, in this case one has to check if $\mathfrak{A} \models \varphi$, where $\| A \|$ is polynomial in n. As we know from the results on the combined complexity of FO, this can be done in PSPACE. Hence, for a vocabulary of bounded arity, the satisfiability problem for $\text{FO}(\exists^*\forall^*)$ is in PSPACE.

We now see an application of this decidability result in database theory. In Chap. 6, we studied conjunctive queries: those of the form $\exists \vec{x} \varphi$, where φ is a conjunction of atomic formulae. We also saw (Exercise 6.19) that containment of conjunctive queries is NP-complete.

Another class of queries often used in database theory is *unions of conjunctive queries*; that is, queries of the form $Q_1 \cup \ldots \cup Q_m$, where each Q_i is a conjunctive query. Can the decidability of containment be extended to union of conjunctive queries? That is, is it decidable whether $Q(\mathfrak{A}) \subseteq Q'(\mathfrak{A})$ for all $\mathfrak{A}$, when Q and Q' are unions of conjunctive queries? We now give the positive answer using the decidability of the Bernays-Schönfinkel class.

Putting all existential quantifiers in front, we can assume without loss of generality that Q is given by $\varphi(\vec{x}) \equiv \exists \vec{y}\, \alpha(\vec{x}, \vec{y})$, and Q' by $\psi(\vec{x}) \equiv \exists \vec{y}\, \beta(\vec{x}, \vec{y})$, where α and β are monotone Boolean combinations of atomic formulae. Our goal is to check whether $\Phi \equiv \forall \vec{x}\, (\varphi(\vec{x}) \rightarrow \psi(\vec{x}))$ is a valid sentence.

Assuming that $\vec{y}$ and $\vec{z}$ are distinct variables, we can rewrite Φ as

$$\forall \vec{x}\, \forall \vec{y}\, \exists \vec{z}\, \big(\neg\alpha(\vec{x}, \vec{y}) \ \vee\ \beta(\vec{x}, \vec{z})\big).$$

We know that Φ is valid iff $\neg\Phi$ is not satisfiable. But $\neg\Phi$ is equivalent to $\exists \vec{x}\, \exists \vec{y}\, \forall \vec{z}\, \big(\alpha \wedge \neg\beta\big)$; that is, to an $\mathrm{FO}(\exists^*\forall^*)$ sentence. This gives us the following.

Proposition 14.4. *Fix a relational vocabulary σ. Let Q and Q' be unions of conjunctive queries over σ. Then testing whether $Q \subseteq Q'$ is decidable in* PSPACE. □

The complexity bound given by the reduction to the Bernays-Schönfinkel class is not the optimal one, but it is not very far off: for a fixed vocabulary σ, the complexity of containment of unions of conjunctive queries is known to be Π_2^p-complete.

We now move to the Ackermann class $\mathrm{FO}(\exists^*\forall\exists^*)$. Again, we have the finite model property.

Theorem 14.5. *Let Φ be an $\mathrm{FO}(\exists^*\forall\exists^*)$ sentence. If Φ is satisfiable, then it has a model whose size is at most exponential in $\|\Phi\|$.* □

Even though the size of the finite model jumps from linear to exponential, the complexity of the decision problem does not get worse, and in fact in some cases the problem becomes easier.

Theorem 14.6. *The satisfiability problem for $\mathrm{FO}(\exists^*\forall\exists^*)$ is* NEXPTIME-*complete. Furthermore, when restricted to sentences that do not mention equality, the problem becomes* EXPTIME-*complete.* □

Finally, we consider finite variable restrictions of FO. Recall that FO^k refers to the fragment of FO that consists of formulae in which at most k distinct variables are used.

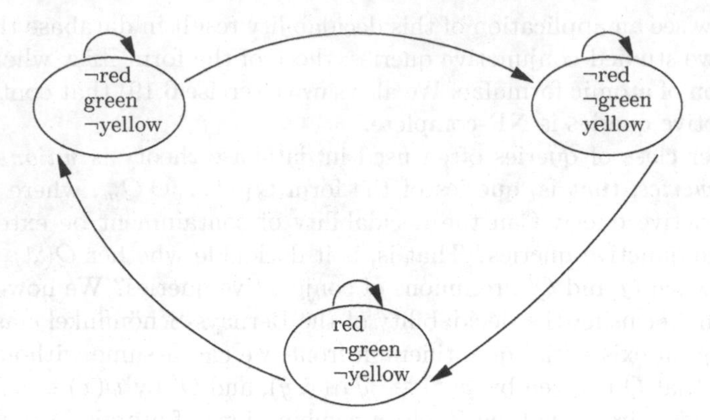

Fig. 14.1. An example of a Kripke structure

Theorem 14.7. FO^2 *has the finite model property: each satisfiable* FO^2 *sentence has a finite model whose size is at most exponential in* $\| \Phi \|$. *Furthermore, the satisfiability problem for* FO^2 *is* NEXPTIME*-complete. The satisfiability problem for* $\mathrm{FO}^k, k > 2$, *is undecidable.* □

14.2 Temporal and Modal Logics

In this section, we look at logics that are used in verifying temporal properties of reactive systems. The finite structure in this case is usually a *transition system*, or a *Kripke structure*. It can be viewed as a labeled directed graph, where the nodes describe possible states the system could be in, and the edges indicate when a transition from one state to another is possible. To describe possible states of the system, one uses a collection of propositional variables, and specifies which of them are true in a given state.

An example of a Kripke structure is given in Fig. 14.1. We have three propositional variables, *red*, *green*, and *yellow*. The states are those in which only one variable is true, and the other two are false. As expected, from a red light one can go to green, from green to yellow, and from yellow to red, and the system can stay in any of these states.

Sometimes edges of Kripke structures are labeled too, but since it is easy to push those labels back into the states, we shall assume that edges are not labeled.

Thus, formally, a Kripke structure, for a finite alphabet Σ, is a finite structure $\mathfrak{K} = \langle S, E, (P_a)_{a \in \Sigma} \rangle$, where S is the set of states, E is a binary relation on S, and for each $a \in \Sigma$, P_a is a unary relation on S, i.e., a subset of S. Since assigning relations P_a can be viewed as labeling states with letters from Σ, we shall also refer to the *labeling function* $\lambda : S \to 2^{\Sigma}$, given by

$$\lambda(s) = \{a \in \Sigma \mid s \in P_a\}.$$

We now define the simplest of the logics we deal with in this section: the *propositional modal logic, ML.* Its formulae are given by the following grammar:

$$\varphi, \psi ::= a \ (a \in \Sigma) \mid \varphi \wedge \psi \mid \neg\varphi \mid \Box\varphi \mid \Diamond\varphi. \tag{14.1}$$

The semantics of ML formulae is given with respect to a Kripke structure $\mathfrak{K}$ and a state s. That is, each formula defines a set of states where it holds. The formal definition of the semantics is as follows:

- $(\mathfrak{K}, s) \models a$, $a \in \Sigma$ iff $a \in \lambda(s)$;
- $(\mathfrak{K}, s) \models \varphi \wedge \psi$ iff $(\mathfrak{K}, s) \models \varphi$ and $(\mathfrak{K}, s) \models \psi$;
- $(\mathfrak{K}, s) \models \neg\varphi$ iff $(\mathfrak{K}, s) \not\models \varphi$;
- $(\mathfrak{K}, s) \models \Box\varphi$ iff $(\mathfrak{K}, s') \models \varphi$ for all s' such that $(s, s') \in E$;
- $(\mathfrak{K}, s) \models \Diamond\varphi$ iff $(\mathfrak{K}, s') \models \varphi$ for some s' such that $(s, s') \in E$.

Thus, $\Box$ is the "for all" modality, and $\Diamond$ is the "there exists" modality: $\Box\varphi$ ($\Diamond\varphi$) means that φ holds in every (in some) state to which there is an edge from the current state.

Notice also that $\Diamond$ is superfluous since $\Diamond\varphi$ is equivalent to $\neg\Box\neg\varphi$.

ML can be translated into FO as follows. For each ML formula φ, we define an FO formula $\varphi^\circ(x)$ such that $(\mathfrak{K}, s) \models \varphi$ iff $\mathfrak{K} \models \varphi^\circ(s)$. This is done as follows:

- $a^\circ \equiv P_a(x)$;
- $(\varphi \wedge \psi)^\circ \equiv \varphi^\circ \wedge \psi^\circ$;
- $(\neg\varphi)^\circ \equiv \neg\varphi^\circ$;
- $(\Box\varphi)^\circ \equiv \forall y \left(R(x, y) \rightarrow \forall x \ (x = y \rightarrow \varphi^\circ(x))\right)$.

For the translation of $\Box\varphi$, we employed the technique of reusing variables that was central in Chapter 11. Thus, φ° is always an FO^2 formula, as it uses only two variables: x and y. Summing up, we obtained the following.

Proposition 14.8. *Every formula of the propositional modal logic* ML *is equivalent to an* FO^2 *formula. Consequently, every satisfiable formula* φ *of* ML *has a model which is at most exponential in* $\|\varphi\|$. $\Box$

The expressiveness of ML is rather limited; in particular, since it is a fragment of FO, it cannot express reachability properties which are of utmost importance in verifying properties of finite-state systems. We thus move to more expressive logics, LTL and CTL.

The formulae of the *linear time temporal logic*, LTL, are given by the following grammar:

$$\varphi, \varphi' \ ::= \ a \ (a \in \Sigma) \ \mid \neg\varphi \mid \ \varphi \wedge \varphi' \ \mid \mathbf{X}\varphi \ \mid \ \varphi \mathbf{U} \varphi'. \tag{14.2}$$

The formulae of the *computation tree logic*, CTL, are given by

$$\begin{array}{ll} \varphi, \varphi' ::= a \ (a \in \Sigma) \mid \neg\varphi & \mid \varphi \wedge \varphi' \quad \mid \\ \quad \mathbf{EX}\varphi & \mid \mathbf{AX}\varphi \mid \mathbf{E}(\varphi \mathbf{U} \varphi') \mid \mathbf{A}(\varphi \mathbf{U} \varphi'). \end{array} \tag{14.3}$$

In both of these logics, we talk about properties of *paths* in the Kripke structure. A path in $\mathfrak{K}$ is an infinite sequence of nodes $\pi = s_1 s_2 \ldots$ such that $(s_i, s_{i+1}) \in E$ for all i. Of course, in a finite structure, some of the nodes must occur infinitely often on a path.

The connective **X** means "next time", or "for the next node on the path". The connective **U** is "until": φ holds until some point where φ' holds. **E** is the existential quantifier "there is a path", and **A** is the universal quantifier: "for all paths".

To give the formal semantics, we introduce a logic that subsumes both LTL and CTL. This logic, denoted by CTL*, has two kinds of formulae: *state formulae* denoted by φ, and *path formulae* denoted by ψ. These are given by the following two grammars:

$$\begin{array}{l} \varphi, \varphi' ::= a \ (a \in \Sigma) \ \mid \ \neg\varphi \mid \varphi \wedge \varphi' \mid \mathbf{E}\psi \ \mid \ \mathbf{A}\psi \\ \psi, \psi' ::= \varphi \mid \ \neg\psi \ \mid \psi \wedge \psi' \mid \mathbf{X}\psi \ \mid \ \psi \mathbf{U} \psi'. \end{array} \tag{14.4}$$

The semantics of a state formula is again given with respect to a Kripke structure $\mathfrak{K}$ and a state s. The semantics of a path formula ψ is given with respect to $\mathfrak{K}$ and a path π in $\mathfrak{K}$. If $\pi = s_1 s_2 s_3 \ldots$, we shall write π^k for the path starting at s_k; that is, $s_k s_{k+1} \ldots$.

Formally, we define the semantics as follows:

- $(\mathfrak{K}, s) \models a$, $a \in \Sigma$ iff $a \in \lambda(s)$;
- $(\mathfrak{K}, s) \models \varphi \wedge \varphi'$ iff $(\mathfrak{K}, s) \models \varphi$ and $(\mathfrak{K}, s) \models \varphi'$;
- $(\mathfrak{K}, s) \models \neg\varphi$ iff $(\mathfrak{K}, s) \not\models \varphi$;
- $(\mathfrak{K}, s) \models \mathbf{E}\psi$ iff there is a path $\pi = s_1 s_2 \ldots$ such that $s_1 = s$ and $(\mathfrak{K}, \pi) \models \psi$;
- $(\mathfrak{K}, s) \models \mathbf{A}\psi$ iff for every path $\pi = s_1 s_2 \ldots$ such that $s_1 = s$, we have $(\mathfrak{K}, \pi) \models \psi$;
- if φ is a state formula, and $\pi = s_1 s_2 \ldots$, then $(\mathfrak{K}, \pi) \models \varphi$ iff $(\mathfrak{K}, s_1) \models \varphi$;
- $(\mathfrak{K}, \pi) \models \psi \wedge \psi'$ iff $(\mathfrak{K}, \pi) \models \psi$ and $(\mathfrak{K}, \pi) \models \psi'$;
- $(\mathfrak{K}, \pi) \models \neg\psi$ iff $(\mathfrak{K}, \pi) \not\models \psi$;
- $(\mathfrak{K}, \pi) \models \mathbf{X}\psi$ iff $(\mathfrak{K}, \pi^2) \models \psi$;

- $(\mathfrak{K}, \pi) \models \psi\mathbf{U}\psi'$ if there exists $k \geq 1$ such that $(\mathfrak{K}, \pi^k) \models \psi'$ and $(\mathfrak{K}, \pi^i) \models \psi$ for all $i < k$.

Note that LTL formulae are path formulae, and CTL formulae are state formulae. LTL formulae are typically evaluated along a single infinite path (hence the name *linear* temporal logic). On the other hand, CTL is well-suited to describe branching processes (hence the name computation *tree* logic). If we want to talk about an LTL formula ψ being true in a given state of a Kripke structure, we shall mean that the formula $\mathbf{A}\psi$ is true in that state.

Some derived formulae are often useful in describing temporal properties. For example, $\mathbf{F}\psi \equiv \mathit{true}\mathbf{U}\psi$, means "eventually", or sometime in the future, ψ holds, and $\mathbf{G}\psi \equiv \neg\mathbf{F}\neg\psi$ means "always", or "globally", ψ holds (*true* itself can be assumed to be a formula in any of the logics: for example, $a \vee \neg a$). Thus, $\mathbf{AG}\psi$ means that ψ holds along every path starting from a given state, and $\mathbf{EF}\psi$ means that along some path, ψ eventually holds.

For the example in Fig. 14.1, consider a CTL formula $\mathbf{AG}(\mathit{yellow} \to \mathbf{AF}\mathit{green})$, saying that if the light is yellow, it will eventually become green. This formula is actually false in the structure shown in Fig. 14.1, since *yellow* can continue to hold indefinitely long due to the loop. However, $\mathbf{AG}\big(\mathit{yellow} \to (\mathbf{AG}\mathit{yellow} \vee \mathbf{AF}\mathit{green})\big)$, saying that either *yellow* holds forever or eventually changes to *green*, is true in that structure.

The main difference between CTL and LTL is that CTL is better suited for talking about branching paths that start in a given node (this is the reason logics like CTL are sometimes referred to as *branching-time* logics), while LTL, on the other hand, is better suited for talking about properties of a single path starting in a given node (and thus one speaks of a *linear-time* logic). For example, consider the CTL formula $\mathbf{AG}(\mathbf{EF}a)$. It says that along every path from a given node, from every node there is a path that leads to a state labeled a. It is known that this formula is not expressible in LTL. The formula $\mathbf{A}(\mathbf{FG}a)$, saying that on every path, starting from some node a will hold forever, is a state formula resulting by applying the $\mathbf{A}$ quantifier to the LTL formula $\mathbf{FG}a$; this formula is not expressible in CTL.

While all the examples seen so far could have been specified in other logics used in this book – for example, MSO or LFP – the main advantage of these temporal logics is that the model-checking problem for them can be solved efficiently. The model-checking problem is to determine whether $(\mathfrak{K}, s) \models \varphi$, for some Kripke structure $\mathfrak{K}$, state s, and a formula φ. The data complexity for CTL* and its sublogics can easily be seen to be polynomial (since CTL* formulae can be expressed in LFP), but it turns out that the situation is much better than this.

Theorem 14.9. *The model-checking problem for* ML, LTL, CTL, *and* CTL* *is fixed-parameter linear. For logics* ML *and* CTL *it can be solved in time* $O(\|\varphi\| \cdot \|\mathfrak{K}\|)$ *and for* LTL *and* CTL*, *the bound is* $2^{O(\|\varphi\|)} \cdot \|\mathfrak{K}\|$. $\square$

We illustrate the idea of the proof for the case of ML. Suppose we have a formula φ and a Kripke structure $\mathfrak{K}$. Consider all the subformulae $\varphi_1, \ldots, \varphi_k$ of φ listed in an order that ensures that if φ_j is a subformula of φ_i, then $j < i$. The algorithm then inductively labels each state s of $\mathfrak{K}$ with either φ_i or $\neg\varphi_i$, depending on which formula holds in that state. For the base case, there is nothing to do since the states are already labeled with either a or $\neg a$ for each $a \in \Sigma$. For the induction, the only nontrivial case is when $\varphi_i \equiv \Box\varphi_j$ for some $j \leq i$. Then for each state s, we check all the states s' with $(s, s') \in E$, and see if all such s' have been labeled with φ_j in the jth step: if so, we label s by φ_i; if not, we label it by $\neg\varphi_i$. This algorithm can be implemented in time $O(\|\varphi\| \cdot \|\mathfrak{K}\|)$.

Next, we look at the connection between temporal and modal logics and other logics for finite structures we have seen. We already mentioned that ML can be embedded into FO^2. What about LTL? We can answer this question for a simple kind of Kripke structures used in Chap. 7: these are structures of the vocabulary $\sigma_\Sigma = (<, (P_a)_{a \in \Sigma})$, used to represent strings.

Theorem 14.10. *Over finite strings viewed as structures of vocabulary* σ_Σ, LTL *and* FO *are equally expressive:* LTL = FO. □

Interestingly, Theorem 14.10 holds for ω-strings as well, but this is outside the scope of this book.

For CTL, one needs to talk about different paths, and hence one should be able to express reachability properties such as "can a state labeled a be reached from a state labeled b"? This suggests a close connection between CTL and logics that can express the transitive closure operator. We illustrate this by means of the following example.

Consider a CTL formula $\mathbf{AF}a$ stating that along every path, a eventually holds. We now express this in a variant of DATALOG. Let (Π_1, T) be the following $\text{DATALOG}_\neg$ program:

$$\begin{array}{lcl} R(x,y) & \text{:-} & \neg P_a(x), E(x,y) \\ R(x,y) & \text{:-} & \neg P_a(z), R(x,z), E(z,y) \end{array}$$

This program computes a subset of the transitive closure: the set of pairs (b, b') for which there is a path $b = b_1, b_2, \ldots, b_{n-1}, b_n = b'$ such that none of the b_i's, $i < n$, is labeled a. Next, we define a program (Π_2, U) that uses R as an extensional predicate:

$$\begin{array}{lcl} U(x) & \text{:-} & R(x,x) \\ U(x) & \text{:-} & \neg P_a(x), E(x,y), U(y) \end{array}$$

Suppose we have an infinite path over $\mathfrak{K}$. Since $\mathfrak{K}$ is finite, it must have a loop. If there is a loop such that $R(x,x)$ holds, then there is an infinite path from x such that $\neg a$ holds along this path. If we have any other path such that $\neg a$ holds along it, then it starts with a few edges and eventually enters a loop in which no node is labeled a. Hence, U is the set of nodes from which there is an infinite path on which $\neg a$ holds. Thus, taking the program (Π_3, Q) given by

$$Q(x) \text{ :- } \neg U(x)$$

we get a program that computes $\mathbf{AF}a$. Notice that this program is *stratified* (for each stratum, the negated predicates are those defined in the previous strata) and *linear* (each intensional predicate appears at most once in the right hand sides of rules). The above translation techniques can be extended to prove the following.

Theorem 14.11. CTL *formulae can be expressed in either of the following:*

- *the linear stratified* DATALOG$_\neg$*;*
- *the transitive closure logic* TRCL. □

Next, we define a fixed point modal logic, called the *μ-calculus* and denoted by CALC$_\mu$, that subsumes LTL, CTL, and CTL*. Consider the propositional modal logic ML, and extend its syntax with propositional variables $x, y, \ldots$, viewed as monadic second-order variables (i.e., each such variable denotes a set of states). Now formulae have free variables. Suppose we have a formula $\varphi(x, \vec{y})$ where x occurs positively in φ. Then $\mu x.\varphi(x, \vec{y})$ is a formula with free variables $\vec{y}$.

To define the semantics of $\psi(\vec{y}) \equiv \mu x.\varphi(x, \vec{y})$ on a Kripke structure $\mathfrak{K}$, assume that each y_i from $\vec{y}$ is interpreted as a propositional variable: that is, a subset Y_i of S consisting of nodes where it holds. Then $\varphi(x, \vec{Y})$ defines an operator $F_\varphi^Y : 2^S \to 2^S$ given by

$$F_\varphi^Y(X) = \{s \in S \mid (\mathfrak{K}, s) \models \varphi(X, \vec{Y})\}.$$

If x occurs positively, then this operator is monotone. We define the semantics of the μ operator by

$$(\mathfrak{K}, s) \models \mu x.\varphi(x, \vec{Y}) \quad \Leftrightarrow \quad s \in \mathbf{lfp}(F_\varphi^Y).$$

Consider, for example, the formula $\mu x.a \vee \Box x$. This formula is true in $(\mathfrak{K}, s)$ if along each path starting in s, a will eventually become true. Hence, this is the CTL formula $\mathbf{AF}a$. In general, every CTL* formula can be expressed in CALC$_\mu$.

Each CALC$_\mu$ formula φ can be translated into an LFP formula $\varphi^\circ(x)$ such that $(\mathfrak{K}, s) \models \varphi$ iff $\mathfrak{K} \models \varphi^\circ(s)$. Furthermore, one can show that CALC$_\mu$ formulae can be translated into MSO formulae as well. Summing up, we have the following relationship between the temporal logics:

$$\text{ML} \subsetneq \left\{ \begin{matrix} \text{LTL} \\ \text{CTL} \end{matrix} \right\} \subsetneq \text{CTL}^* \subsetneq \text{CALC}_\mu \subsetneq \left\{ \begin{matrix} \text{LFP} \\ \text{MSO} \end{matrix} \right\}.$$

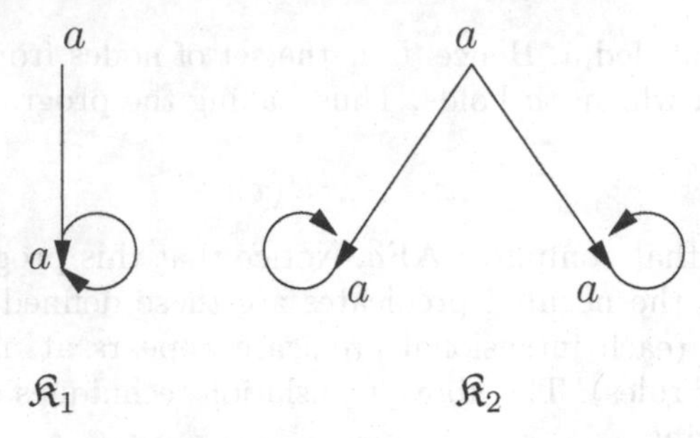

Fig. 14.2. Bisimulation equivalence

In the μ-calculus, it is common to use both least and greatest fixed points. The latter are definable by $\nu x.\varphi(x) \equiv \neg\mu x.\neg\varphi(\neg x)$, assuming that x occurs positively in φ. Notice that negating both φ and each occurrence of x in it ensures that if x occurs positively in φ, then it occurs positively in $\neg\varphi(\neg x)$, and hence the least fixed point is well-defined. Using the greatest and the least fixed points, the formulae of the μ-calculus can be written in the alternating style so that negations are applied only to propositions. We shall denote the fragment of the μ-calculus that consists of such alternating formulae with alternation depth at most k by CALC_μ^k.

Theorem 14.12. *The complexity of the model-checking problem for* CALC_μ^k *is* $O(\|\varphi\| \cdot \|\mathfrak{K}\|^k)$. □

Since CALC_μ can be embedded into LFP, its data complexity is polynomial. The combined complexity is known to be in NP ∩ coNP. Furthermore, CALC_μ has the finite model property: if φ is a satisfiable formula of CALC_μ, then there is a Kripke structure $\mathfrak{K}$ of size at most exponential in φ such that $(\mathfrak{K}, s) \models \varphi$ for some $s \in S$.

Finally, we present another way to connect temporal logics with other logics seen in this book. Since logics like CALC_μ talk about temporal properties of paths, they cannot distinguish structures in which all paths agree on all temporal properties, even if the structures themselves are different. For example, consider the structures $\mathfrak{K}_1$ and $\mathfrak{K}_2$ shown in Fig. 14.2. Even though they are different, all the paths realized in these structures are the same: an infinite path on which every node is labeled a. CALC_μ cannot see the difference between them, although these structures are easily distinguished by the FO sentence "There is a node with two distinct successors."

One can formally capture this notion of indistinguishability using the definition of *bisimilarity*. Let $\mathfrak{K} = \langle S, E, (P_a)_{a\in\Sigma}\rangle$ and $\mathfrak{K}' = \langle S', E', (P'_a)_{a\in\Sigma}\rangle$. We say that $(\mathfrak{K}, s)$ and $(\mathfrak{K}', s')$ are *bisimilar* if there is a binary relation $R \subseteq S \times S'$ such that

- $(s, s') \in R$;

- if $(u, u') \in R$, then $P_a(u)$ iff $P'_a(u')$, for all $a \in \Sigma$;
- if $(u, u') \in R$ and $(u, v) \in E$, then there is $v' \in S'$ such that $(v, v') \in R$ and $(u', v') \in E'$;
- if $(u, u') \in R$ and $(u', v') \in E'$, then there is $v \in S$ such that $(v, v') \in R$ and $(u, v) \in E$.

A property of Kripke structures is *bisimulation-invariant* if whenever it holds in $(\mathfrak{K}, s)$, it also holds in every $(\mathfrak{K}', s')$ which is bisimilar to $(\mathfrak{K}, s)$. As we have seen, even FO can express properties which are not bisimulation-invariant, but CALC_μ and its sublogics only express bisimulation-invariant properties.

The following result shows how to use bisimulation-invariance to relate temporal logics and other logics seen in this book.

Theorem 14.13.
- *The class of bisimulation-invariant properties expressible in* FO *is precisely the class of properties expressible in* ML.
- *The class of bisimulation-invariant properties expressible in* MSO *is precisely the class of properties expressible in* CALC_μ. □

14.3 Constraint Satisfaction and Homomorphisms of Finite Models

Constraint satisfaction problems are problems of the following kind. Suppose we are given a set V of variables, a finite domain D where the variables can take values, and a set of constraints C. The problem is whether there exists an assignment of values to variables that satisfies all the constraints.

Each constraint in the set C is specified as a pair $(\vec{v}, R)$ where $\vec{v}$ is a tuple of variables from V, of length n, and R is an n-ary relation on D. The assignment of values to variables is then a mapping $h : V \to D$. Such a mapping satisfies the constraint $(\vec{v}, R)$ if $h(\vec{v}) \in R$.

For example, satisfiability of certain propositional formulae can be viewed as a constraint satisfaction problem. Consider the MONOTONE 3-SAT problem. That is, we have a CNF formula $\varphi(x_1, \ldots, x_m)$ in which every clause is either $(x_i \vee x_j \vee x_k)$, or $(\neg x_i \vee \neg x_j \vee \neg x_k)$. Consider the constraint satisfaction problem where $V = \{x_1, \ldots, x_n\}$, $D = \{0, 1\}$, and for each clause $(x_i \vee x_j \vee x_k)$ we have a constraint $((x_i, x_j, x_k), \{0,1\}^3 - \{(0,0,0)\})$, and for each clause $(\neg x_i \vee \neg x_j \vee \neg x_k)$ we have a constraint $((x_i, x_j, x_k), \{0,1\}^3 - \{(1,1,1)\})$. Then the resulting constraint satisfaction problem (V, D, C) has a solution iff φ is satisfiable.

There is a nice representation of constraint satisfaction problems in terms of the existence of a certain homomorphism between finite structures.

Suppose we are given a constraint satisfaction problem $\mathcal{P} = (V, D, C)$. Let $R_1^{\mathcal{P}}, \ldots, R_l^{\mathcal{P}}$ list all the relations mentioned in C. Let $\sigma_{\mathcal{P}} = (R_1, \ldots, R_l)$. We define two $\sigma_{\mathcal{P}}$-structures as follows:

$$\begin{aligned}\mathfrak{A}_{\mathcal{P}} &= \langle V, \{\vec{v} \mid (\vec{v}, R_1^{\mathcal{P}}) \in C\}, \ldots, \{\vec{v} \mid (\vec{v}, R_m^{\mathcal{P}}) \in C\}\rangle \\ \mathfrak{B}_{\mathcal{P}} &= \langle D, R_1^{\mathcal{P}}, \ldots, R_m^{\mathcal{P}}\rangle.\end{aligned}$$

Then

$$\begin{aligned}&\mathcal{P} \text{ has a solution} \\ \Leftrightarrow\ &\text{there exists a homomorphism } h : \mathfrak{A}_{\mathcal{P}} \to \mathfrak{B}_{\mathcal{P}}.\end{aligned}$$

Thus, the constraint satisfaction problem is really the problem of checking whether there is a homomorphism between two structures. We thus use the notation

$$\text{CSP}(\mathfrak{A}, \mathfrak{B}) \quad \Leftrightarrow \quad \text{there exists a homomorphism } h : \mathfrak{A} \to \mathfrak{B}.$$

To see another example, let K_m be the clique on m elements. Then $\text{CSP}(G, K_m)$ holds iff G is m-colorable.

The constraint satisfaction problem can easily be related to conjunctive query evaluation. Suppose we have a vocabulary σ that consists only of relation symbols, and a σ-structure $\mathfrak{A}$. Let $A = \{a_1, \ldots, a_n\}$. We define the Boolean conjunctive query $\text{CQ}_{\mathfrak{A}}$ as

$$\text{CQ}_{\mathfrak{A}} \equiv \exists x_1 \ldots \exists x_n \bigwedge_{R \in \sigma} \quad \bigwedge_{(a_{i_1}, \ldots, a_{i_m}) \in R^{\mathfrak{A}}} R(x_{i_1}, \ldots, x_{i_m}).$$

Proposition 14.14. $\text{CSP}(\mathfrak{A}, \mathfrak{B})$ *is true iff* $\mathfrak{B} \models \text{CQ}_{\mathfrak{A}}$. □

If $\mathcal{C}$ and $\mathcal{C}'$ are two classes of structures, then we write $\text{CSP}(\mathcal{C}, \mathcal{C}')$ for the class of problems $\text{CSP}(\mathfrak{A}, \mathfrak{B})$ where $\mathfrak{A} \in \mathcal{C}$ and $\mathfrak{B} \in \mathcal{C}'$. We use All for the class of all finite structures.

The m-colorability example shows that CSP(All, All) contains NP-hard problems. Furthermore, each problem in CSP(All, All) can be solved in NP: given $\mathfrak{A}$ and $\mathfrak{B}$, we simply guess a mapping $h : A \to B$, and check, in polynomial time, if it is a homomorphism between $\mathfrak{A}$ and $\mathfrak{B}$. Thus, CSP(All, All) is NP-complete.

This naturally leads to the following question: under what conditions is $\text{CSP}(\mathcal{C}, \mathcal{C}')$ tractable?

We first answer this question in the setting suggested by the examples of MONOTONE 3-SAT and m-colorability. In both of these examples, we were interested in the problem of the form $\text{CSP}(\text{All}, \mathfrak{B})$; that is, in the existence of a homomorphism into a fixed structure. This is a very common class of constraint satisfaction problems. We shall write $\text{CSP}(\mathfrak{B})$ for $\text{CSP}(\text{All}, \mathfrak{B})$. Thus, the first question we address is when $\text{CSP}(\mathfrak{B})$ can be guaranteed to be tractable.

All problems of the form $\text{CSP}(\mathfrak{B})$ whose complexity is known fall into two categories: they are either tractable, or NP-complete. This is a real dichotomy: if PTIME $\neq$ NP, there are NP problems which are neither tractable nor NP-complete. In fact, it has been conjectured that for any $\mathfrak{B}$, the problem CSP($\mathfrak{B}$) is either tractable, or NP-complete. In general, this conjecture remains unproven, but some partial solutions are known. For example:

Theorem 14.15. *For every $\mathfrak{B}$ with $|B| \leq 3$, CSP($\mathfrak{B}$) is either tractable, or NP-complete.* ☐

Moreover, for the case of $|B| = 2$ (so-called Boolean constraint satisfaction problem), one can classify precisely for which structures $\mathfrak{B}$ the corresponding problem CSP($\mathfrak{B}$) is tractable.

For more general structures $\mathfrak{B}$, one can use logical definability to find some fairly large classes that guarantee tractability.

If one tries to think of a logic in which CSP($\mathfrak{B}$) can be expressed, one immediately thinks of MSO. Indeed, suppose that the universe of $\mathfrak{B}$ is $\{b_0, \ldots, b_{n-1}\}$. Then the MSO sentence characterizing CSP($\mathfrak{B}$) is of the form

$$\exists X_0 \ \ldots \ \exists X_{n-1} \ \Psi,$$

where Ψ is an FO sentence stating that, on a structure $\mathfrak{A}$ expanded with n sets $X_0, \ldots, X_{n-1}$, the sets X_i form a partition of A, and the map defined by sending all elements of X_i into b_i, for $i = 0, \ldots, n-1$, is a homomorphism from $\mathfrak{A}$ to $\mathfrak{B}$.

However, while in many cases MSO is tractable, in general it is not suitable to establish tractability results without putting restrictions on a class of structures $\mathfrak{A}$, since MSO can express NP-complete problems.

To express CSP($\mathfrak{B}$) in a tractable logic, we instead consider the negation of CSP($\mathfrak{B}$): that is,

$$\neg\text{CSP}(\mathfrak{B}) = \{\mathfrak{A} \mid \text{there is no homomorphism } h : \mathfrak{A} \to \mathfrak{B}\}.$$

If $\mathfrak{A} \in \neg$CSP($\mathfrak{B}$) and $\mathfrak{A}$ is a substructure of $\mathfrak{A}'$, then $\mathfrak{A}' \in \neg$CSP($\mathfrak{B}$). This monotonicity property suggests that for some $\mathfrak{B}$, the class $\neg$CSP($\mathfrak{B}$) could be definable in a rather expressive tractable monotone language such as Datalog. If this were the case, then CSP($\mathfrak{B}$) would be tractable as well.

Trying to express $\neg$CSP($\mathfrak{B}$) in Datalog may be a bit hard, but it turns out that instead one could attempt to express $\neg$CSP($\mathfrak{B}$) in a richer infinitary logic.

Theorem 14.16. *For each $\mathfrak{B}$, the problem $\neg$CSP($\mathfrak{B}$) is expressible in* Datalog *iff it is expressible in $\exists\mathcal{L}^{\omega}_{\infty\omega}$.* ☐

Thus, one general way of achieving tractability is to show that the negation of the constraint satisfaction problem is expressible in the existential fragment of the very rich finite variable logic $\mathcal{L}^{\omega}_{\infty\omega}$.

Moving back to the general problem CSP($\mathcal{C}, \mathcal{C}'$), one may ask whether CSP($\mathcal{C}, \mathcal{C}'$) is tractable whenever CSP($\mathcal{C}, \mathfrak{B}$) is tractable for all $\mathfrak{B} \in \mathcal{C}'$. The answer to this is negative: for each fixed graph G, the problem $\text{CSP}(\{K_m \mid m \in \mathbb{N}\}, G)$ is tractable, but $\text{CSP}(\{K_m \mid m \in \mathbb{N}\}, \text{All})$ is not. However, for the class of structures above, a uniform version of the tractability result can be shown.

Theorem 14.17. *Let $\mathcal{C}_{\text{DATALOG}^k}$ be the class of structures $\mathfrak{B}$ such that $\neg\text{CSP}(\mathfrak{B})$ is expressible by a* DATALOG *program that uses at most k distinct variables. Then $\text{CSP}(\text{All}, \mathcal{C}_{\text{DATALOG}^k})$ is in* PTIME. □

Yet another tractable restriction uses the notion of *treewidth* encountered in Chap. 6. If we let $\mathcal{TW}_k$ be the class of graphs of treewidth at most k, then one can show that $\neg\text{CSP}(\mathcal{TW}_k, \mathfrak{B})$ is expressible in DATALOG (in fact, in the k-variable fragment of DATALOG). Hence, $\text{CSP}(\mathcal{TW}_k, \mathfrak{B})$ is tractable.

In fact, this can be generalized as follows. We call two structures $\mathfrak{A}$ and $\mathfrak{B}$ *homomorphically equivalent* if there exist homomorphisms $h : \mathfrak{A} \to \mathfrak{B}$ and $h' : \mathfrak{B} \to \mathfrak{A}$. Let $\mathcal{HTW}_k$ be the class of all structures homomorphically equivalent to a structure in $\mathcal{TW}_k$.

Theorem 14.18. $\text{CSP}(\mathcal{HTW}_k, \text{All})$ *can be expressed in* LFP *(in fact, using at most $2k$ variables) and consequently is in* PTIME. □

Thus, definability results for fixed point and finite variable logics describe rather large classes of tractable constraint satisfaction problems.

14.4 Bibliographic Notes

A comprehensive survey of decidable and undecidable cases for the satisfiability problem is given in Börger, Grädel, and Gurevich [25]. It describes both the Bernays-Schönfinkel and Ackermann classes, and proves complexity bounds for them. The finite model property for FO^2 is due to Mortimer [184]; the complexity bound is from Grädel, Kolaitis, and Vardi [100]. The Π_2^p-completeness of containment of unions of conjunctive queries is due to Sagiv and Yannakakis [211].

There are a number of books and surveys in which temporal and modal logics are described in detail: van Benthem [240], Clarke, Grumberg, and Peled [37], Emerson [64, 65], Vardi [246]. Theorem 14.10 is from Kamp [141]. Abiteboul, Herr, and Van den Bussche [2] showed that Kamp's theorem no longer holds if one moves from strings to arbitrary structures. It is also known that for the translation from LTL to FO, three variables suffice (i.e., over strings, LTL equals FO^3, see, e.g., Schneider [214]), but two variables do not suffice (as shown by Etessami, Vardi, and Wilke [69]). The example of expressing a CTL property in DATALOG is from Gottlob, Grädel, and Veith [93], and Theorem 14.11 is from [93] and Immerman and Vardi [136]. Equivalence of bisimulation-invariant FO and modal logic is from van Benthem [240], and the corresponding result for MSO and CALC$_\mu$ is from Janin and Walukiewicz [138]; for a related result about CTL*, see Moller and Rabinovich [183].

Constraint satisfaction is a classical AI problem (see, e.g., Tsang [235]). The idea of viewing constraint satisfaction as the existence of a homomorphism between two structures is due to Feder and Vardi [77]. They also suggested using expressibility in DATALOG as a tool for proving tractability, and formulated the dichotomy conjecture. Theorem 14.15 is due to Schaefer [213] (for $|B| = 2$) and Bulatov [28] (for $|B| = 3$). The existence of complexity classes between PTIME and NP-complete, mentioned before Theorem 14.15, is due to Ladner [159]. Other results in that section are from Kolaitis and Vardi [156] and Dalmau, Kolaitis, and Vardi [48]. The converse of Theorem 14.18 was proved recently by Grohe [112].

References

1. S. Abiteboul, K. Compton, and V. Vianu. Queries are easier than you thought (probably). In *ACM Symp. on Principles of Database Systems*, 1992, ACM Press, pages 23–32.
2. S. Abiteboul, L. Herr, and J. Van den Bussche. Temporal connectives versus explicit timestamps to query temporal databases. *Journal of Computer and System Sciences*, 58 (1999), 54–68.
3. S. Abiteboul, R. Hull, and V. Vianu. *Foundations of Databases*, Addison-Wesley, 1995.
4. S. Abiteboul, M. Y. Vardi, and V. Vianu. Fixpoint logics, relational machines, and computational complexity. *Journal of the ACM*, 44 (1997), 30–56.
5. S. Abiteboul and V. Vianu. Fixpoint extensions of first-order logic and datalog-like languages. In *Proc. IEEE Symp. on Logic in Computer Science*, 1989, pages 71–79.
6. S. Abiteboul and V. Vianu. Computing with first-order logic. *Journal of Computer and System Sciences*, 50 (1995), 309–335.
7. J.W. Addison, L. Henkin, A. Tarski, eds. *The Theory of Models*. North-Holland, 1965.
8. F. Afrati, S. Cosmadakis, and M. Yannakakis. On datalog vs. polynomial time. *Journal of Computer and System Sciences*, 51 (1995), 177–196.
9. A. Aho and J. Ullman. The universality of data retrieval languages. In *Proc. ACM Symp. on Principles of Programming Languages, 1979*, ACM Press, pages 110–120.
10. M. Ajtai. Σ_1^1 formulae on finite structures. *Annals of Pure and Applied Logic*, 24 (1983), 1–48.
11. M. Ajtai and R. Fagin. Reachability is harder for directed than for undirected graphs. *Journal of Symbolic Logic*, 55 (1990), 113–150.
12. M. Ajtai, R. Fagin, and L. Stockmeyer. The closure of monadic NP. *Journal of Computer and System Sciences*, 60 (2000), 660–716.
13. M. Ajtai and Y. Gurevich. Monotone versus positive. *Journal of the ACM*, 34 (1987), 1004–1015.

14. G. Asser. Das Repräsentantenproblem im Prädikatenkalkül der Ersten Stufe mit Identität. *Zeitschrift für Mathematische Logik und Grundlagen der Mathematik*, 1 (1955), 252–263.
15. D.A.M. Barrington, N. Immerman, C. Lautemann, N. Schweikardt, and D. Thérien. The Crane Beach conjecture. In *IEEE Symp. on Logic in Computer Science, 2001*, pages 187–196.
16. D.A.M. Barrington, N. Immerman, and H. Straubing. On uniformity within NC^1. *Journal of Computer and System Sciences*, 41 (1990), 274–306.
17. J. Barwise. On Moschovakis closure ordinals. *Journal of Symbolic Logic*, 42 (1977), 292–296.
18. J. Barwise and S. Feferman, eds. *Model-Theoretic Logics*. Springer-Verlag, 1985.
19. M. Benedikt and L. Libkin. Relational queries over interpreted structures. *Journal of the ACM*, 47 (2000), 644–680.
20. M. Benedikt and L. Libkin. Tree extension algebras: logics, automata, and query languages. In *IEEE Symp. on Logic in Computer Science*, 2002, pages 203–212.
21. M. Benedikt, L. Libkin, T. Schwentick, and L. Segoufin. Definable relations and first-order query languages over strings. *Journal of the ACM*, 50 (2003), 694–751.
22. A. Blass, Y. Gurevich, and D. Kozen. A zero-one law for logic with a fixed-point operator. *Information and Control*, 67 (1985), 70–90.
23. A. Blumensath and E. Grädel. Automatic structures. In *IEEE Symp. on Logic in Computer Science*, 2000, pages 51–62.
24. H. Bodlaender. A linear-time algorithm for finding tree-decompositions of small treewidth. *SIAM Journal on Computing*, 25 (1996), 1305–1317.
25. E. Börger, E. Grädel, and Y. Gurevich. *The Classical Decision Problem*. Springer-Verlag, 1997.
26. V. Bruyère, G. Hansel, C. Michaux, and R. Villemaire. Logic and p-recognizable sets of integers. *Bulletin of the Belgian Mathematical Society*, 1 (1994), 191–238.
27. J.R. Büchi. Weak second-order arithmetic and finite automata. *Zeitschrift für Mathematische Logik und Grundlagen der Mathematik*, 6 (1960), 66–92.
28. A. Bulatov. A dichotomy theorem for constraints on a three-element set. *IEEE Symp. on Foundations of Computer Science*, 2002, pages 649–658.
29. S. R. Buss. First-order proof theory of arithmetic. In *Handbook of Proof Theory*, Elsevier, Amsterdam, 1998, pages 79–147.
30. J. Cai, M. Fürer, and N. Immerman. On optimal lower bound on the number of variables for graph identification. *Combinatorica*, 12 (1992), 389–410.
31. P.J. Cameron. The random graph revisited. In *Eur. Congr. of Mathematics, Vol. 1*, Progress in Mathematics, Birkhäuser, 2001, pages 267–274.
32. A. Chandra and D. Harel. Computable queries for relational databases. *Journal of Computer and System Sciences*, 21 (1980), 156–178.
33. A. Chandra and D. Harel. Structure and complexity of relational queries. *Journal of Computer and System Sciences*, 25 (1982), 99–128.

34. A. Chandra and P. Merlin. Optimal implementation of conjunctive queries in relational data bases. In *ACM Symp. on Theory of Computing*, 1977, pages 77–90.
35. C.C. Chang and H.J. Keisler. *Model Theory.* North-Holland, 1990.
36. O. Chapuis and P. Koiran. Definability of geometric properties in algebraically closed fields. *Mathematical Logic Quarterly*, 45 (1999), 533–550.
37. E. Clarke, O. Grumberg, and D. Peled. *Model Checking.* The MIT Press, 1999.
38. H. Comon, M. Dauchet, R. Gilleron, F. Jacquemard, D. Lugiez, S. Tison, and M. Tommasi. *Tree Automata: Techniques and Applications.* Available at `www.grappa.univ-lille3.fr/tata`. October 2002.
39. S.A. Cook. The complexity of theorem-proving procedures. In *Proc. ACM Symp. on Theory of Computing*, 1971, ACM Press, pages 151–158.
40. S.A. Cook. *Proof complexity and bounded arithmetic.* Manuscript, Univ. of Toronto, 2002.
41. S.A. Cook and Y. Liu. A complete axiomatization for blocks world. In *Proc. 7th Int. Symp. on Artificial Intelligence and Mathematics*, January, 2002.
42. S. Cosmadakis. Logical reducibility and monadic NP. In *Proc. IEEE Symp. on Foundations of Computer Science*, 1993, pages 52–61.
43. S. Cosmadakis, H. Gaifman, P. Kanellakis, and M. Vardi. Decidable optimization problems for database logic programs. In *ACM Symp. on Theory of Computing*, 1988, pages 477–490.
44. B. Courcelle. Graph rewriting: an algebraic and logic approach. In *Handbook of Theoretical Computer Science, Vol. B*, North-Holland, 1990, pages 193–242.
45. B. Courcelle. On the expression of graph properties in some fragments of monadic second-order logic. In [134], pages 33–62.
46. B. Courcelle. The monadic second-order logic on graphs VI: on several representations of graphs by relational structures. *Discrete Applied Mathematics*, 54 (1994), 117–149.
47. B. Courcelle and J. Makowsky. Fusion in relational structures and the verification of monadic second-order properties. *Mathematical Structures in Computer Science*, 12 (2002), 203–235.
48. V. Dalmau, Ph. Kolaitis, and M. Vardi. Constraint satisfaction, bounded treewidth, and finite-variable logics. *Proc. Principles and Practice of Constraint Programming*, Springer-Verlag LNCS 2470, 2002, pages 310–326.
49. A. Dawar. A restricted second order logic for finite structures. *Logic and Computational Complexity*, Springer-Verlag, LNCS 960, 1994, pages 393–413.
50. A. Dawar, K. Doets, S. Lindell, and S. Weinstein. Elementary properties of finite ranks. *Mathematical Logic Quarterly*, 44 (1998), 349–353.
51. A. Dawar and Y. Gurevich. Fixed point logics. *Bulletin of Symbolic Logic*, 8 (2002), 65-88.
52. A. Dawar and L. Hella. The expressive power of finitely many generalized quantifiers. *Information and Computation*, 123 (1995), 172–184.
53. A. Dawar, S. Lindell, and S. Weinstein. Infinitary logic and inductive definability over finite structures. *Information and Computation*, 119 (1995), 160–175.

54. A. Dawar, S. Lindell, and S. Weinstein. First order logic, fixed point logic, and linear order. In *Computer Science Logic*, Springer-Verlag LNCS Vol. 1092, 1995, pages 161–177.
55. L. Denenberg, Y. Gurevich, and S. Shelah. Definability by constant-depth polynomial-size circuits. *Information and Control*, 70 (1986), 216–240.
56. M. de Rougemont. Second-order and inductive definability on finite structures. *Zeitschrift für Mathematische Logik und Grundlagen der Mathematik*, 33 (1987), 47–63.
57. G. Dong, L. Libkin, and L. Wong. Local properties of query languages. *Theoretical Computer Science*, 239 (2000), 277–308.
58. R. Downey and M. Fellows. *Parameterized Complexity*. Springer-Verlag, 1999.
59. D.-Z. Du, K.-I. Ko. *Theory of Computational Complexity*. Wiley-Interscience, 2000.
60. H.-D. Ebbinghaus and J. Flum. *Finite Model Theory*. Springer-Verlag, 1995.
61. H.-D. Ebbinghaus, J. Flum, and W. Thomas. *Mathematical Logic*. Springer-Verlag, 1984.
62. A. Ehrenfeucht. An application of games to the completeness problem for formalized theories. *Fundamenta Mathematicae*, 49 (1961), 129–141.
63. T. Eiter, G. Gottlob, and Y. Gurevich. Existential second-order logic over strings. *Journal of the ACM*, 47 (2000), 77–131.
64. E.A. Emerson. Temporal and modal logic. In *Handbook of Theoretical Computer Science, Vol. B*, North-Holland, 1990, pages 995–1072.
65. E.A. Emerson. Model checking and the mu-calculus. In [134], pages 185–214.
66. H. Enderton. *A Mathematical Introduction to Logic*. Academic-Press, 1972.
67. P. Erdös and A. Rényi. Asymmetric graphs. *Acta Mathematicae Academiae Scientiarum Hungaricae*, 14 (1963), 295–315.
68. K. Etessami. Counting quantifiers, successor relations, and logarithmic space. *Journal of Computer and System Sciences*, 54 (1997), 400–411.
69. K. Etessami, M.Y. Vardi, and T. Wilke. First-order logic with two variables and unary temporal logic. *Information and Computation*, 179 (2002), 279–295.
70. R. Fagin. Generalized first-order spectra and polynomial-time recognizable sets. In *Complexity of Computation*, R. Karp, ed., *SIAM-AMS Proceedings*, 7 (1974), 43–73.
71. R. Fagin. Monadic generalized spectra. *Zeitschrift für Mathematische Logik und Grundlagen der Mathematik*, 21 (1975), 89–96.
72. R. Fagin. A spectrum hierarchy. *Zeitschrift für Mathematische Logik und Grundlagen der Mathematik*, 21 (1975), 123–134.
73. R. Fagin. Probabilities on finite models. *Journal of Symbolic Logic*, 41 (1976), 50–58.
74. R. Fagin. Finite-model theory — a personal perspective. *Theoretical Computer Science*, 116 (1993), 3–31.
75. R. Fagin. Easier ways to win logical games. In [134], pages 1–32.
76. R. Fagin, L. Stockmeyer, and M.Y. Vardi. On monadic NP vs monadic co-NP. *Information and Computation*, 120 (1994), 78–92.

77. T. Feder and M.Y. Vardi. The computational structure of monotone monadic SNP and constraint satisfaction: a study through datalog and group theory. *SIAM Journal on Computing*, 28 (1998), 57–104.
78. T. Feder and M.Y. Vardi. Homomorphism closed vs. existential positive. *IEEE Symp. on Logic in Computer Science*, 2003, pages 311–320.
79. S. Feferman and R. Vaught. The first order properties of products of algebraic systems. *Fundamenta Mathematicae*, 47 (1959), 57–103.
80. J. Flum, M. Frick, and M. Grohe. Query evaluation via tree-decompositions. *Journal of the ACM*, 49 (2002), 716–752.
81. J. Flum and M. Grohe. Fixed-parameter tractability, definability, and model-checking. *SIAM Journal on Computing* 31 (2001), 113–145.
82. J. Flum and M. Ziegler. Pseudo-finite homogeneity and saturation. *Journal of Symbolic Logic*, 64 (1999), 1689–1699.
83. H. Fournier. Quantifier rank for parity of embedded finite models. *Theoretical Computer Science*, 295 (2003), 153–169.
84. R. Fraïssé. Sur quelques classifications des systèmes de relations. *Université d'Alger, Publications Scientifiques, Série A*, 1 (1954), 35–182.
85. M. Frick and M. Grohe. The complexity of first-order and monadic second-order logic revisited. In *IEEE Symp. on Logic in Computer Science*, 2002, pages 215–224.
86. M. Furst, J. Saxe, and M. Sipser. Parity, circuits, and the polynomial-time hierarchy. *Mathematical Systems Theory*, 17 (1984), 13–27.
87. H. Gaifman. Concerning measures in first-order calculi. *Israel Journal of Mathematics*, 2 (1964), 1–17.
88. H. Gaifman. On local and non-local properties, *Proc. Herbrand Symp., Logic Colloquium '81*, North-Holland, 1982.
89. H. Gaifman and M.Y. Vardi. A simple proof that connectivity is not first-order definable. *Bulletin of the EATCS*, 26 (1985), 43–45.
90. F. Gécseg and M. Steinby. Tree languages. In *Handbook of Formal Languages, Vol. 3.* Springer-Verlag, 1997, pages 1–68.
91. F. Gire and H. K. Hoang. A more expressive deterministic query language with efficient symmetry-based choice construct. In *Logic in Databases, Int. Workshop LID'96*, Springer-Verlag, 1996, pages 475–495.
92. Y.V. Glebskii, D.I. Kogan, M.A. Liogon'kii, and V.A. Talanov (Ю. В. Глебский, Д. И. Коган, М. И. Лиогонький, В. А. Таланов). Range and degree of realizability of formulas in predicate calculus (Объём и доля выполнимости формул исчисления предикатов). *Kibernetika (Кибернетика)*, 2 (1969), 17–28.
93. G. Gottlob, E. Grädel, and H. Veith. Datalog LITE: a deductive query language with linear time model checking. *ACM Transactions on Computational Logic*, 3 (2002), 42–79.
94. G. Gottlob and C. Koch. Monadic datalog and the expressive power of languages for Web information extraction. *Journal of the ACM*, 51 (2004), 74–113.
95. G. Gottlob, Ph. Kolaitis, and T. Schwentick. Existential second-order logic over graphs: charting the tractability frontier. In *IEEE Symp. on Foundations*

of Computer Science, 2000, pages 664–674.
96. G. Gottlob, N. Leone, and F. Scarcello. The complexity of acyclic conjunctive queries. *Journal of the ACM*, 48 (2001), 431–498.
97. E. Grädel. Capturing complexity classes by fragments of second order logic. *Theoretical Computer Science*, 101 (1992), 35–57.
98. E. Grädel and Y. Gurevich. Metafinite model theory. *Information and Computation*, 140 (1998), 26–81.
99. E. Grädel, Ph. Kolaitis, L. Libkin, M. Marx, J. Spencer, M.Y. Vardi, Y. Venema, S. Weinstein. *Finite Model Theory and its Applications*. Springer-Verlag, 2004.
100. E. Grädel, Ph. Kolaitis, and M.Y. Vardi. On the decision problem for two-variable first-order logic. *Bulletin of Symbolic Logic*, 3 (1997), 53–69.
101. E. Grädel and G. McColm. On the power of deterministic transitive closures. *Information and Computation*, 119 (1995), 129–135.
102. E. Grädel and M. Otto. Inductive definability with counting on finite structures. *Proc. Computer Science Logic*, 1992, Springer-Verlag, pages 231–247.
103. R.L. Graham, B.L. Rothschild and J.H. Spencer. *Ramsey Theory*. John Wiley & Sons, 1990.
104. E. Grandjean. Complexity of the first-order theory of almost all finite structures. *Information and Control*, 57 (1983), 180–204.
105. E. Grandjean and F. Olive. Monadic logical definability of nondeterministic linear time. *Computational Complexity*, 7 (1998), 54–97.
106. M. Grohe. *The structure of fixed-point logics*. PhD Thesis, University of Freiburg, 1994.
107. M. Grohe. Fixed-point logics on planar graphs. In *IEEE Symp. on Logic in Computer Science*, 1998, pages 6–15.
108. M. Grohe. Equivalence in finite-variable logics is complete for polynomial time. *Combinatorica*, 19 (1999), 507–532.
109. M. Grohe. The parameterized complexity of database queries. In *ACM Symp. on Principles of Database Systems*, 2001, ACM Press, pages 82–92.
110. M. Grohe. Large finite structures with few L^k-types. *Information and Computation*, 179 (2002), 250–278.
111. M. Grohe. Parameterized complexity for the database theorist. *SIGMOD Record*, 31 (2002), 86–96.
112. M. Grohe. The complexity of homomorphism and constraint satisfaction problems seen from the other side. In *IEEE Symp. on Foundations of Computer Science*, 2003, pages 552–561.
113. M. Grohe and T. Schwentick. Locality of order-invariant first-order formulas. *ACM Transactions on Computational Logic*, 1 (2000), 112–130.
114. M. Grohe, T. Schwentick, and L. Segoufin. When is the evaluation of conjunctive queries tractable? In *ACM Symp. on Theory of Computing*, 2001, pages 657–666.
115. S. Grumbach and J. Su. Queries with arithmetical constraints. *Theoretical Computer Science*, 173 (1997), 151–181.

116. Y. Gurevich. Toward logic tailored for computational complexity. In *Computation and Proof Theory*, M. Richter et al., eds., Springer Lecture Notes in Mathematics, Vol. 1104, 1984, pages 175–216.
117. Y. Gurevich. Logic and the challenge of computer science. In *Current trends in theoretical computer science*, E. Börger, ed., Computer Science Press, 1988, pages 1–57.
118. Y. Gurevich, N. Immerman, and S. Shelah. McColm's conjecture. In *IEEE Symp. on Logic in Computer Science*, 1994, 10–19.
119. Y. Gurevich and S. Shelah. Fixed-point extensions of first-order logic. *Annals of Pure and Applied Logic*, 32 (1986), 265–280.
120. W. Hanf. Model-theoretic methods in the study of elementary logic. In [7], pages 132–145.
121. L. Hella. Logical hierarchies in PTIME. *Information and Computation*, 129 (1996), 1–19.
122. L. Hella, Ph. Kolaitis, and K. Luosto. Almost everywhere equivalence of logics in finite model theory. *Bulletin of Symbolic Logic*, 2 (1996), 422–443.
123. L. Hella, L. Libkin, and J. Nurmonen. Notions of locality and their logical characterizations over finite models. *Journal of Symbolic Logic*, 64 (1999), 1751–1773.
124. L. Hella, L. Libkin, J. Nurmonen, and L. Wong. Logics with aggregate operators. *Journal of the ACM*, 48 (2001), 880–907.
125. W. Hodges. *Model Theory*. Cambridge University Press, 1993.
126. J. Hopcroft and J. Ullman. *Introduction to Automata Theory, Languages, and Computation*. Addison-Wesley, 1979.
127. R. Hull and J. Su. Domain independence and the relational calculus. *Acta Informatica*, 31 (1994), 513–524.
128. N. Immerman. Upper and lower bounds for first order expressibility. *Journal of Computer and System Sciences*, 25 (1982), 76–98.
129. N. Immerman. Relational queries computable in polynomial time (extended abstract). In *ACM Symp. on Theory of Computing*, 1982, ACM Press, pages 147–152.
130. N. Immerman. Relational queries computable in polynomial time. *Information and Control*, 68 (1986), 86–104.
131. N. Immerman. Languages that capture complexity classes. *SIAM Journal on Computing*, 16 (1987), 760–778.
132. N. Immerman. Nondeterministic space is closed under complementation. *SIAM Journal on Computing*, 17 (1988), 935–938.
133. N. Immerman. *Descriptive Complexity*. Springer-Verlag, 1998.
134. N. Immerman and Ph. Kolaitis, eds. *Descriptive Complexity and Finite Models*, Proc. of a DIMACS workshop. AMS, 1997.
135. N. Immerman and E. Lander. Describing graphs: a first order approach to graph canonization. In *Complexity Theory Retrospective*, Springer-Verlag, Berlin, 1990.
136. N. Immerman and M.Y. Vardi. Model checking and transitive-closure logic. In *Proc. Int. Conf. on Computer Aided Verification*, Springer-Verlag LNCS 1254,

1997, pages 291–302.

137. D. Janin and J. Marcinkowski. A toolkit for first order extensions of monadic games. *Proc. of Symp. on Theoretical Aspects of Computer Science*, Springer-Verlag LNCS vol. 2010, Springer Verlag, 2001, 353–364.
138. D. Janin and I. Walukiewicz. On the expressive completeness of the propositional mu-calculus with respect to monadic second order logic. In *Proc. of CONCUR'96*, Springer-Verlag LNCS 1119, 1996, pages 263–277.
139. D.S. Johnson. A catalog of complexity classes. In *Handbook of Theoretical Computer Science, Vol. A*, North-Holland, 1990, pages 67–161.
140. N. Jones and A. Selman. Turing machines and the spectra of first-order formulas. *Journal of Symbolic Logic*, 39 (1974), 139–150.
141. H. Kamp. *Tense logic and the theory of linear order.* PhD Thesis, University of California, Los Angeles, 1968.
142. P. Kanellakis, G. Kuper, and P. Revesz. Constraint query languages. *Journal of Computer and System Sciences*, 51 (1995), 26–52.
143. C. Karp. Finite quantifier equivalence. In [7], pages 407–412.
144. M. Kaufmann and S. Shelah. On random models of finite power and monadic logic. *Discrete Mathematics*, 54 (1985), 285–293.
145. B. Khoussainov and A. Nerode. *Automata Theory and its Applications.* Birkhäuser, 2001.
146. S. Kleene. Arithmetical predicates and function quantifiers. *Transactions of the American Mathematical Society*, 79 (1955), 312–340.
147. Ph. Kolaitis. Languages for polynomial-time queries – an ongoing quest. In *Proc. 5th Int. Conf. on Database Theory*, Springer-Verlag, 1995, pages 38–39.
148. Ph. Kolaitis. On the expressive power of logics on finite models. In [99].
149. Ph. Kolaitis and J. Väänänen. Generalized quantifiers and pebble games on finite structures. *Annals of Pure and Applied Logic*, 74 (1995), 23–75.
150. Ph. Kolaitis and M.Y. Vardi. The decision problem for the probabilities of higher-order properties. In *ACM Symp. on Theory of Computing*, 1987, pages 425–435.
151. Ph. Kolaitis and M.Y. Vardi. 0-1 laws and decision problems for fragments of second-order logic. *Information and Computation*, 87 (1990), 301–337.
152. Ph. Kolaitis and M.Y. Vardi. Infinitary logic and 0-1 laws. *Information and Computation*, 98 (1992), 258–294.
153. Ph. Kolaitis and M.Y. Vardi. Fixpoint logic vs. infinitary logic in finite-model theory. In *IEEE Symp. on Logic in Computer Science*, 1992, pages 46–57.
154. Ph. Kolaitis and M.Y. Vardi. On the expressive power of Datalog: tools and a case study. *Journal of Computer and System Sciences*, 51 (1995), 110–134.
155. Ph. Kolaitis and M.Y. Vardi. 0-1 laws for fragments of existential second-order logic: a survey. In *Proc. Mathematical Foundations of Computer Science*, Springer-Verlag LNCS 1893, 2000, pages 84–98.
156. Ph. Kolaitis and M.Y. Vardi. Conjunctive-query containment and constraint satisfaction. *Journal of Computer and System Sciences*, 61 (2000), 302–332.
157. B. Kuijpers, J. Paredaens, and J. Van den Bussche. Topological elementary equivalence of closed semi-algebraic sets in the real plane. *Journal of Symbolic*

Logic, 65 (2000), 1530–1555.

158. G. Kuper, L. Libkin, and J. Paredaens, eds. *Constraint Databases*. Springer-Verlag, 2000.
159. R.E. Ladner. On the structure of polynomial time reducibility. *Journal of the ACM*, 22 (1975), 155–171.
160. R.E. Ladner. Application of model theoretic games to discrete linear orders and finite automata. *Information and Control*, 33 (1977), 281–303.
161. C. Lautemann, N. Schweikardt, and T. Schwentick. A logical characterisation of linear time on nondeterministic Turing machines. In *Proc. Symp. on Theoretical Aspects of Computer Science*, Springer-Verlag LNCS 1563, 1999, pages 143–152.
162. C. Lautemann, T. Schwentick, and D. Thérien. Logics for context-free languages. In *Proc. Computer Science Logic 1994*, Springer-Verlag, 1995, pages 205–216.
163. J.-M. Le Bars. Fragments of existential second-order logic without 0-1 laws. In *IEEE Symp. on Logic in Computer Science*, 1998, pages 525–536.
164. J.-M. Le Bars. The 0-1 law fails for monadic existential second-order logic on undirected graphs. *Information Processing Letters*, 77 (2001), 43–48.
165. D. Leivant. Inductive definitions over finite structures. *Information and Computation* 89 (1990), 95–108.
166. L. Libkin. On counting logics and local properties. *ACM Transactions on Computational Logic*, 1 (2000), 33–59.
167. L. Libkin. Logics capturing local properties. *ACM Transactions on Computational Logic*, 2 (2001), 135–153.
168. L. Libkin. Embedded finite models and constraint databases. In [99].
169. L. Libkin and L. Wong. Query languages for bags and aggregate functions. *Journal of Computer and System Sciences*, 55 (1997), 241–272.
170. L. Libkin and L. Wong. Lower bounds for invariant queries in logics with counting. *Theoretical Computer Science*, 288 (2002), 153–180.
171. S. Lindell. An analysis of fixed-point queries on binary trees. *Theoretical Computer Science*, 85 (1991), 75–95.
172. A.B. Livchak (А. Б. Ливчак). Languages for polynomial-time queries (Языки для полиномиальных запросов). In *Computer-based Modeling and Optimization of Heat-power and Electrochemical Objects (Расчёт и Оптимизация Теплотехнических и Електрохимических Объектов с Помощью ЭВМ)*, Sverdlovsk, 1982, page 41.
173. J. Lynch. Almost sure theories. *Annals of Mathematical Logic*, 18 (1980), 91–135.
174. J. Lynch. Complexity classes and theories of finite models. *Mathematical Systems Theory*, 15 (1982), 127–144.
175. R.C. Lyndon. An interpolation theorem in the predicate calculus. *Pacific Journal of Mathematics*, 9 (1959), 155–164.
176. J. Makowsky. Model theory and computer science: an appetizer. In *Handbook of Logic in Computer Science, Vol. 1*, Oxford University Press, 1992.

177. J. Makowsky. Algorithmic aspects of the Feferman-Vaught Theorem. *Annals of Pure and Applied Logic*, 126 (2004), 159–213.
178. J. Makowsky and Y. Pnueli. Arity and alternation in second-order logic. *Annals of Pure and Applied Logic*, 78 (1996), 189–202.
179. J. Marcinkowski. Achilles, turtle, and undecidable boundedness problems for small datalog programs. *SIAM Journal on Computing*, 29 (1999), 231–257.
180. O. Matz, N. Schweikardt, and W. Thomas. The monadic quantifier alternation hierarchy over grids and graphs. *Information and Computation*, 179 (2002), 356–383.
181. G.L. McColm. When is arithmetic possible? *Annals of Pure and Applied Logic*, 50 (1990), 29–51.
182. R. McNaughton and S. Papert. *Counter-Free Automata.* MIT Press, 1971.
183. F. Moller and A. Rabinovich. On the expressive power of CTL. In *IEEE Symp. on Logic in Computer Science*, 1999, pages 360-369.
184. M. Mortimer. On language with two variables. *Zeitschrift für Mathematische Logik und Grundlagen der Mathematik*, 21 (1975), 135–140.
185. Y. Moschovakis. *Elementary Induction on Abstract Structures.* North-Holland, 1974.
186. F. Neven. Automata theory for XML researchers. *SIGMOD Record*, 31 (2002), 39–46.
187. F. Neven and T. Schwentick. Query automata on finite trees. *Theoretical Computer Science*, 275 (2002), 633–674.
188. J. Nurmonen. On winning strategies with unary quantifiers. *Journal of Logic and Computation*, 6 (1996), 779–798.
189. J. Nurmonen. Counting modulo quantifiers on finite structures. *Information and Computation*, 160 (2000), 62–87.
190. M. Otto. A note on the number of monadic quantifiers in monadic Σ^1_1. *Information Processing Letters*, 53 (1995), 337–339.
191. M. Otto. *Bounded Variable Logics and Counting: A Study in Finite Models.* Springer-Verlag, 1997.
192. M. Otto. Epsilon-logic is more expressive than first-order logic over finite structures. *Journal of Symbolic Logic*, 65 (2000), 1749–1757.
193. M. Otto and J. Van den Bussche. First-order queries on databases embedded in an infinite structure. *Information Processing Letters*, 60 (1996), 37–41.
194. C. Papadimitriou. A note on the expressive power of Prolog. *Bulletin of the EATCS*, 26 (1985), 21–23.
195. C. Papadimitriou. *Computational Complexity.* Addison-Wesley, 1994.
196. C. Papadimitriou and M. Yannakakis. On the complexity of database queries. *Journal of Computer and System Sciences*, 58 (1999), 407–427.
197. J. Paredaens, J. Van den Bussche, and D. Van Gucht. First-order queries on finite structures over the reals. *SIAM Journal on Computing*, 27 (1998), 1747–1763.
198. A. Pillay and C. Steinhorn. Definable sets in ordered structures. III. *Transactions of the American Mathematical Society*, 309 (1988), 469–476.

199. E. Pezzoli. Computational complexity of Ehrenfeucht-Fraïssé games on finite structures. *Computer Science Logic 1998*, Springer-Verlag, LNCS 1584, pages 159–170.
200. B. Poizat. Deux ou trois choses que je sais de L_n. *Journal of Symbolic Logic*, 47 (1982), 641–658.
201. B. Poizat. *A Course in Model Theory: An Introduction to Contemporary Mathematical Logic*. Springer-Verlag, 2000.
202. M. Rabin. Decidability of second-order theories and automata on infinite trees. *Transactions of the American Mathematical Society*, 141 (1969), 1–35.
203. R. Rado. Universal graphs and universal functions. *Acta Arithmetica*, 9 (1964), 331–340.
204. N. Robertson and P. Seymour. Graph minors V. Excluding a planar graph. *Journal of Combinatorial Theory, Series B*, 41 (1986), 92–114.
205. N. Robertson and P. Seymour. Graph minors XIII. The disjoint paths problem. *Journal of Combinatorial Theory, Series B*, 63 (1995), 65–110.
206. J. Robinson. Definability and decision problems in arithmetic. *Journal of Symbolic Logic*, 14 (1949), 98–114.
207. E. Rosen. Some aspects of model theory and finite structures. *Bulletin of Symbolic Logic*, 8 (2002), 380–403.
208. E. Rosen and S. Weinstein. Preservation theorems in finite model theory. In *Logic and Computational Complexity*, Springer-Verlag LNCS 960, 1994, pages 480–502.
209. J. Rosenstein. *Linear Orderings*. Academic Press, 1982.
210. B. Rossman. Successor-invariance in the finite. In *IEEE Symp. on Logic in Computer Science*, 2003, pages 148–157.
211. Y. Sagiv and M. Yannakakis. Equivalences among relational expressions with the union and difference operators. *Journal of the ACM*, 27 (1980), 633–655.
212. V. Sazonov. Polynomial computability and recursivity in finite domains. *Elektronische Informationsverarbeitung und Kybernetik*, 16 (1980), 319–323.
213. T. Schaefer. The complexity of satisfiability problems. In *Proc. 10th Symp. on Theory of Computing*, 1978, pages 216–226.
214. K. Schneider. *Verification of Reactive Systems*. Springer-Verlag, 2004.
215. T. Schwentick. On winning Ehrenfeucht games and monadic NP. *Annals of Pure and Applied Logic*, 79 (1996), 61–92.
216. T. Schwentick. Descriptive complexity, lower bounds and linear time. In *Proc. of Computer Science Logic*, Springer-Verlag LNCS 1584, 1998, pages 9–28.
217. T. Schwentick and K. Barthelmann. Local normal forms for first-order logic with applications to games and automata. In *Proc. 15th Symp. on Theoretical Aspects of Computer Science (STACS'98)*, Springer-Verlag, 1998, pages 444–454.
218. D. Seese. The structure of models of decidable monadic theories of graphs. *Annals of Pure and Applied Logic*, 53 (1991), 169–195.
219. D. Seese. Linear time computable problems and first-order descriptions. *Mathematical Structures in Computer Science*, 6 (1996), 505–526.

220. O. Shmueli. Decidability and expressiveness of logic queries. In *ACM Symp. on Principles of Database Systems*, 1987, ACM Press, pages 237–249.

221. M. Sipser. *Introduction to the Theory of Computation*. PWS Publishing, 1997.

222. L. Stockmeyer. The complexity of decision problems in automata and logic. PhD Thesis, MIT, 1974.

223. L. Stockmeyer. The polynomial-time hierarchy. *Theoretical Computer Science*, 3 (1977), 1–22.

224. L. Stockmeyer and A. Meyer. Cosmological lower bound on the circuit complexity of a small problem in logic. *Journal of the ACM*, 49 (2002), 753–784.

225. H. Straubing. *Finite Automata, Formal Logic, and Circuit Complexity*. Birkhäuser, 1994.

226. R. Szelepcsényi. The method of forced enumeration for nondeterministic automata. *Acta Informatica*, 26 (1988), 279–284.

227. V.A. Talanov and V.V. Knyazev (В. А. Таланов, В.В. Князев). The asymptotic truth value of infinite formulas (Об асимптотическом значении истинности бесконечных формул). *Proc. All-Union seminar on discrete mathematics and its applications (Материалы Всесоюзного Семинара по Дискретной Математике и её Приложениям)*, Moscow State University, Faculty of Mathematics and Mechanics, 1986, pages 56–61.

228. R. Tarjan and M. Yannakakis. Simple linear-time algorithms to test chordality of graphs, test acyclicity of hypergraphs, and selectively reduce acyclic hypergraphs. *SIAM Journal on Computing*, 13 (1984), 566–579.

229. A. Tarski. *A Decision Method for Elementary Algebra and Geometry*. Univ. of California Press, 1951. Reprinted in *Quantifier Elimination and Cylindrical Algebraic Decomposition*, B. Caviness and J. Johnson, eds. Springer-Verlag, 1998, pages 24–84.

230. J. Thatcher and J. Wright. Generalized finite automata theory with an application to a decision problem of second-order logic. *Mathematical Systems Theory*, 2 (1968), 57–81.

231. W. Thomas. Classifying regular events in symbolic logic. *Journal of Computer and System Sciences*, 25 (1982), 360–376.

232. W. Thomas. Logical aspects in the study of tree languages. In *Proc. 9th Int. Colloq. on Trees in Algebra and Programming (CAAP'84)*, Cambridge University Press, 1984, pages 31–50.

233. W. Thomas. Languages, automata, and logic. In *Handbook of Formal Languages, Vol. 3*, Springer-Verlag, 1997, pages 389–455.

234. B. A. Trakhtenbrot (Б.А. Трахтенброт). The impossibilty of an algorithm for the decision problem for finite models (Невозможность алгоритма для проблемы разрешимости на конечных классах). *Doklady Academii Nauk SSSR (Доклады Академии Наук СССР)*, 70 (1950), 569–572.

235. E. Tsang. *Foundations of Constraint Satisfaction*. Academic Press, 1993.

236. G. Turán. On the definability of properties of finite graphs. *Discrete Mathematics*, 49 (1984), 291–302.

237. J. Väänänen. Generalized quantifiers. *Bulletin of the EATCS*, 62 (1997), 115–136.

238. J. Väänänen. Unary quantifiers on finite models. *Journal of Logic, Language and Information*, 6 (1997), 275–304.

239. J. Väänänen. *A Short Course in Finite Model Theory*. University of Helsinki. 44pp. Available at `www.math.helsinki.fi/logic/people/jouko.vaananen`.

240. J. van Benthem. *Modal Logic and Classical Logic*. Bibliopolis, 1983.

241. D. Van Dalen. *Logic and Structure*. Springer-Verlag, 1994.

242. J. Van den Bussche. Constraint databases: a tutorial introduction. *SIGMOD Record*, 29 (2000), 44–51.

243. L. van den Dries. *Tame Topology and O-Minimal Structures*. Cambridge University Press, 1998.

244. M.Y. Vardi. The complexity of relational query languages. In *Proc. ACM Symp. on Theory of Computing*, 1982, 137–146.

245. M.Y. Vardi. On the complexity of bounded-variable queries. In *ACM Symp. on Principles of Database Systems*, ACM Press, 1995, pages 266–276.

246. M.Y. Vardi. Why is modal logic so robustly decidable? In [134], pages 149–183.

247. H. Vollmer. *Introduction to Circuit Complexity*. Springer-Verlag, 1999.

248. A.J. Wilkie. Model completeness results for expansions of the ordered field of real numbers by restricted Pfaffian functions and the exponential function. *Journal of the American Mathematical Society*, 9 (1996), 1051–1094.

249. M. Yannakakis. Algorithms for acyclic database schemes. In *Proc. Conf. on Very Large Databases*, 1981, pages 82–94.

250. M. Yannakakis. Perspectives on database theory. In *IEEE Symp. on Foundations of Computer Science*, 1995, pages 224–246.

List of Notation

Index

Name Index

Monographs in Theoretical Computer Science · An EATCS Series

K. Jensen
Coloured Petri Nets
Basic Concepts, Analysis Methods
and Practical Use, Vol. 1
2nd ed.

K. Jensen
Coloured Petri Nets
Basic Concepts, *Analysis Methods*
and Practical Use, Vol. 2

K. Jensen
Coloured Petri Nets
Basic Concepts, Analysis Methods
and *Practical Use,* Vol. 3

A. Nait Abdallah
The Logic of Partial Information

Z. Fülöp, H. Vogler
Syntax-Directed Semantics
Formal Models Based
on Tree Transducers

A. de Luca, S. Varricchio
**Finiteness and Regularity
in Semigroups and Formal Languages**

E. Best, R. Devillers, M. Koutny
Petri Net Algebra

S.P. Demri, E. S. Orłowska
**Incomplete Information:
Structure, Inference, Complexity**

J.C.M. Baeten, C.A. Middelburg
Process Algebra with Timing

L.A. Hemaspaandra, L. Torenvliet
Theory of Semi-Feasible Algorithms

E. Fink, D. Wood
Restricted-Orientation Convexity

Zhou Chaochen, M. R. Hansen
Duration Calculus
A Formal Approach to Real-Time
Systems

M. Große-Rhode
**Semantic Integration
of Heterogeneous Software
Specifications**

Texts in Theoretical Computer Science · An EATCS Series

J. L. Balcázar, J. Díaz, J. Gabarró
Structural Complexity I

M. Garzon
Models of Massive Parallelism
Analysis of Cellular Automata and Neural Networks

J. Hromkovič
Communication Complexity and Parallel Computing

A. Leitsch
The Resolution Calculus

G. Păun, G. Rozenberg, A. Salomaa
DNA Computing
New Computing Paradigms

A. Salomaa
Public-Key Cryptography
2nd ed.

K. Sikkel
Parsing Schemata
A Framework for Specification and Analysis of Parsing Algorithms

H. Vollmer
Introduction to Circuit Complexity
A Uniform Approach

W. Fokkink
Introduction to Process Algebra

K. Weihrauch
Computable Analysis
An Introduction

J. Hromkovič
Algorithmics for Hard Problems
Introduction to Combinatorial Optimization, Randomization, Approximation, and Heuristics
2nd ed.

S. Jukna
Extremal Combinatorics
With Applications in Computer Science

P. Clote, E. Kranakis
Boolean Functions and Computation Models

L. A. Hemaspaandra, M. Ogihara
The Complexity Theory Companion

C.S. Calude
Information and Randomness.
An Algorithmic Perspective
2nd ed.

J. Hromkovič
Theoretical Computer Science
Introduction to Automata, Computability, Complexity, Algorithmics, Randomization, Communication and Cryptography

A. Schneider
Verification of Reactive Systems
Formal Methods and Algorithms

S. Ronchi Della Rocca, L. Paolini
The Parametric Lambda Calculus
A Metamodel for Computation

Y. Bertot, P. Castéran
Interactive Theorem Proving and Program Development
Coq'Art: The Calculus of Inductive Constructions

L. Libkin
Elements of Finite Model Theory